Extension Management Strategies for Sustainable Agriculture

Opportunities and Challenges

NIPA® GENX ELECTRONIC RESOURCES & SOLUTIONS P. LTD.
New Delhi-110 034

www.nipaers.com
Browse, Search, Read & Buy...
Print Books eChapters
eBooks Publishers
Forthcoming Subject Catalogues
New Titles
Online Resources on:
✓ Current Affairs
✓ Reasoning, Logic & Aptitude
✓ Competitive Examinations
✓ English Language Lab
✓ Writing & Pronunciation Tools
✓ Personality Development
✓ Online Programmes for Professionals
✓ Interview Preparation
Pay Using
PhonePe Google Pay paytm PayPal UPI CCAvenue

Extension Management Strategies for Sustainable Agriculture

Opportunities and Challenges

Edited by

J. Vasantha Kumar

Dean, Faculty of Agriculture (Retd.) and
Registrar, Annamalai University (Retd.)
Annamalai University, Annamalai Nagar
Chidambaram, Tamil Nadu

T. Rathakrishnan

Professor, Agricultural Extension
Agricultural College and Research Institute
Madurai – 625 104, Tamil Nadu

H. Philip

Director of Extension Education I/C
Tamil Nadu Agricultural University
Coimbatore – 641 003, Tamil Nadu

P.P. Murgan

Professor and Head
Department of Agricultural Extension
Agricultural College and Research Institute
Mudurai – 625 104, Tamil Nadu

and

M. Jagadeesan

Assistant Professor, Agricultural Extension
PRO, Tamil Nadu Agricultural University
Coimbatore – 641 003, Tamil Nadu

NIPA® GENX ELECTRONIC RESOURCES & SOLUTIONS P. LTD.

New Delhi-110 034

NIPA® GENX ELECTRONIC
RESOURCES & SOLUTIONS P. LTD.

101,103, Vikas Surya Plaza, CU Block
L.S.C. Market, Pitam Pura, New Delhi-110 034
Ph : +91 11 27341616, 27341717, 27341718
E-mail: newindiapublishingagency@gmail.com
www: www.nipabooks.com

For customer assistance, please contact
Phone: + 91-11-27 34 17 17
Fax: + 91-11- 27 34 16 16
E-Mail: feedbacks@nipabooks.com

ISBN: 978-81-19103-76-8

Composed and Designed by NIPA®.

Preface

Extension science with respect to Agricultural and allied fields has undergone a metamorphosis over a period of several decades in India and across the globe. It has been vibrant, dynamic and aids in bringing about remarkable changes in the livelihood of peasants across India. Any science which fell in line with the pace of change of the system in which it is operating, ready to accept challenges and evolve pragmatic strategies, models, policies can thrive and contribute to the humankind. Hence, it is imperative for extension science to change according to changing needs of its clientele.

Climate change has brought in tremendous changes in the way the agriculture is being practiced. Fragmentation and shrinking of land holdings, water and labour shortage, changing food habits of people, shift towards high value horticultural crops, enhanced thrust for quality aspects and value addition of agricultural commodities and other plethora of factors resulted in 'Agriculture' being viewed as 'Agribusiness'. In such a scenario, Indian farmers are craving for a different type of Extension system which could deliver a variety of information through innovative modes.

In response to the shift in farmer's requirement, several Central and State Government interventions were put in place and their results were mixed. National level Institutional arrangements like ATMA, ATIC and state specific arrangements like Krishi Bhavans of Kerala, DATTS of Andhra Pradesh etc., brought significant change in technology delivery. Group based Extension, Market led Extension, ICT enabled Extension, Participatory Extension, Youth and Women Inclusive Extension, Extension plus are some of the conspicuous shifts observed in Extension approaches. Several Extension Service Providers like Central, State departments, State Agricultural Universities, Private Firms, NGOs have been trying different approaches and models of technology delivery to farmers from time to time. The contours of Agricultural technology transfer across the globe has been realigned with the Public Extension is slowly replaced by several models. We need to identify Good Extension Practices (GEP) as that of Good

Agricultural Practices which can be thoroughly debated and replicated in appropriate locations.

The management of this change contemplated through several new approaches and models of Extension holds the key. Hence, a multidimensional and honest stocktaking of what has happened so far and what is required to be done in terms of Extension Research and Field Extension is the need of the hour. This warrants deliberation of higher order which resulted in creation of new knowledge and cross learning. The Extension scholars and researchers need to share their unique experiences in research and communicate the concepts that they have developed and experimented. This sharing will enrich the Extension science in one hand and the scientists who share will get a research network for themselves.

Keeping the noble cause of dissemination of research findings in Extension science in mind, this edited book has been prepared through collection of precious research articles from eminent Extension scientists and scholars across India. They were addressed to contribute papers on Institutional arrangements for technology delivery, innovative Extension approaches, break through methodologies in Extension research, the reach and effectiveness of ICT tools in technology delivery, youth and gender inclusive extension strategies and Extension strategies for changing scenario of Agriculture.

The authors are grateful to the Extension scientists and scholars across India who has taken strenuous efforts in preparing their valuable research work into research article and send them with utmost sincerity. Every paper in this book is unique in terms of contemporary relevance, content and the rigour with which they have been written.

The authors have greater confidence that the papers in this book would definitely make an impact among Extension scientists and post graduate scholars of Extension in Agriculture, Veterinary and Fishery Sciences. The inputs that this book offers will be of much helpful to the researchers, administrators and policy makers. From the research findings of the selected research papers of this book, the Extension researchers can sharpen their future research strategies, the Extension administrators and policy makers in the Government can get to know about the impact and constraints of several institutional mechanisms of technology delivery which will facilitate them in evidence based policy making.

The authors place on record their sincere thanks to New India Publishing Agency who have come forward to bring out this useful compilation of research papers.

Date: 28 January, 2016 **Editors**

Contents

List of Contributors

A. Anitha Pauline
Dept of Agricultural Extension
Tamil Nadu Agricultural University
Coimbatore - 641 003

A. Janaki Rani
Assistant Professor (Agrl. Extension)
Dept. of Human Development
HSC&RI
Madurai

Akoijam Christina
Department of Agricultural Extension,
College of Agriculture, Vellayani
Kerala Agricultural University
Kerala

A. Latha
Agricultural Research Station
Kerala Agricultural University
Mannuthy, Kerala

A. Prabhuraj
Professor (Agricultural Entomology)
College of Agriculture
University of Agricultural Sciences, Raichur
Karnataka

A. Sakeer Husain
Institutional Co-ordinator
Centre for e-learning
Kerala Agricultural University
Thrissur, Kerala

Anup Prakash Uppadhyay
Assistant Professor
Communication & Extension Management
Indian Institute of Forest Management
Bhopal, Madhya Pradesh

Ambily Varughese
Subject Matter Specialist
CARD Krishi Vigyan Kendra
Pathanamthitta, Kerala

Anbarasan, P.
Department of Agricultural Extension.
Vanavarayar Institute of Agriculture
Manakadavu, Pollachi
Tamil Nadu

Arulnathan, N.
Department of Animal Nutrition
Veterinary College and Research Institute
Tirunelveli, Tamil Nadu

B. Narayanaswamy
Principal Scientists
Division of Extension and Training
Indian Institute of Horticultural Research
Bangalore -560089

B. Shanmugasundaram
Associate Professor (Agrl Extension)
Regional Agricultural Research Station
(RARS)
Mele Pattambi, Palakkad District
Kerala 679 306

B. Swaminathan
Department of Agricultural Economics
Tamil Nadu Agricultural University
Coimbatore - 641 003

B. Vijayabhinandana
Scientist
DATT Centre
Visakhapattinam, Andhra Pradesh

B.J. Pandian
Director
WTC, Tamil Nadu Agricultural University
Coimbatore - 641 003
Tamil Nadu

Chellapandian, M.
Department of Animal Nutrition
Veterinary College and Research Institute
Tirunelveli, Tamil Nadu

C. Karthikeyan
Agricultural Extension
Tamil Nadu Agricultural University
Coimbatore - 641 003, Tamil Nadu

C. Kathiresan
Principal Technical Officer
Centre for Development of Advanced Computing (C-DAC)
Hyderabad, Telangana - 500 030

C. Karpagam
Senior Scientist (Agricultural Extension)
Sugarcane Breeding Institute (ICAR)
Coimbatore, Tamil Nadu

Chitra Parayil
Assistant Professor (Agrl. Economics)
Regional Agricultural Research Station (RARS)
Mele Pattambi, Palakkad
Kerala - 679 306

D. Jebapreetha
Research Associate (Agril. Extension)
Home Science College & Research Institute
Madurai – 625104

D. Puthira Prathap
Principal Scientist (Agricultural Extension)
Sugarcane Breeding Institute (ICAR)
Coimbatore, Tamil Nadu

D. Mahalakshmi
Department of Fisheries Extension
Fisheries College and Research Institute
Thoothukudi - 628 008

D. Sreenivas
Krishi Vigyan Kendra, Mamnoor
Warangal, Telangana - 506 166

D. Sekhar
Scientist
DATT Centre
Visakhapattinam

G.V. Suneel Kumar
Krishi Vigyan Kendra
Darsi – 523 247
Andhra Pradesh

G.M. Hiremath
Assistant Professor (Agricultural Economics)
College of Agriculture
University of Agricultural Sciences
Raichur, Karnataka

G. N. Maraddi
College of Agriculture
University of Agricultural Sciences
Raichur–584 102, Karnataka

G. Arul Oli
Department of Fisheries Extension
Fisheries college and Research Institute
Thoothukudi – 628 008, Tamil Nadu

Guru
Department of Agronomy
Tamil Nadu Agricultural University
Coimbatore, Tamil Nadu

Gopika, M.H.
Assistant Technology Manager (ATM)
ATMA Project
Department of Agriculture Hoskote Taluk
Bangalore Rural District- 562 114

Hinduja. N.A.
Department of Agricultural Extension
College of Agriculture
Vellayani, Kerala Agricultural University
Thiruvananthapuram – 695 522, Kerala

H. Philip
Director of Extension Education
Tamil Nadu Agricultural University
Coimbatore - 641 003

Hidetoshi Miyazaki
Tamil Nadu Agricultural University and
Research Institute for Humanity and Nature
Tamil Nadu

Ishimoto Yudai
Tottori University, Japan

J. Tulasiram
Department of Agricultural Extension
Education
College of Agriculture
University of Agricultural Sciences
Raichur, Karnataka-584 102

J.V. Patil
Director
Directorate of Sorghum Research
Hyderabad, Telangana-500 030

J.K. Das
Department of Agricultural Extension
BCKV, Mohanpur, Nadia
West Bengal-741 252

Jain, P. K.
Assist Professor
School of Agriculture
IGNOU, New Delhi

Jayasree Datta
Department of Agricultural Extension
BCKV, Mohanpur, Nadia
West Bengal-741 252

Jaliya, M. K.
Vocational Instructor in Agriculture
CVHSS, Ramavarmapuram
Thrisssur – 680631, Kerala

K. Senthil Kumar
PG Research Scholor
Department of Fisheries Extension
Fisheries College and Research Institute
Thoothukudi – 628 008

K. Dhanasree
Scientist
DATT Centre, Visakhapattinam

Koichi Fujita
Center for Southeast Asian Studies
Kyoto University, Japan

K.N. Krishnakumar
Professor
Communication & Extension Management
Indian Institute of Forest Management
Bhopal, Madhya Pradesh

K. Veerabhadran
Department of Fisheries Extension,
Fisheries College and Research Institute
Thoothukudi – 628 008

K.R. Murugan
Associate Professor & Head
Department of Women's Studies
Alagappa University, Karaikudi, Tamil Nadu

Kumari Manisha
Research Fellow
Fellow Program in Management (FPM)
Faculty of Communication & Extension
Management
Indian Institute of Forest Management
Bhopal, Madhya Pradesh

M.R. Hegde
Principal Scientists
Division of Extension and Training
Indian Institute of Horticultural Research
Bangalore -560 089

M. Israel Thomas
Associate Professor (Agrl. Extension)
Krishi Vigyan Kendra
Kerala Agricultural University
Pattambi - 679 306, Kerala

M. Ramasubramanian
Assistant Professor
Department of Agricultural Extension and
Rural Sociology Agricultural College and
Research Institute, Madurai

Muniandi Jegadeesan
Tamil Nadu Agricultural University and
Research Institute for Humanity and Nature
Tamil Nadu

M. Senthilkumar
Assistant Professors & Scientist Incharge
TN IAMWARM at KVK and
Agrl. Research Station
Vrinjipuram, Tamil Nadu

M.C. Narayanan Kuttty
Associate Director Research
Regional Agricultural Research Station (RARS)
Mele Pattambi, Palakkad District
Kerala - 679 306

Miyazaki Hidetoshi
Research Institute for Humanity and Nature
Japan

M. Jegadeesan
Center for Southeast Asian Studies
Kyoto University, Japan

M. Sundaramari
Centre for Indigenous Knowledge in Agriculture
F.Ag. & A.H., Gandhigram Rural Institute
(Deemed University)
Gandhigram, Dindigul

Manoj P. Samuel
Scientist
Indian Institute of Horticultural Research
Bangalore – 560 089
Karnataka

Meti, S.K.
Department of Agricultural Extension
College of Agriculture
University of Agricultural Sciences
Raichur, Karnataka - 584 102

N. Jayakumar
Agricultural Extension
Adhiparasakthi Agricultural College
Kalavai, Vellore

Nagesh
College of Agriculture
University of Agricultural Sciences
Raichur – 584 102, Karnataka

N. Kishore Kumar
Department of Agricultural Extension
College of Agriculture
Vellayani, Kerala Agricultural University
Thiruvananthapuram – 695 522
Kerala

Nathan Kanuma Taremwa
Research Scholar
Gandhigram Rural Institute -Deemed University, GRI-DU
Gandhigram, Dindigul District,
Chinnalapatti, Tamil Nadu - 624 302

O. Sarada
SMS (Extn.), KVK
Darsi – 523 247, Andhra Pradesh

P.P. Murugan
Department of Agricultural Extension & Rural Sociology
AC& RI, TNAU, Madurai
Tamil Nadu

P. Sethuraman Sivakumar
Senior Scientist
Central Tuber Crops Research Institute
(ICAR-CTCRI)
Sreekariyam, Thiruvananthapuram - 695 017
Kerala, India

Pandiyan, M.
Professor & Head, ARS & KVK
Vrinjipuram, Vellore - 632 104, Tamil Nadu

Paulpandi, V.K.
Professor and Head, KVK
TNAU, Aruppukottai - 626 107
Tamil Nadu

P. Sandya Rani
Krishi Vigyan Kendra
Darsi – 523247, Andhra Pradesh

P. Murali
Scientist (Agricultural Economics)
Sugarcane Breeding Institute (ICAR)
Coimbatore, Tamil Nadu

P. Mooventhan
National Research Centre on Meat, (ICAR)
Hyderabad, Telangana - 500 092

P.S. Preman
Agricultural Research Station
Kerala Agricultural University
Mannuthy, Kerala

Preethi Priya Dharshini
Assistant Professor (Agrl. Extension)
Department of Human Development
HSC&RI, Madurai

P. Venkatesan
Senior Scientist (Agril. Extn.)
NAARM
Hyderabad - 500 030

P. Marimuthu
Programme Coordinator
CENDECT KVK, Kamatchipuram (S.O)
Theni District - 625 520

P.B. Pradeep Kumar
Scientist
DATT Centre
Visakhapattinam

P.R. Deshmukh
Department of Extension Education
College of Agriculture
Vasantrao Naik
Marathawada Krishi Vidyapeeth
Parbhani, Maharashtra

P. Sivaraj
Department of Agricultural Extension
Tamil Nadu Agricultural University
Coimbatore

R.R. Chapke
Senior Scientist
Directorate of Sorghum Research
Hyderabad, Rajendra Nagar
Hyderabad, Telangana - 500 030

R.K. Naika
Professor of Sericulture
UAS (B) College of Sericulture
Chinthamani - 563 125, Karnataka

Ramakrishnan, K
Assistant Professor (Agrl.Extn)
KVK, TNAU
Aruppukottai - 626 107
Tamil Nadu

R. Karthik
Farm Manager (Horticulture)
KVK, TNAU, Aruppukottai - 626 107
Tamil Nadu

R. Ganesan
Professor
(Agrl. Extension) Retd.
TNAU, Coimbatore - 641 003

R. Prakash
Professor and Head
Department of Agricultural Extension
College of Agriculture, Vellayani
Kerala Agricultural University
Thiruvananthapuram - 695 522
Kerala

Rexlin Selvin
Professor (Agril. Extension)
Agriculture College, Madurai
Tamil Nadu

R. Sureshverma
Department of Agricultural Extension
Telangana State Agricultural University
Rajendranagar, Hyderabad
Telangana - 500 030

R. Rajasekaran
Department of Agricultural Extension
Tamil Nadu Agricultural University
Coimbatore, Tamil Nadu

R. Sendilkumar
(Agrl. Extension)
Kerala Agricultural University, CCBM
Vellanikkara, Thrissur
Kerala

R. Venkattakumar
Senior Scientist (Agril. Extn.)
Indian Institute of Horticultural Research
Bangalore - 560 089, Karnataka

R.P. Kadam
Department of Extension Education
College of Agriculture
Vasantrao Naik
Marathawada Krishi Vidyapeeth
Parbhani, Maharashtra

Robert C.P.
Subject Matter Specialist,
CARD Krishi Vigyan Kendra
Pathanamthitta, Kerala

Sathish
Ph.D, Scholar
Department of Agricultural Extension
College of Agriculture
University of Agricultural Sciences, Raichur
Karnataka

Sheela, P
Ph.D., Sholar
Department of Food Science and Nutrition
Home Science College and Research Institute
Madurai

Suganthi, A.
Assistant Professors
(Agril. Extension)
KVK, Virinjipuram, Vellore - 632 104

S.R. Padma
Assistant Professor (Agrl.Extension)
AC&RI, Killikulam
Tutilorin Dist. Vallanad - 628 252
Tamil Nadu

S. Senthil Vinayagam
Professor and Principal Scientist
(Agril. Extension)
ESM Division, NAARM
Hyderabad - 500 030

Senthilkumar, M.
Assistant Professors at KVK & ARS
Vrinjpuram, Vellore - 632 104
Tamil Nadu

Sendilkumar, R.
Associate Professor (Agrl. Extension)
Kerala Agricultural University
Thrissur -680 656

Saju George
Senior Scientist
Division of Extension and Training
Indian Institute of Horticultural Research
Bangalore -560 089

Sreenivasulu, S.
Ph.D. Scholar
School of Agriculture
IGNOU, New Delhi

Srividya Rani, N
Department of Agricultural Extension
S.V. Agricultural College, Tirupati
Andhra Pradesh

S.V. Prasad
Department of Agricultural Extension
S.V. Agricultural College
Tirupati, Andhra Pradesh

Sakunthalai, A.
Professor (Agrl. Extension)
AEC&RI, Kumulur- 621 712, Trichy

Sency Mathew
Subject Matter Specialist
CARD Krishi Vigyan Kendra
Pathanamthitta, Kerala

Shyla Joseph
Agricultural Research Station
Kerala Agricultural University
Mannuthy, Kerala

Shipin, P.
2. PG Student, MBA (ABM)
Kerala Agricultural University
CCBM, Vellanikkara, Thrissur - 680 656
Kerala

Sindhu Sadanandan
Subject Matter Specialist
CARD Krishi Vigyan Kendra
Pathanamthitta, Kerala

Suresh, C.
Department of Animal Nutrition
Veterinary College and Research Institute
Tirunelveli, Tamil Nadu

Shantha Govind
Department of Agricultural Extension
Annamalai University
Annamalainagar – 608 002, Tamil Nadu

S.N. Bhat
College of Agriculture
University of Agricultural Sciences
Raichur–584 102, Karnataka

Shely Mary Koshy
College of Agriculture
Vellayani, Trivandram, Kerala

Thimmaiyah Ranganathan
Professor of Agriculture,
Gandhigram Rural Institute -Deemed University, GRI-DU
Gandhigram, Dindigul District
Chinnalapatti, Tamil Nadu

T. Rathakrishnan
Department of Agricultural Extension and Rural Sociology
AC&RI, Madurai
Tamil Nadu

T. Lakshmi
Department of Agricultural Extension
S.V. Agricultural College, Tirupati
Andhra Pradesh

Tanaka Ueru
Research Institute for Humanity and Nature
Japan

Thirumeignanam, D.
Department of Animal Nutrition
Veterinary College and Research Institute
Tirunelveli, Tamil Nadu

T.P. Sastry
Professor & University Head (Rtd.)
S.V. Agriculture College
ANGRAU, Tirupati

T. Rajula Shanthy
Principal Scientist (Agricultural Extension)
Sugarcane Breeding Institute (ICAR)
Coimbatore, Tamil Nadu

Tapas Kumar Mondal
Department of Agricultural Extension
BCKV, Mohanpur, Nadia
West Bengal - 741 252

U. Jaikumaran
Agricultural Research Station
Kerala Agricultural University
Mannuthy, Kerala

Umetsu Chieko
Nagasaki University Japan
457-4 Motoyama Kamigamo Kita-ku
Kyoto - 603 8047 Japan

V. G. Dhanakumar
Director
Indian Institute of Plantation Management
Jnana Bharathi Campus, P.O. Malathalli
Bangalore - 560 056

V. Venkatasubramaniam
Senior Scientist (Agricultural Extension)
Sugarcane Breeding Institute (ICAR)
Coimbatore, Tamil Nadu

V. Ravichandran
Professor (Agrl. Extension)
Dept of AE&RS, Tamil Nadu Agricultural University, Coimbatore - 641 003

Venkatapirabu
Professor
Tamil Nadu Agricultural University
Coimbatore - 641003, Tamil Nadu

Waheeda Munaver
Research Scholar, MANAGE
Directorate of Sorghum Research
Hyderabad
Telangana - 500 030

Yamashita Megumi
Tokyo University of Agriculture and Technology
Japan

Extension Management Strategies for Sustainable Agriculture, pp 1-12

New India Publishing Agency, New Delhi, India

1

Designing Value and Talent Management Process in National Extension System

V.G. Dhanakumar

The field of talent management continues to mature from agri-business to extension service, but it is also being impeded both by the scale of debates taking place. The field continues to agree that talent management "adds value" to the organisation such as extension system. To understand the development scenarios of strategic talent management, we need a "theory of value" to understand the elements of talent management which are the most valuable for business-led extension service (BLESS) effectiveness. "Theory of Value" is a generic term which encompasses all the theories within economics that although explain the exchange value (or) price of goods and services. Alternatively, "Value of Theory" encompasses a range of approaches in extension system to understand how, why, and to what degree humans should value things, whether the thing is person, idea, object, (or) anything else for effective functioning of strategic business-led extension service to sustain its clients system.

The purpose of talent management is to ensure that extension unit has the right talent with the right skills at the right time. However, what may have been the right skills in the past may not be all that is required in the features of future. Many things change presenting new challenges and requiring new skills. The extension personnel who was able to solve yesterday's problem is not necessarily the right person to solve tomorrow's problem. Having identified the future agri-business environment and business strategy of the system, it is necessary for management to identify what the success criteria will be for clients in the future. As described above, Talent Management is a natural evolution of a HR. Additionally, Talent Management is a "forward-looking" function. Not only should talent management improve organization's flexibility and performance, it should give us the information and tools to plan for growth, change, acquisitions, and critical new production and service initiatives. To accomplish this goal, a HR can provide value by focusing on five key areas: ensuring extension stability, emphasizing extensionist brand and reputation, developing integrated talent strategies, supporting multilevel accountability, getting involved in talent management initiative and offering opportunities for career and personal development.

While the challenges facing extension unit are critical, however, if addressed proactively they are surmountable. Over the next decade, extension system will be confronted with multiple fundamental agri-business and talent threats, and opportunities. Many are paying close attention to how this crucial extension system confronts these challenges, through Talent Management System (TMS) within Return on Talent (ROT) perspectives as indicated both the past and future perspectives of extension approaches in Fig.1:

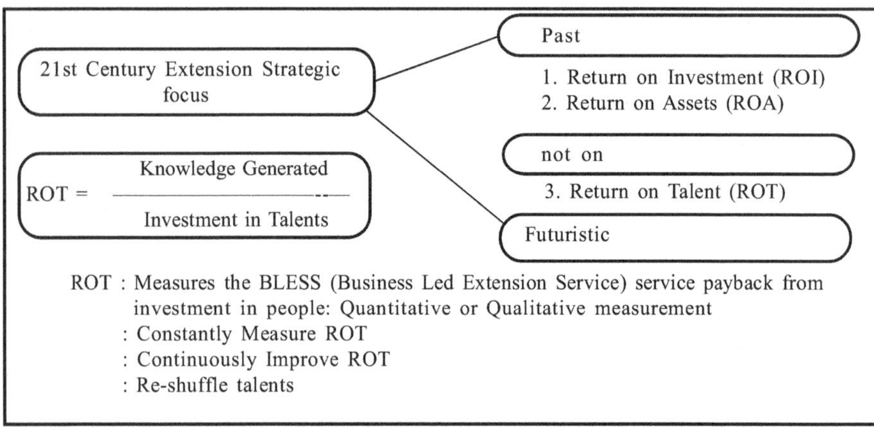

Fig. 1: Talent Management Systems (TMS) Focus in Extension

1. An Understanding on Talent Management

Talent management is "the systematic attraction, identification, development, engagement / retention and deployment of those individuals with high potential who are of particular value to extension system, either in view of their "high potential" for the future or because they are fulfilling agri-business / operation-critical roles. Moreover, we need to re-consider the meaning of talent management in extension system. An equally important consideration is how to integrate talent management with other processes such as planning, supply chain management, resource management, learning and development, performance management etc.

Numerous authors are revealed more complex understanding when referring to the term of "talent", as the sum of a person's abilities, or his or her intrinsic gifts, skills, knowledge, experience, intelligence, judgement, attitude, character, and drive. It also includes his or her ability to learn or to grow (Michaels *et al.*, 2001). In global markets, talent is considered an "asset of the organisation". Despite the pressures of the current economic downturn, equality, diversity and talent remain high priority. In fact, it is at such difficult times that extension organisations most need to invest in their people over return on talents. Talent management is one of the most important process of extension organizational success, more important than money, market share, management team or the rest of the organization's resource. We can state that the importance of an extension is influenced by the success in talent management and in managing human capital. Talent management should focus its value creation on extension, research, clients, L&SCM and administrators' potential and, most important, on the client's future performance, if given proper development of skills and increased responsibility in order to achieve the extension's mission. Now-a-days organizations are focused on utilizing talent as unique source of competitive advantage (Collings and Mellahi, 2009).

Therefore, within extension system, it is very important to show the differences in application between Agriculture Human Resource Management (AHRM) and Strategic Talent Management (STM), because these are two processes that can be easily confused: the first difference that we have noted is that the talent management system is integrated across all aspects of human resource management. There are important connections between talent management and several main activities of the agriculture human resource management. Strategic Talent Management is a process that ensures the optimal development of personnel for each extension units. On the other hand, agriculture human resource management is more focused on the development of people. Talent management is viewed as a strategic process and mandatory for the global strategy. The fundamental components are particularly important to ensure that talent management is justified within a whole extension system's approach to learning & development and where there are clear pathways for succession planning leading towards extension processes in agri-business. The quality of extension organization is the quality of the extensionists it possesses; that is why the development of the culture of learning within extension and development of ability, skill and proficiency is very important.

A diverse clients, that fully reflects our increasingly diverse society, is essential to our capability to meet the challenges of the future. Failure to make such investment could mean, we lose people – the talented people – and lose along with them much of the richness and diversity of our future talent. An expectation of an employer and employee perspective with reference to talent retention and management is shown in the table.

From an employer perspective? Talent management can bring significant agribusiness benefits:	From an individual perspective? When we invest in talented people they are more likely to:
• Ensure that the leadership of our organisation is rich and diverse;	• Engage with their work and be more effective;
• Build a high performance workplace;	• Have a good opinion of their superiors
• Encourage a culture of learning and development; Engage people;	• Have stronger feelings of personal and professional growth and accomplishment;
• Ensures value for money through targeting talent spend and ensuring talent is coherently managed;	• Feel valued and important to the success of the extension service.

Source: Modified version from Nurturing Talent – Research report from Cranfield School of Management and learn direct Business 2008

2. Value Chain Analysis (VCA) and Management

The value chain describes the full range of activities which are required to bring an agri-based product or service from conception, through the different phases of production and delivery to final consumers. Value Chain Analysis (VCA) in extension activities that take place in an agri-business and relates them to an analysis of the competitive strength of the business. Value Chain can also be defined as stages of value addition in a process of deriving products by a single or several extension units. It is a string of companies or players working together to delight market demands for a particular agri-product or service.

1. Primary Activities: those that are directly concerned with creating and delivering a product (e.g. production management components in rice or tea commodity); and
2. Support Activities: which whilst they are not directly involved in production, may increase effectiveness or efficiency (e.g. agricultural, human resource management e.g. talent). It is rare for an agri-business to undertake all primary and support activities.

Value Chain Analysis is one way of identifying which activities are best undertaken by an agri-business in an extension system and which are best provided by others ("out sourced"). However, in order to perform, extension system need to make sure that they have chosen the right partners across Logistics and Supply Chain Management (L&SCM) in creating and sharing value. In order to create success and sharing the value, the extensionists must have the same vision and values and the capacity to create value in extension system. It also indicates that extension system and their chain partners won't use power against each other to get a bigger share of this value. Value chain members of extension can provide customers and each other with the product required in the right quantities, in the desired form, with the right paper work, at the right place, at the right time, and at the optimum cost possible.

Through value chain management, the value created by one member (e.g. client) of a chain is shared – and – therefore, the benefits that chain members receive comes in multiples. Through value chain management, extensionists get benefits not only from the value they create, but also from the value that their partners (processors / traders / suppliers) create. Using the value chain approach, agri-business goes beyond simple across-the-chain cuts and attempts to lower cost and improve efficiency within each value-creating process. For e.g. an extension system might negotiate lower costs of process inputs such as raw materials (or) evaluate make-or-buy options. Value can be defined as indicated below in three ways within agribusiness extension system:

- Look at the *content* of talent management (what is valuable, who values what and where the value resides);
- Think about the *different targets for whom value is being created* by talent management (farming agri-business owners, R&D, extension, chain partners, stakeholders such as society or nation, states, customers, employees or important employee segments); or
- Look at the *process by which value is created or generated* by talent management (the mechanisms that allow value, once created by the talent system, to actually be captured and exploited beyond the individual).

Once the concept of talent management is understood, we need to focus on four crucial value-driven processes that must result in designing a valuable talent management function within extension system from scratch, then combine all the following four processes indicated below and manage them within a single framework shown in Fig. 2.

The four talent value systems need to ensure that the extension organization has viable options in any negotiations that take place between actors in the talent system. Organizations need mechanisms to capture, assimilate and re-create captured know-how and commoditize that know-how, if they are not to be overly dependent on the provider of talent.

Besides, there are two types of value at the extension system level, viz. 'use value' and 'exchange' value. Use value refers to the specific quality of a new extension assignment, task, product or service as perceived by system / clients

- Ability to build and acquire talent outweigh alternative investments to in extension
- Balanced perceived use value between network.
- Structured portfolio of talent resources farming community and supplier of talent,
- Creation of " new use value" in agri-business led extension clusters exploit agri-business opportunities through renewal, governance of talent augmenting capabilities by and combining individual skills adapting current
- Limiting value appropriation or leakage to
- Estimates of talent lifetime value, costs of lost talent, accelerated failure of effectiveness of strategic execution in extension relationships and talent-disengagement

- Unique position of the extensionists in a professional individual (e.g. R&D, extension, NGOs) network
- Bundling into new capabilities to Consideration of reputational issues with tacit of extension knowledge.
- Structural management activities and competitors practices improving the efficiency and
- Creating organizational learning

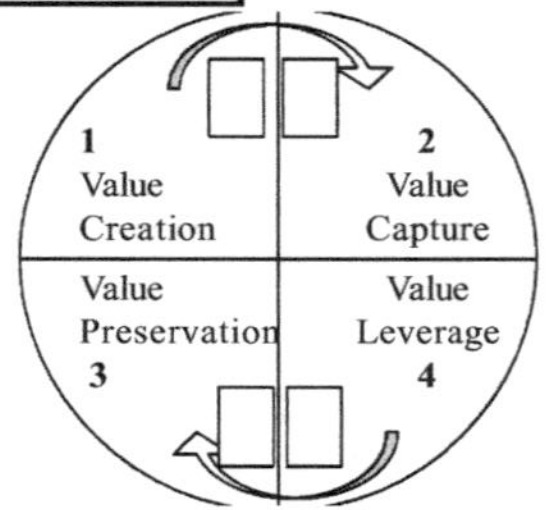

- Design and maintenance of effective
- Limiting value appropriation or leakage to
- Estimates of talent lifetime value, costs of lost talent, accelerated failure of effectiveness of strategic execution in extension relationships and talent-disengagement

- Amplifying talent through renewal, governance of talent augmenting capabilities by and combining individual skills adapting current
- Consideration of reputational issues with tacit of extension knowledge.
- Structural management activities and competitors practices improving the efficiency and
- Creating organizational learning

Fig. 2: Four Value-Driven Processes within a Talent Management Framework

Source: Modified version from CPHR-White Paper 14/01 May 2014

in relation to their requirement. The perceived use value of talent management is also directly related to senior extension executives' perceptions (policy matters) of the value they expect by investing in talent management strategies and these perceptions vary from one system to another. On the other hand, 'exchange value' refers to either the monetary amount realized at a certain point in time, when the exchange of the new task, good, service or product takes place. Therefore, institutionalization, value management and commercialization within extension system is the systematic effect to provide profitable agri-business and to meet customer needs and expectations. It is therefore, necessary that efforts to understand the producers' to adopt valuebased group approach for the sustainability of the extension system through value chain analysis.

3. Analysis of the Talent Management Process in Extension through Value Creation

The purpose of analyzing the value chain of an extension system's talent management process is to identify how each agri-business activity within the process contributes to the overall competitive strategy and where there may be efficiencies and cost savings to be gained within the process. An agri-business unit may create a cost advantage through value chain analysis by reducing the cost of the individual (or) group based agri-business activities within the value chain or by reconfiguring the value chain. The following five steps in the value chain analysis should facilitate to create a value:

1. ***Identify or clarify the overall competitive strategy:*** It may mean revisiting the extension system's vision or mission or reiterating extensionist short-term and long-term goals, ensuring that everyone within all units in the extension is on the same page.
2. ***Activity Analysis:*** Identify each activity undertaken in the talent management process-recruiting, retaining and growing human capital. This can be accomplished by conducting a brainstorming session or an agribusiness process review with key individuals who are responsible for carrying out these activities and those who are impacted by these activities.
3. ***Value Analysis:*** For each activity, assess the value that the activity brings to the process.
4. ***Evaluation:*** Evaluate the pros and cons of making changes to the process. We must also consider how the changes may affect other departments within the organization and evaluate the overall impact of proposed changes
5. ***Planning:*** Design and implement a strategic plan to reduce the costs of the activities within the value chain or the cost to reconfigure the value chain.

4. Approaches to implement Talent Management in Extension System (TMES)

The way to approach talent management depends significantly upon the extension that is doing it. It may be an implementation for some; it may legitimately have a narrower focus for others. The key thing is for it to be able to pose, and then answer, the really hard questions about a proposed agri-business led extension strategy. In order for it to do this, it has developed the following three stage process within

"BLESS" perspective to help to anticipate what this means for the extension. This process of initiation in extension system is described below in Table 1.

Table 1: Extension System Process

Define	1.	Understanding strategic extension talent management requirements
	2.	Develop an integration framework
	3.	Map role and competency requirements to the integration framework
Measure	1.	Design an assessment strategy and approaches
	2.	Measure individual strengths and determine gaps
	3.	Assess individual, group and organisation implications
Realise	1.	Closing the talent gap through improved attraction, retention, development and engagement
	2.	Identify and validate key agri-business success metrics to review and maintain commitment

Source: Modified version from SHL Approach to Integrated Talent Management, 2007

As indicated in table 1, competency management is the process through which the strategic talent requirements of the extension are defined in behavioural terms and then integrated into the full extensionists' lifecycle to create a talent architecture.

Further, Figure 3, shows how the agri-business environment provides the context within which an extension decides what results it needs to achieve. The decision made about the success criteria in turn have implications for the behaviours. It is necessary to encourage if those results are to be achieved. Having identified the behaviours, it is then necessary to see whether the people in extension organisation have these competencies or whether they need to be developed.

In general, competencies are based on observable and behavioural language from some form of competency model. These behaviours make assessment and evaluation easy. Organisation that have their own competency framework need to be clear with themselves about the potentials pitfalls of their competency framework.

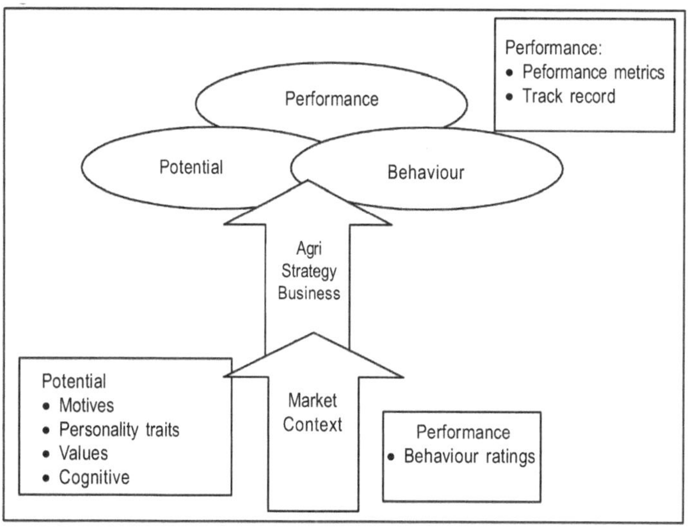

Fig. 3: Extension Performance at work model

Source: Modified version from SHL Approach to Integrated Talent Management, 2007

5 Norms for Implementation of Talent Management

The strategic talent management in extension has received a remarkable degree of both practitioners and academic interest over the last two decades for effective implementation. The talent management approach should be conceptualized and defined in terms of effective talent management at the grassroot level. The talent management should be implemented within the following framework of extension system.

1. ***People Approach:*** In extension system, talent management should focus on a categorization of talented people across extension, research, farming, supplier, marketing, distribution and up in the value chain of agri-business. For example, in an extension system, an organizational effectiveness and design perspective, we need to identify talent and management strategic sourcing for unique individual talent, and thereby managing such talent into collective extension system capability to ensure the knowledge is turned into performance. An elite extensionist generate sustainable value for a particular system.

2. ***Practices Approach:*** Talent Management in the presence of key AHRM practices should be effectively implemented under this category. The client based approach outlined above argue that differentiator for high performing should adopt much of the thinking implicit in AHRM perspective. These component practices in the field-client-system and organization network which have to be more than just a string of HR programme practices and

processors. It must form part of a broader BLESS system and a core agri-business process matched to the organization strategy.

3. ***Position Approach:*** Talent management should identify the pivotal positon across the value chain system. This component argues that the focus of extension organization effectiveness really lies in the management of key positions across the extension systems, including clients' component rather than in key people. Through this a system identification of key positions which definitely contribute to extension's sustainable competitive advantage accompanied by the development of a talent pool of high potential and high performing incumbents to fill these roles.
4. ***Strategic Pools Approach:*** Talent management as an internal talent pool and succession planning is essential for high performance of the extension system. Under this unit, measurement of the impact that human capital (talent seen as either an individual elite or collective of particular skill group or otherwise segmented population) on the ability to execute an extension's agri-business process in terms of impact of operational outcome. Under this approach, talent management draws upon supply chain management theory in order to embrace planning, supply & demand and inability to forecast away uncertainty and plan years into the future.

For the administrators and policy-makers of extension system and those well versed in the above four approaches, it is advised to spare a maximum capacity building and talent management architecture to make projected assumption to meet the requirement of the system.

6. Essential Practices of a Talent Multiplier

A new management or human resources practice is grabbing headlines and clamouring the extension system's attention. Extension with the best clients based financial performance achieve those results not by chasing, but by adhering to sound clients management practices that are as old as agri-business itself. These practices enable leading extension to create a "talent multiplier", that is they enable extension to generate superior results from their clients and stakeholders. The concept of talent multiplier is central to an extension's performance. But even though extensionists' have long recognised these practices as both successful and fundamental, many do not implement them. This looks at the specific activities that constitute the practices strongly related to production, productivity and financial success. In the face of this talent shortage, how to achieve long term talent management goals by creating a talent management strategy- a blueprint or master plan that helps to identify, evaluate, develop and keep individuals who can bring the special and sometimes unique skills need to help team succeed.

An effective way to understand the broad concept of talent management en-route to developing own talent management strategy is to recognize its basic components: extension system management, performance management, learning management, clients' management, succession planning, and social-business networking. Once the talent has gained an understanding of each area and have developed a strategy, then it is time for consistent, active engagement in each area to maximize the quality and length of the extensionist life cycle for up-stream talent.

7. Conclusion

In an attempt to analyse return on investment in extension, we need to design more strategic extension talent systems and to do this, we need to co-opt thinking of other management fields such as 'strategy as BLESS'. By surfacing the assumption made in four approaches, talent management as a categorization of people, the presence of AHRM practices, an identification of pivotal positions, and human capital and planning around strategically important internal talent pools brings an implicit client about the value of talent management. In an agri-business environment, and rapidly changing agri-business scenario, it is necessary for an extension system to make a difference in its needs to possess a highly qualified and motivated personnel. The demand for competent extensionists is high especially for key positions; therefore, extension systems are expected to be a continuous search; for the best personnel which result in competitiveness of most talent people. The policy of developing and applying talent management in ICAR, SAU, SDA, NGO and Farming Community is different from other, mainly because it is in direct correspondence with its necessary strategy and objective. Within agri-business led extension service, the concept of talent management can be adopted and applied in many different ways. But its main advantage remains; in creating positive effects and performance for all stakeholders through value creation.

References

Axelrod, B. Handfield-Jones, H. and Michaels, E. (2001). *The War of Talent,* Harvard Business School Press, Boston.

Cheese, P. and Green, W. (2008). "Where will all the talent come from?" *The Journal of High-Performance Business. (http://www.accenture.com/usen/outlook/pages/outlook-journal-2008-global-talent-strategy.aspx, [accessed March 16th, 2012]).*

Collings, D. and Mellahi, K. (2009). "Strategic Talent Management: A Review and Research Agenda". *Human Resource Management Review,* Vol. 19, No.4, *Pp. 304-331.*

Dhanakumar, V.G (2005). Agri-business led Extension Service (ABES): India's Experience within WTO-SPS-TBT-AOA Scenario of Extension System. *Strategic Business Alliances (SBA) for the Promotion of Participatory Extension Approach* (PEA), Proceedings of 17th ESEE Seminar held at Izmir, Turkey, 30th August to 3rd September, pp. 181- 185 & 325 – 328.

Dhanakumar, V.G. (2006). Integrating a Value in Supply Chain Management (SCM) for Agri-Business Led Extension Service (ABES) in India, *International Journal of Extension Education*, Vol. No. 2 (December). Pp. 21-33.

Dhanakumar, V.G (2008). Micro level Institution for agri-business Led extension Service. National seminar on Current Agrarian Issues. *International Extension Forum*, Tamil Nadu Agriculture University, Coimbatore (December).

Dhanakumar, V.G. (2010). "Business Led Extension Services (BLESS): Global Perspective", *National seminar on extension management reforms – initiatives and impact*, December 11th & 12th, edited by H. Philip and et. al. Society of Extension Education, TN.

Nurturing Talent. (2008). Research Report from Cranfield School of Management and Learndirect Business. *Practitioner Guide: How to Create a Talent Management Strategy that Reflects Diversity.*

Paul Sparrow, Lilian Otaye. Hba Makram. (2014). CPHR. *How Should We Value Talent Management?* White paper 14/01 (May).

SHL. (2007). *Talent Management Issues and Observations.* SHL Whitepaper Series: SHL Guidelines for Best Practice in the Use of Job Analysis Techniques, SHL Guidelines for Best Practice in the Use of Assessment and Development Centres, Surrey, KT7 ONE. *The Five Essential Practices of a Talent Multiplier*. (2007). Business Strategy Series. 8(5:358-364.

Extension Management Strategies for Sustainable Agriculture, pp 13-31

New India Publishing Agency, New Delhi, India

2

Collaborative Institutional Support Systems for Effective Customer Oriented Technology Delivery and Extension Service

K.N. Krishnakumar, Anup Prakash Uppadhyay and Kumari Manisha

1. Prelude

Indian agriculture has experienced spectacular changes in the recent period, manifesting large-scale commercialization and diversification. They broadly include cultivation of new crops and varieties, increase in the share of area under cash crops, large scale spread of livestock activities and fisheries, pursuance of high tech in the areas of aquaculture, bio-technology, horticulture, processing etc. India currently is the second largest food producer. If we look critically the resources available with the country we can say it has the potential to be the world's number one producing over 600 million tons of food products. However, the demand for agricultural products has been increasing at a faster rate as a result of influences of a large number of factors like growing population, rising income level of people, inelastic demand for food items, dynamic economic policy, consumer subsidies, public distribution system, minimum support prices and above all, changing tests and preferences of consumers.Under the new scenario, the most important thing is to provide the extension support to the farmers, especially in the market oriented agribusiness. The farm scientists and the experts in agribusiness in general, and the extension professionals in particular, are expected to help and guide the communities engaged in farming. The time has come to take into consideration the consumers' point of view.

Though we have a strong network of extension professionals from national to village level, sophisticated means of communication for quick and effective transfer of technologies, excellent institutional support for training the human resource and above all, millions of innovative farmers ready to cooperate with us in our endeavor to promote sustainable agribusiness,in the changing circumstances, the extension system will have to be still broad based. Extension organization should make effort to promote diversification at farm level and value addition for international market.This situation requires for reorientation of extension system to promote agri-based enterprises with public – private participation and sustainable development through community empowerment and participation.

1. Lead paper presented in the National Seminar on "Extension Management Strategies for Sustainable Agriculture-Challenges and Opportunities" organized by Extension Education Society at AC&RI, TNAU, Madurai during 12-13 December 2014.
2. Professor, Communication & Extension Management, Indian Institute of Forest Management, Bhopal.
3. Assistant Professor, Communication & Extension Management, Indian Institute of Forest Management, Bhopal.
4. Research Fellow, Fellow Program in Management (FPM), Faculty of Communication &Extension Management, Indian Institute of Forest Management, Bhopal.

In order to achieve this, the extension service has to come out of its traditional mode of over dependence on government support system and transformed to be inclusive of different institutional support systems including that of government, non-government, private agencies, corporate sectors etc.

2. Extension Service for Effective Technology Delivery

As the World Bank defines, extension service is a process that helps farmers become aware of improved technologies and adopt them in order to improve their efficiency, income and welfare. Technology adoption depends on a number of factors that include the relationship between extension service activities and changes in the attitude of farmers, the skills to use technologies and improved practices, farmers' access to information and availability of technology, input supply, and other support mechanisms. Technology delivery is the main activity of the extension program in India, as in many developing countries. In a agrarian country like India technology delivery in extension system ranges from information on new technologies, expert advise, infrastructural support, infrastructural services (roads, bridges, check dams, etc.), exposure to performance of technology (demonstrations), Identify problems and opportunities, problem solving, to market information and many more.

The major institutional supporters involved in establishing a creative partnership in providing the extension services include the government, private sector, corporate sector, civil society, NGOs, etc. The exact role of these agencies in supporting the technology delivery to the client system varies depending on the national and regional situations, but they have helped to make the agenda more responsive to the needs of farming communities, including the output marketing services. Also the emerging institutional approach to research and extension emphasizes the empowerment of rural communities and adopts a demand-led approach, using information from various sources, including the private sector.

In terms of institutional support to technology delivery to the farming communities through participatory extension service, two emerging trends v.i.z. 'Public-Private Partnership' i.e the PPP model and the 'Corporate Social Responsibility', shortly known as CSR are increasingly observed across different agrarian regions of the countries. The main objective of the study is to explore the level of effectiveness of these two types of institutional support systems in technology delivery and extension service for overall community empowerment and to find out how such systems affects performance, and also to what extent the expectations of the beneficiaries of extension services are fulfilled. This paper reviews two case studies pertaining to the extension service rendered by ATMA promoting agri-business enterprises and LUPIN Welfare Foundation promoting rural women empowerment through SHGs, both representing the two different type of institutional support systems v.i.z. 'Public-Private Partnership' and the 'Corporate Social Responsibility' respectively. It distinguishes strategies that have been employed to strengthen both the supply and demand sides of service provision in the area of agricultural extension.

3. Objective of the Study

3.1 To study the institutional and extension support provided by ATMA to Agri-based enterprises through public-private partnership.

3.2 To study the institutional and extension support provided by LUPIN-CSR initiative to empower women and enhance their communication behavior through SHG activates.

4. Public –private partnership - An institutional & extension support system provided by ATMA to promote agri-business enterprises: Case study from Udham Singh Nagar district of Uttarakhand state

4.1 Institutional & Extension support under ATMA

ATMA is a society of key stakeholders involved in agricultural activities for sustainable agricultural development in the district of Udham Singh Nagar of Uttarakhand state. It is a focal point for integrating research and extension activities and decentralizing day to day management of the public Agricultural Technology System (ATS). It is a registered society responsible for technology dissemination at the district level. As a society, it would be able to receive and expend project funds, entering into contracts & agreements and maintaining revolving accounts that can be used to collect fees and thereby recovering operating cost. The major focus is to develop interface at various levels, develop research-extension linkage and to explore and utilize the Agri-based enterprise opportunity. The ATMA at district level would be increasingly responsible for all the technology dissemination activities. It would have linkage with all the line departments, research organizations, non-governmental organizations and agencies associated with agricultural development in the district, such as ZRS or substations, KVKs and the key line departments of agriculture, animal husbandry, horticulture and fisheries etc. would become constituent members of ATMA.

Agribusiness as a concept includes all operations involved in the manufacture and distribution of farm supplies. These are production activities on farm, storage, processing, and distribution of farm commodities. In the modern context, on-farm activities are called 'farming sector' of agriculture and off-farm activities are called as 'agribusiness sector' (Satpathy and Nayak, 2001). Most of the Indian farmers are unorganized, illiterate and live in perpetual debt. According to Anjan, 2001, Indian farmers view agriculture and business as two separate sectors. So there is need to create awareness about agribusiness and to promote it in the country.

Realizing the importance of scope of agri-business in India and in order to promote the same for the benefit of farming communities, the National Agricultural Technology Project (NATP) came with a new strategy of public private partnership extension. Keeping in view the needs to strengthen the research-extension linkage, farm planning linkage with credit institutions, NGOs and private institutions with the main objective of integrated approach, the project devised Agricultural Technology Management Agency (ATMA) implemented in each NATP district. This initiative aims to bring research, extension, service organizations, farmers clubs and farmers together to jointly evolve a new research extension strategy which is demand driven, location specific and multi disciplinary team approach. Hence with this back up an attempt has been made to explore the strategies followed by ATMA with Special Reference to Agri-Based Enterprise in Udham Singh Nagar District of Uttarakhand state.

4.2 Field data collection

The primary data presented here was collected from village by using PRA techniques. While conducting the PRA, the co-operation from resource poor to resource rich farmers and landless villagers were obtained. Due consideration was also given to the participation of women members of village. Detailed field observation was made to understand field based issues related to Agri based enterprises and strategically support being given by ATMA and FAC members. The secondary data were collected from various government and semi government agencies and departments situated in U.S. Nagar, MANAGE, ATMA and PRA survey in the district during 2010-11.

4.3 Extension strategies of ATMA

Based upon the analysis, problems and opportunities, relevant and feasible institutional support and extension strategies of ATMA to promote Agri-preneurship in the district have been explored. The strategies have been classified under nine major groups as indicated below.

4.3.1 Diversification and Intensification of Existing Farming System

Low spread and low yield of crops are undesirable and need diversification. Such crops are either to be substituted or their yield level has to be increased by technological and input intervention. High spread and high yield crops should be encouraged and horizontal expansion of such crops needs to be considered. ATMA through intensive campaigning suggesting farmers that the high spread and high yield crops are to be encouraged and value addition, agro processing and marketing of such crops need to be improved. The less remunerations enterprises need diversification and existing farming systems require further intensification to increase the production level.

4.3.2 Improvement in Productivity and Income of Existing Enterprise and Commodities Within Existing Farming System

ATMA's Strategies are to reduce the yield gap by improving productivity and income of existing enterprises. The farmers are inclined towards higher production and profit with minimum risk. Various extension strategies to overcome the technological gaps and reduce anticipated risk are applied by ATMA. Some of worth mentioning are technical consultancy provided by university experts, private extension agencies, input agencies, NGOs and linking farmers with line department officials and other private agencies including CSR institutions and micro finance institutions.

4.3.3 Natural Resource Management for ensuring Sustainability in Production and Productivity

During the field study it was observed that unabated soil erosion and degradation of natural resource base poses a serious threat to sustainable crop production. Problems related to soil and water management IPM and INM have been identified and strategies suggested conserving the natural resources and regenerating them. Besides, farmer awareness campaigns are also organized by Block technical team of ATMA in collaboration with agricultural university, government, non-government and private organizations.

4.3.4 Human Resource Development

Human resource development in terms of capacity development of key-stakeholders in the development of agri-business enterprises being carried out through ATMA agency through following components.

i. Increasing techno-managerial capacity of farming community on Farming system approach.

ii. Acquaintance to gap in technical, managerial and organizational aspect.

iii. Training on importance of follow-up action and feed back.

iv. Organizing training to extension personnel on participatory research, extension techniques, behavioral science and inter personal skills.

v. Training to extension personnel in training needs assessment.

vi. Regular interface between G.B., FAC, NGOs with BTT, FIGs and Research Scientists.

vii. Training on IT and cyber extension.

viii. Skill up gradation training for grass root level workers.

ix. Specialized training course for NGOs, farmer, farm women and private - para-technicians.

x. Organizing need based training program for extension functionaries of line department, NOOs, private-para-technicians.

4.3.5 Promotion of Community Organizations

Group approach is the cornerstone of the restructured extension mechanism. A major component of extension services was found to be the mobilization of the community into farmers groups — Farmers Interest Groups (FIGs), Farmer Organizations (FOs) and SHGs. Farmers organization linked with Panchayat through existing statutory institutional arrangements. FOs are supported directly through public funds and are involved in the planning, implementation, monitoring and feed back of the program. The ultimate aim is for FOs to internalize extension services for its members and provide backward (input, credit, technology) and forward linkage (post harvest facilities, market, value addition) in a vertically integrated arrangement.

4.3.6 Sustainability of the Agri based Projects

Most of the projects do not sustain after the funds are withdrawn. Cost sharing by the farmers and building a revolving fund by ATMA is ensured for project suitability after the fund flow is withdrawn. In addition, contingency plan is also under consideration for the sustainability of project.

Sustainability of the project achieved through

- Realistic cast recovery for agro-services offered by ATMA.
- Cost sharing by farmers on sustainable issue like soil and water. conservation and NRM.
- Building up revolving fund by collecting cost for supply of inputs like biofertilizer, planting material, and soil testing.
- Opening Agri-clinics and providing consultancy on payment.

Sustainability of extension system through

- Linkages of extension services provided by Govt./SAU/ NGOs and private companies to improve quality of message and avoid multiplicity.
- Establishing of strong linkage between research, extension farmers and services providing agencies.
- Involving farming community in decision making and delivering mechanism in agriculture extension.
- Handing over the extension system of the farming community in the long run so as to operate an extension system.
- Encourage farmers to share the cost of critical extension services and inputs to reduce the burden on public extension etc.

4.3.7 Media Support and use of Information Technology

All the block level FarmInformation and Advisory Centers in Udham Singh Nagar district are electronically linked to district, state and national institutions. ATMA supports the use of electronic linkage so that marketing, research, extension and farming communities are linked to each other at local, national and global market. IFFCO, ITC etc. through their ICT based kiosk system are also supporting farmers for better and timely access of information. ATMA is constantly taking feed back from the farmers regarding their services with the help of FIGs and FOs.

4.3.8 Promotion to NGOs and Private organizations

Some of ATMA funds have been utilized to support NGOs. A systematic training, capacity building and technical back stopping mechanism would be supported though ATMA funds to provide extension services. Strengthening of NGOs in mobilizing communities into farmer organization FOs/FIGs/Market Associations is effectively utilized and it is further explored by ATMA to make it more meaningful to support Agri based enterprises in Udham Singh Nagar district.

4.3.9 Empowerment of Farm Women

Farm women contribute significantly to agriculture production, livestock production and house hold food security in the district. Their skill can be improved to reduce drudgery and to improve the quality of work through need-based training. Formation of women groups into SHGs or FIGs is also considered essential for farm women. There are some agri based enterprises in Udham Singh Nagar district which are already practiced by women but their scope is limited to household only. Hence ATMA is deriving a broad based long term action plan to support farm women groups to come forward and establish themselves as Agri-preneurs with the active support of NGOs and private agencies, including agri-clincics.

4.4 Conclusion

Agribusiness can be promoted through integration of agricultural activities of production, processing marketing and export/import under a single institutional frame work. In the present scenario of globalization agribusiness has a characteristic role for Indian economy to integrate with world economy. Most of the problems stated by the respondents can be easily bridged with the help of organization such as ATMA. This district where the study had been conducted has huge potential for agri-based enterprise. This district has important position in agriculture production not only in state of Uttarakhand but also on country level. The reform in extension system was felt necessary since long time, because it was primarily top down in nature and implemented without participation of key-stakeholders like farmers, extension workers, concerned departments, NGOs and private agencies which include CSR organizations. The feedback system was also very weak and real field problems were not brought before the researchers and planners. Despite efforts of the government, the vast extension system has not been able to provide solution to the farmers of the post WTO (GATT era). Farmers want to become entrepreneur. They are keen to invest in diversified agriculture. Farmers are willing to change their cropping patterns. They are adopting diversified agriculture with invent of open market, new avenues are open for agri-based enterprises. This is only one facet of the coin. Another side is problems related with these agri-based enterprises and diversified agriculture. In order to further improve the functioning of ATMA and the strategy being implemented, respondents suggested for involving all the stakeholders in preparation of SREP (Strategic Research and Extension Plan), and making technology support services more demand driven and location specific. It was found that generation and dissemination of multiple technological options, conducting training programs and demonstrations along with input and information support could prove to be some of the major intervention which can not only motivate farmers to take up Agri-entrepreneurship but also will help in sustaining the established one. It was evident that strengthening research and farmer linkages is the need of the hour as well as need for strong R-E linkages at planning, implementation and follows up stages. More representation of farmers groups and private enterprises including CSR agencies and micro finance groups in ATMA Governing Board and Management Committee will enhance its viability and efficiency in terms of providing institutional support for productive technology delivery at farmer's door steps.

5. Corporate Social Responsibility (CSR)- an emerging institutional & extension support system for rural women empowerment: Case of women SHGs in Raisen district of Madhya Pradesh state promoted by LUPIN Human Welfare & Research Foundation (LHWRF)

5.1 Changing Scenario-glancing at Success Examples

Shubh Desna lives in Obdullaganj block of Raisen district. Shubh's husband was a mason and his earnings were just Rs. 250 per day, if he got any work at all. This was the situation four years earlier. Then the Lupin Foundation entered in her life. It guided her and other women to form the Satyam women self-help group (SHG) in March, 2009 and thereafter her life changed gradually. Now

Shubh is the proud owner of a centering business and recently bought a second hand mixture plant and is leading a respectful life. After six months of the SHG formation, Shubh Desna got her first loan of Rs 3,000 from the Indian Bank under the SHG loan scheme to start the centering business and repaid the loan in just ten months. She took a second loan of Rs. 11,000 to purchase more material for the centering business and after repaying it, took a third loan of Rs 25,000 to invest more in this business. With a fourth loan of Rs 31,000, her family bought a second hand mixing plant besides continuing the centering work. Shubh Desna's family now earns Rs 6000-7000 per month and is thinking of doing some charity by contributing money to install a hand pump in her village because there always remains acute shortage of water there.

Jyoti Baia resident of Rajalwadi rural area earlier was involved in agriculture in her small piece of farm to earn two square meals for the family. Her husband used to work as daily wage driver. He usually got work for only few days in a month. For remaining days of months, he remained idle. Family's monthly income was merely Rs.1000-1500. This income did not suffice for even two square meals for the family, so educating children was beyond their imagination. One day Jyoti came to the Lupin HWRF office and explained her predicament. Lupin suggested to her to form a self help group, through which she could do some activity. After the formation of Shivam SHG and once she become a member of SHG organization, she was provided with a loan of Rs 0.06 lakhs at first, consequently she took second loan of Rs 0.12 lakhs, then third loan of Rs 0.40 lakhs and then Rs 0.66 lakhs, which she used to start up a Kirana shop and for running a truck business by taking truck on rent for her husband. The financial support from the bank and the couple's enterprising nature has enabled them to earn Rs. 6000-7000 per month. In the village six women's SHGs have been formed through her help. Jyoti arranges meetings of the groups regularly and keeps account of all banking for these groups. She inspires other members of SHGs to get socially and economically empowered. Other women in her village aspire to start economic activities on the line of Jyoti so that they can not only help their family but also feel economically empowered to bring-up their children better.Her life mission now is to help poor women of the village. Jyoti's hard labor and dedication has brought happiness to the family and her example is paving the way for many other families to come up in life through self effort.

The above stories from Lupin library not only represent the changing picture of rural women in Madhya Pradesh by help of LHWRF but also the success of Lupin foundation in its initiative of empowering rural women. Lupin Foundation has touched all the issues of social development by formation of women Self Help Groups in each of its adopted villages. It has linked them with banks for financial assistance and provided the women folk, with technical knowledge on how to start income generation activities.

5.2 Lupin-CSR Initiative to empower rural through SHGs

Lupin Human Welfare & Research Foundation, as a pioneer CSR initiative, was set up on October 2, 1988 with the objective of providing an alternative model of rural development in the country, which is sustainable, replicable and ever evolving. Initiating the programme of rural development within a small number of 35 villages, LHWRF has now succeeded in revitalizing, revamping and recreating life in 2,200 villages in Rajasthan, Madhya Pradesh, Maharashtra, Uttarakhand, Uttar Pradesh and Gujarat States of India, which has led to LHWRF emerging as one of the largest NGOs in the country.

The Foundation has been successful in making a big difference in the development of poverty-ridden villages, and especially in the life of the poorest of the poor and empowerment of a large number of women in these areas. Till date, Lupin has formed more than 4,000 SHG groups with a total membership of 50,000 in Rajasthan, Madhya Pradesh, Maharashtra, Uttarakhand, Uttar Pradesh and Gujarat States of India. All the members of SHG groups are insured against natural and accidental deaths. Nearly 20,000 women have been self employed through these financial resources of Rashtriya Mahila Kosh, Banks and inter-loaning of SHGs. In Madhya Pradesh this foundation is working in Raisen district and Vidisha District with 26 Gram Panchayats covering 84 villages for poverty elimination of 3500 families.

5.3 Framework of institutional support provided by LUPIN-CSR initiative for empowering women

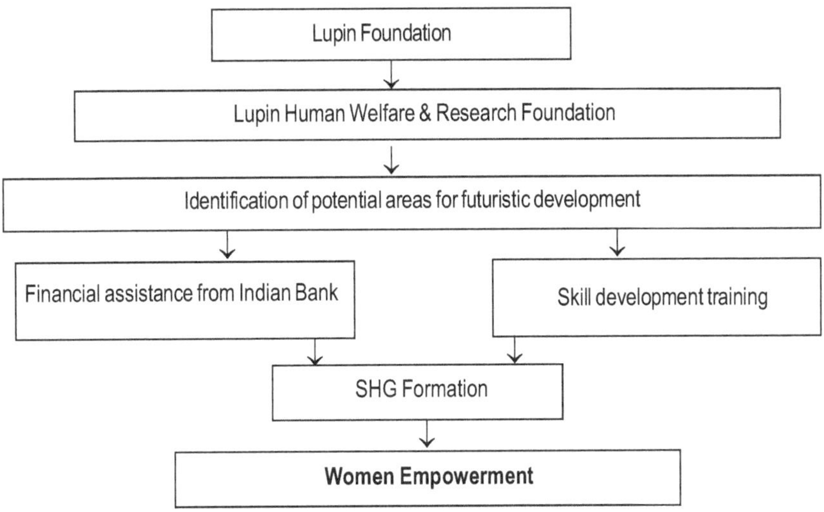

Fig. 1: Framework for empowering women

Fig. 2: Systematic working framework

5.4 Development communication – an effective tool of women empowerment

The term "development communication" was first coined in 1972 by Nora C. Quebral, who defines the field as ,"the art and science of human communication linked to a society's planned transformation from a state of poverty to one dynamic socio-economic growth that makes for greater equality and the larger unfolding of individual potentials." (Quebral, 1973/72).

The overall rural women empowerment is to be achieved through two aspects social and economic empowerment. Development communication acts as the bridge between the gaps to empower rural women. Communication can promote changes in attitudes and social behavior and help communities to identify sustainable opportunities and development solutions that are within their reach. In addition to putting development planners in a position to respect women's traditional knowledge, communication processes can improve the management and effectiveness of new social organizations and institutions, ensuring that they provide services to women in a participatory and democratic manner. Media and communication technologies can also help rural women to exchange experiences and learn from each other. They can be powerful tools for advising women about new ideas and practices and improving training. The majority of rural women are illiterate and live in remote, isolated areas where access to information, lack of transportation, a scarcity of trainers, and cultural and language differences are common problems.However these problems can be overcome through the use of appropriate communication technologies. Development communication approach through SHG formation for empowering rural women leads to:

- **Economic empowerment**- generating livelihood options and making them economically sound and independent.
- **Social empowerment**- improvement in their status through skill development, education, leadership, organizational process and so on.

5.5 Method adopted

The main objective of this study was to study influence of SHG on empowerment through better communication behavior of rural women who are the members of SHGs. So, a matched group experiment was designed in order to study before-after behavior of rural women in SHGs. In this type of experiment each condition uses different participants, but they are matched in terms of certain characteristics, e.g. sex, age, intelligence etc. So, two groups of women identical in all respect except the membership of SHG were identified i.e. Non-SHG and SHG. This categorization was done in order to study the before-after behavior because of two things; first it's convenient nature secondly to control situations when a particular member was unable or struggle to recall her previous status. Further the SHG group was subdivided into three categories for assessment of change in behavior with period of membership.

5.5.1 Data collection

Primary data were collected for Impact analysis through individual interviewing with SHG and Non-SHG members and Lupin and Indian Bank officials and observation and discussion. The Secondary data were collected through internet sources, library journals and documents provided by Lupin.

5.5.2 Respondents

Convenient sampling method was followed to select both SHG and non-SHG members. Multi-stage process was adopted for selection of audience for data collection. At the first stage, two categories SHG members and Non-SHG members having same characteristics in all aspects except membership to some SHG were identified as per data made available by Lupin Human Welfare & Research Foundation. Thus the sample size was constituted by:

- Non –SHG members: 30
- SHG members: 30

At the second stage, SHGs formed at different periods under following categories were identified:

- Newly formed SHGs (less than a year) (6)
- SHGs formed between 1 & 3years (19)
- Old SHGs formed between 3 & 5years (5)

At the third stage in consultation with Lupin Human Welfare & Research Foundation Sultanpur, Obdullaganj, Berkheri, Salamatpur blocks nearby Bhopal were selected where LHWRF was in active working mode for primary data collection. The responses were recorded using personal interviewing method with sections related their demographic information, SHG information, and communication behavior. From the 72 total responses, only 60 complete responses were selected which included 30 SHG members and 30 non-SHG members.

5.6 Major Findings

5.6.1 Description of communication behavior observed from the respondents –members and non members of SHGs

The women who were SHG members were observed to be highly forthcoming in their communication behavior, within their households as well as outside, including the officials from NGOs, govt. organizations, banks etc. The result of the personal interview with respondents- members and non members of SHGs is shown in Table 1.

Table 1: Types and nature of communication behavior of the selected SHG and non SHG members within as well as outside their households/local social settings

	Communication behavior	
Communication pattern	Members	Non-members
Within family	• More frequently with children and husband (76.66%) • Financial discussion, health and education (83.33%) • Regular talks (73.33%)	• Less frequent communication with male member (33.34%) • Discussion on Family issues almost all cases • Occasionally on serious matters discussions takes place (20%)
With other households	• Mainly with female members (46.66%) • Discussions related to SHG and social problems (46.66%) • Occasionally (30%)	• Only with female members (23.33%) • Discussion on Family issues (26.6%) • During social or family gatherings (23.3%)
With officials and leaders	• Interaction is less because of Parda system (46.66%) • Problems related to SHG and village (46.66%) • Faces less hesitation while interaction (46.66%)	• Very less (16.66%) • Hesitation while presentation and interaction with outsiders is prohibited
Within SHG	• Plays important role in decisions of SHG (86.6%) • Problems related to SHG, finance, family (almost all cases) • Almost every week (56.66%)	NA
With Lupin foundation	• More friendly interaction (all cases) • Repayment of loans and related issues (60%) • Monthly (all cases)	NA

From the above Table it could be inferred that the pattern of communication behavior of SHG members was more frequent and constructive within household as well as outside. It could also be seen that the communication of SHG members with male members was also more frequent as compared to non-SHG members. The SHG members were also found to be more aware and active in financial matters, health and education. The level of awareness and knowledge of SHG members was not just confined to family issues but of surrounding and their neighborhoods also. The communication behavior of SHG members, in general was found to be more cosmopolite as compared to non SHG members since the above results clearly revealed that the SHG members had frequent communication with outsiders such as officials of government and Lupin foundation and elected representatives on varieties of issues.

5.6.2 Impact of duration of membership on communication behavior of the SHG and non SHG members

The results revealed the duration or longevity of the membership had definite influence in shaping the nature and type of communication behavior of the rural women. The result is shown in Table 2.

From the above Table 1, it could be inferred that the pattern of communication behavior of SHG members is more frequent and constructive within household as well as outside with two influencing factors that were new age members of newly formed SHGs with new ideas and more literacy level in new age members. It could also be seen that the communication of SHG members with male members was also more frequent in case of young members comprising new SHGs. The communication behavior of SHG members, in general was found to be more cosmopolite over years of membership since the above results clearly revealed that the SHG members had frequent communication with outsiders such as officials of government and Lupin foundation and elected representatives on varieties of issues.

5.6.3 Constraints faced in communication by SHG and Non-SHG members

The responses of the selected SHG and non SHG members with respect to the constraints faced by them in communicating at various levels i.e. within their own households, with neighbors, within SHG and outside were listed in the following Table 2. The results revealed that there was still immense scope for improvement in the communication pattern of the SHG members.

Table 2: Impact of duration of membership on communication behavior of the selected SHG and non SHG members within as well as outside their households/local social settings (Total respondents: 30 SHG members)

Communication pattern	Communication behavior		
	Newly formed SHGs (less than a year) (n=6)	SHGs formed between 1 & 3years (n=19) (n=5)	Old SHGs formed between 3 & 5 years
Within family	• More frequently with children and husband (83.33%) • Financial discussion, health and education (66.66%) • Regular talks (83.33%)	• More frequently with children and husband (73.68%) • Financial discussion, health and education (84.21%) • Regular talks (68.41%)	• More frequently with children and husband (80.00%) • Financial discussion, health and education (80%) • Regular talks (80%)
With other households	• Mainly with female members (33.33%) • Discussions related to SHG and social problems (33.33%) • Occasionally (33.33%)	• Mainly with female members (52.63%) • Discussions related to SHG and social problems (52.63%) • Occasionally (31.57%)	• Mainly with female members (40%) • Discussions related to SHG and social problems (40%) • Occasionally (20%)
With officials and leaders	• Less because of Parda system (83.33%) • Problems related to SHG and village (83.33%) • Faces less hesitation while interaction (83.33%)	• Less because of Parda system (52.63%) • Problems related to SHG and village (52.63%) • Faces less hesitation while interaction (52.63%)	• Less because of Parda system (40%) • Problems related to SHG and village (40%) • Faces less hesitation while interaction (40%)
Within SHG	• Plays important role in decisions of SHG (83.33%) • Problems related to SHG, finance, family (almost all cases) • Almost every week (66.67%)	• Plays important role in decisions of SHG (89.47%) • Problems related to SHG, finance, family (almost all cases) • Almost every week (42.10%)	• Plays important role in decisions of SHG (80%) • Problems related to SHG, finance, family (almost all cases) • Almost every week (100%)
With Lupin foundation	• More friendly interaction(all cases) • Repayment of loans and related issues (50%) • Monthly (all cases)	• More friendly interaction(all cases) • Repayment of loans and related issues (63.15%) • Monthly (all cases)	• More friendly interaction (all cases) • Repayment of loans and related issues (60%) • Monthly (all cases)

Table 3: Constraints faced in communication as expressed by SHG and Non-SHG members

	Constraints faced in communication	
Communication pattern	Members	Non-members
Communication behavior within family	Hesitation while communicating with elders due to traditional respect	Male domination in decision making and not allowing in participating in decision making
Communication with other household	Communication was prohibited to large extent because of Parda system	Mobility outside is prohibited
Communication with officials and leaders	Due to own hesitation and parda system as well	Restricted by family
Communication within SHG	Minimum participation by regular members due to the domination of position holders in decision making	NA
Communication with Lupin foundation	Only position holders communicate	NA

5.7 Findings and conclusion

This study mainly finds out the influence of SHG membership on communication behavior of rural women which is an indicator of social empowerment. Non SHG members were considered for comparison. The non SHG members selected were similar in their socio- economic profile except membership like SHG members. Hence, the responses with respect to communication behavior of the non SHG members has been considered equivalent to the expected responses of the SHG members before joining SHG.

- Qualitative as well as quantitative improvement was found in interacting with the outsiders by SHG members. Majority of them were found to be bold enough in expressing their issues in the presence of outsiders- officials and elected representative.
- Majority of rural women who had joined SHG were found to be more empowered both knowledge wise and skill improvement. This showed the strengthening of the educative and informative functions of SHG.
- Considering the above aspect, the educational and skill development programs could be combined and imparted through local and traditional music and skits for disseminating development and social messages.
- This study revealed the type of changes that had emerged on the basis of the longevity of membership as well.

- As all the SHG members were employed, they were empowered both economically and socially. The change in the communication behavior represented the change in their mind set as well as empowerment.
- Apart from several positive influence of SHG membership over communication behavior, some shortcomings such as lack of proper documentation of activities of SHGs, lack of sufficient exposure to information media like computer, internet, and limited availability of market for handicraft items were also observed.
- This comparative analysis of communication pattern of SHG and non SHG members clearly revealed that the SHG membership and the duration or longevity of the membership had definite influence in shaping the nature and type of communication behavior of the rural women.

References

Anjan, N., Raveendaran, N. and Ramaswamy, C. (2001). Impact of WTO on Tamil Nadu Agriculture and Agro based products. Paper presented at National Seminar on Implications of WTO on agriculture in India conducted by CARDS, TNAU, Coimbatore on November 26-27.

Babbie, E. (1989). The Practice of Social Research. Belmont: Wordsworth.

Gupta, V. (2004). Communication, development and civil society : essays on social development and civil society. New Delhi: Concept Publishing Company.

Kaul, V. (2011). Development Communication in India: Prospects, Issues and Trends. Global Media Journal – Indian Edition, 2249-5835.

Lupin. (2008). Lupin Human Welfare & Research Foundation (LHWRF). Retrieved March- April 18, 2013, from Lupin world: http://www.lupinworld.com/socialinitiatives.htm

McClelland, D. C. (1961). The Achieving Society. New York: The Macmillan Comapny.

Quebral, N. C. (1973/72). What Do We Mean by 'Development Communication? International Development Review , 25–28.

Satpathy, C. and Nayak, N. (2001). Self Help Group and Agri-based opportunity (ed. B.S. Hansra and K. Vijayaragavan). 153-162.

Wikimedia Foundation, I. (n.d.). *Wikipedia*. Retrieved May;June 28, 2012, from Wikipedia: http:/ /en.wikipedia.org

Extension Management Strategies for Sustainable Agriculture, pp 33-40
© 2016, Editors, T. Rathakrishnan, H. Philip, P.P. Murgan and Jagadeesan
New India Publishing Agency, New Delhi, India

3

Negotiating Reality: A Pragmatic Approach for Conducting Quality Extension Research

P. Sethuraman Sivakumar

Though the extension education was originated as a "field approach" aimed to enhance the life quality of rural people, it has evolved into a scientific discipline that relies heavily on empirical research to develop and test its "field-oriented" approaches and methods. Since its birth, the extension research has passed through several phases and a range of research paradigms like community mobilisation, farmer participatory approaches, technology assessment and refinement, impact assessment, Information and Communication Technologies (ICTs) aided development, etc. Despite its diversity and wide scope, the extension research is still regarded a low status by other disciplines and its utility for agricultural development is being questioned. The research managers, academic coordinators and development administrators view the extension as a "Field work" and extension specialists are often burdened with field extension activities, leaving little time to conduct quality research. Besides, the extension researchers are also criticised by their counterparts for our inability to produce quality research output, resulting low impact factors for extension journals. In this context, there is a need for objective analysis of extension research arena as a whole, to derive a realistic assessment of the state of extension research in India. There is also a need to reorient the extension research to produce quality output i.e. publications in high-rated International journals, to restore the prestige of our discipline. This paper will provide an overview of the state of extension research in India. Various research traditions and emerging paradigms will be discussed. Finally it will provide "easy-to-do" methods for conducting systematic and impact-driven extension research within the time and resources available at our workplaces.

Introduction

Research is a backbone of any professional discipline that is aimed to bring out a significant and positive change in the society to foster long-term development. For a field discipline like extension, the research plays crucial role in standardizing the intervention models to bring out an equitable development in the society. Since its birth as a simple field appraisal, the extension research has acquired several dimensions to cater to the emerging needs of the agricultural science. As a part of a "Field-oriented" professional discipline, the extension research differs significantly from other social science research in terms of its content and methods. The extension research is an "applied problem solving" enquiry, conducted in a complex environment - created by the interplay of natural, social and biological forces. Despite its complexity and cumbersome nature, extension research is often criticized by other agricultural discipline scientists, research managers and policy makers as a "non-performing discipline". Failure of extension is often cited as a chief reason for poor technology adoption by the farmers and the role of extension research is questioned. The extension scientists and academicians are often under pressure to deliver significant and tangible outputs to restore the prestige of our discipline. Though there is no "magic wand" to deliver output in a short time, a systematic but pragmatic approach can be tried to create deliverables in a relatively short span of time.

Historical Foundations of Extension Research

The foundations for extension research were laid out in the 1960s by the scientists in ICAR and SAU system. Extension research methods of 1960-1980s were drawn mostly from the psychology and sociology, with a strong psychometric background. The leading extension science journals published very high quality research papers with few of them attracted International accolades. The quantitative research paradigm dominated by multivariate statistical modelling, has brought a new scientific outlook for the field extension. Several scales and indices like socio-economic status (Trivedi, 1963), economic motivation thescale (Supe, 1969), etc were developed during this period. Teacher-made knowledge tests and several scales pertaining to different aspects of extension were also developed in the later stages. Since beginning, the classical test paradigm is being used for scale construction. With the advent of participatory methods, Participatory Rural Appraisal (PRA) has dominated the extension research. Studies on training - need analysis and effectiveness, knowledge, perception and attitude assessment on specific aspects, communication / information processing behaviour, adoption behaviour, constraint analysis, farmer involvement in extension programmes, gender mainstreaming and empowerment, and job satisfaction and productivity of extension personnel dominated the extension research for several decades. Emergence of Information and Communication Technology (ICT), entrepreneurship, organizational management and impact assessment paradigms in the early 1990s has redefined the extension research in the 1990s-2000s.

Methods and Tools used in Extension Research

The current extension research is mostly non-experimental or descriptive in nature. It consists of (i) field studies (exploratory/hypothesis-testing), (ii) expost-facto research, (iii) survey research, (iv) contentand readability/ comprehension analysis, (v) case study, and (vi) ethnographic studies (MANAGE, 2007). These methods employ a variety of data collection methods and tools, to generate the data. These methods and tools are mostly borrowed from cultural anthropology, sociology, cognitive psychology, social psychology, educational psychology, marketing and computer science and engineering. The data generated by extension researchers is mostly ordinal or interval in nature, which can't be analyzed through parametric statistical analyses. Thus, extension researchers depend heavily on descriptive statistics (frequencies, percentages, mean, standard deviation *etc*), non-parametric inferential methods (chi-square, Friedman ANOVA, *etc*), and parametric inferences (correlation and regression). However, there is wide spread misuse of statistical tools among extension researchers who employ inappropriate statistical methods without understanding the nature of data. In the recent years, there is an increasing emphasis on using participatory methods and tools. Though participatory methods are good for exploratory research, they hardly result in quality research publication.

Research Quality in Extension

Quality of social research is judged by the scientificprocess encompassing all aspects of study design. It pertains to the judgment regarding the matchbetween the methods and questions, selection of subjects,measurement of outcomes, and protection againstsystematic bias, non-systematic bias, and inferential error (Boaz & Ashby, 2003; Lohr, 2004). In specific terms, the research quality, whether primarily designed to gather quantitative orqualitative data, should reflect adequate objectivity,internal validity, external validity, reliability, rigor,open-mindedness, and honest and thorough reporting (Shavelson & Towne, 2002;Wooding & Grant, 2003).

A publication in the peer-reviewed journals of International or National repute is a primary indicator of research quality in extension research. Though several Indian Journals like Indian Journal of Extension Education (ISEE, New Delhi), Indian Research Journal of Extension Education (SEE, Agra), Journal of Extension Education (TNAU) among others follow a rigorous peer-review process, their impact factor as determined by the National Academy of Agricultural Research Management (NAAS), New Delhi is very low (<5) (NAAS, 2014). This trend is also similar for extension journals of International repute. With this limitation, the extension scientists and students face innumerable difficulties in applying for awards given by ICAR and other reputed societies.

The reasons for decline in research quality may be due to following reasons

- Confusion over conducting rigorous scientific research as against the need for practical problem-solving in the field. Since extension is perceived as a "Field-work" aims to solve a problem by applying the fruits of science, conducting systematic, objective empirical research is often considered as a unproductive exercise. On the other hand, the results of extension research are also not perceived tangible or significant by the research managers leading to frustration among extension professionals.
- Most of the research methods used today is borrowed from other disciplines. Though the research practice in these disciplines witnessed a sea change in the last two decades with state-of-art techniques, the extension researchers continue to use out-dated methods from these disciplines. Though the extension research themes became diverse with wide scope, the research methods continue to be old and obsolete. This phenomenon resulted in poor quality research as indicated by stereotypic publications and duplicating results.
- The extension professionals are still following the Classical Test Theory for construction of tests and scales while the Item-Response Theory (IRT) has witnessed an exponential growth and widely used in many social science research works in Western countries. In general, classical test theory can't differentiate the examinee characters from the test character, while the IRT help to achieve this. Emergence of multivariate models like Structural Equation Modeling, have revolutionized the social science research, but extension research is yet to use these models for behavioural inquiry.
- Over emphasis on qualitative research methods leads to serious questions on the reproducibility of research. Though PRA, Focus Groups and other ethnographic methods can generate descriptive data, it is not generalizable across populations. The lack of reproducibility of scientific extension research undermines public confidence in extension and leads to the stereo typed findings with huge wastage of resources.

Improving Extension Research Quality

In general, extension research is conducted in three basic settings i.e. (i) research project in a research Institute or college, (ii) student project in a college or research Institute, (iii) field research in a Krishi Vigyan Kendra, (iv) field research in a NGO, (v) field research in a state extension department, (vi) field research in a agri-input/ marketing/ credit agencies, (vii). However, the number and volume of extension research vary among these settings. While the first three settings have a compulsive environment to deliver quality research output, the pressure to deliver is low among others. With extension is perceived as an "action-oriented" discipline by research managers, there is a pressure on the extension scientists/ academicians to get involved in the field-oriented programs like Front-Line Demonstrations, field days, field visits, trainings and exhibitions resulting little time for conducting quality extension research. The extension professionals' lack of time and resources and lack of guidance on new developments in social science research result in poor quality publications. In this context, there is a need to look into probable ways of delivering quality research output without disturbing routine schedule in the office.

A Pragmatic Approach

Recognising the limitations of present extension professional set-up, a pragmatic approach is proposed for the following actors in the extension research system.

i. A researcher who is burdened with field extension

ii. A academician overloaded with teaching

iii. A student who is under compulsion to complete his PhD in prescribed time

iv. A KVK SMS who is burdened with FLDs and trainings

v. A field worker who has little time to complete household survey.

Here we go!

This approach assumes that the extension professionals are engaged in the field extension based on their mandate and demands of their bosses. Any research conducted on these activities like exhibitions, Front-Line Demonstrations etc. will be beneficial to those involved.

a) Research can be conducted in any mandated activity

Any mandated activity can produce quality research output. If someone is continuously engaged in teaching/ training/ exhibitions, they can conduct research on their routine activity. In case of exhibitions, high quality research were conducted on the contextual model of learning – learning in a free-choice settings (Falk and Storksdieck, 2005), exhibit labelling and visitor concept development (Falk, 1997) and impact of prior knowledge on learning at an exhibition (Falkand Adelman, 2003). Method demonstrations can be researched using traditional media/ method comparison techniques (Crouch et al., 2004). Skill learning research (Scully and Newell, 1985) can be conducted for farm machinery and other equipment studies. Motor skill learning is still under researched in extension and the research methods in exercise sciences can help. Since most extension research follows a convenient sampling pattern, a purposeful sampling can be sufficient for these studies.

b) Extension research is multi-disciplinary in nature. There are many multi-disciplinary journals that publish extension research

Extension research is not a stand-alone phenomenon. It is multi-disciplinary in nature and helps extension science to grow stronger through sustained supply of vital elements (concept, tools, techniques and methods). If a scientist is posted in a crop or animal / aspect specific Institute, it is good to get the mastery over the mandated thing before starting any empirical enquiry. Application of a social science concept to biological or engineering science will produce quality outputs. E.g. Participatory plant breeding, sensory evaluation of food products, ergonomic studies on farm implements and tools. There are several engineering or biology or chemistry or other discipline journals that publish extension research. An exhaustive list can be found in Agricultural Extension in South Asia Website. (Avl in: http://www.crispindia.org/Where%20we%20can%20publish%20 extension%20research%20-%20Final%20Note%20%281%29.pdf)

c) Collaborative work is the essence of research quality

A great team can deliver wonderful outputs. A research team can produce a synergy of unique talents of each member which could result in high quality publications. Innovative research thinking, rigor in research design and method, effective data collection, intelligent analysis and interpretation and scholarly language are few essential ingredients for quality publications. Try to identify and work with the people who possess these abilities. In a research team, each one deliver a part of the work that are pooled together to produce quality output. For example, a participatory evaluation of a FLD could be split into several parts like selecting research design and instruments (extension scientist), attribute ranking and impact indicator data collection (research fellow), yield estimation and cost-benefit ratio (agronomist), sensory evaluation for quality (home scientist/ post-harvest engineer), data analysis and interpretation (statistician) and paper writing (anyone who has mastery of English language).

d) Mastery of multivariate statistics is a necessity

With the advent of multivariate statistical methods, the social science research has witnessed a radical transformation. Traditionally, the data analysis is performed after collecting the data. These multivariate methods are not merely tools of data analysis, but they form integral part of every stage in data collection. For example, structural equation modeling, which is a combination of correlations, confirmatory factor analysis, path analysis and goodness-of-fit tests, provides detailed steps in constructing an attitude scale beginning from collection of statements to assessing reliability and validity. It is an integrated tool where all the analyses related to constructing an attitude scale are inbuilt and reduce the researchers' effort and time considerably. Conjoint analysis is another statistical method used for quantifying farmers' preferences of a particular technology. It estimates the farmers' "perceived utility" of a particular varietal attribute, and provides us with an ideal variety combining the preferred varietal attributes in a systematic way. If someone wishes to conduct a farmers attribute preference study, they can directly employ the steps in conjoint analysis for conducting research. Limited dependent models like logistic regression can be used to assess the effect of nominal and ordinal independent variables on a nominal dependent variable. For example, adoption studies (adopted/ not adopted) or acceptance of a food product (accepted/ not accepted) etc can be well-researched using these models. Multi-nominal and ordinal regressions can add one more category i.e. partially adopted into the logistic model.

Conclusion

Extension research needs a face-lift to increase its impact and restore its credibility among research managers and other discipline scientists. Extension is a unique discipline among social sciences as it has evolved from "field practices" that are implemented to improve the quality of life of rural communities. So, research "output" in extension should contribute to improve the efficiency of an existing practice. The perspective taken here is that extension science is an applied, problem-oriented field, and 'scientific knowledge' (which includes theory) within extension should exhibit both 'scientific rigor' and the 'applied perspective' of the extension work. A pragmatic approach consistent with the mandate of the organization will help to produce quality output. More detailed version of these methods can be found in the AESA Blog http://aesa-gfras.net/Resources/file/Blog % 204 %2 0 Enhancing % 20 the % 20 potential %20 of % 20 quality % 20 videos % 20 for % 20 farmers.pdf.

References

Boaz, A. and Ashby, D. (2003). Fit for purpose? Assessing research quality for evidence based policy and practice. London: ESRC UK Centre for Evidence Based Policy and Practice.

Crouch, C. H., A. P. Fagen, J. P. Callan and E. Mazur. (2004). Classroom demonstrations: Learning tools or entertainment? American Journal of Physics, 72(6):835–838.

Falk, J.H. (1997). Testing a museum exhibition design assumption: Effect of explicit labeling of exhibit clusters on visitor concept development. Science Education. 81(6):679 – 687.

Falk, J.H. and Storksdieck, M. (2005). " Using the Contextual Model of Learning to Understand Visitor Learning from a Science Center Exhibition. Science Education, 89:744-778.

Falk, JH and Adelman (2003). Investigating the impact of prior knowledge, experience and interest on aquarium visitor learning. Journal of Research in Science Teaching, 40:163-176.

Lohr, K. N. (2004). Rating the strength of scientific evidence: Relevance for quality improvement programs. International Journal for Quality in Health Care, 16(1):9–18.

NAAS (2014). NAAS score from scientific journals 2013. Avl in: http://naasindia.org/documents/journal2013.pdf

Sculy, D.M. and Newel, K.M. (1985). Observational learning and the acquisition of motor skils: Toward a visual perception perspective. Journal of Human Movement Studies, 11:169-186.

Shavelson, R. J. and Towne, L. (Eds.) (2002). Scientific research in education. Washington, DC: National Research Council, National Academy Press.

Sokoloff, D.R and Thornton, R.K. (1997). Using Interactive Lecture Demonstrations to Create an Active Learning Environment, Physics Teacher, 35, 340.

Supe,S.V. (1969). Factors related to different degrees of rationality in decision making among farmers. Ph.D. thesis (unpub.), Division of Agril. Extension, I.A.R.I., New Delhi.

Trivedi, G.(1963). Measurement and analysis of socio-economic status of rural families, Ph.D. Thesis, Division of Agril. Extension, I.A.R.I., New Delhi.

Wooding, S. and Grant, J. (2003). Assessing research: The researchers' view. Cambridge, England: RAND Europe.

Extension Management Strategies for Sustainable Agriculture, pp 41-51

New India Publishing Agency, New Delhi, India

4

Review of Extension Approaches in India-Prospects and Challenges

S.Senthil Vinayagam, R.R.Chapke, J.V. Patil and Waheeda Munaver

Farmers require a diverse range of information to support their farm enterprises. Information is needed not only on best practices and technologies for crop production, but also for post -harvest aspects. Agricultural Extension is an essential pillar for research and development in agriculture. Despite a wide range of reform initiatives in agricultural extension in India in the past decades, the coverage of, access to, and quality of information provided to marginalized and poor farmers is uneven. Approaches to agricultural extension in India and worldwide continue to evolve. Since the Green Revolution the agricultural extension, with its focus on increasing production via technology transfer, has adopted decentralized, participatory, and demand driven approaches in which accountability is geared toward the users (Kokate, 2009). While the call for demand driven agricultural extension has existed for several decades now, new modes of reaching out to farmers could have significant impact in India, as they might better reflect the local information needs of farmers. The diverse nature of the Indian subcontinent, with its wide variety of agro climatic regions and broad range of socioeconomic conditions in the rural population, calls for agricultural extension approaches that are context and situation-specific (Singh and Sharma, 2004).

In India, the role of agricultural extension in improving agricultural growth is today being recognized with increasing investment. India's 10th and 11th five-year plans emphasize agricultural extension as a key to increasing agricultural growth by reducing the yield gap in farmer fields, and therefore stress the need to strengthen agricultural extension in India (Planning Commission 2001, 2005,

2006). However, despite the renewed interest and investment in agricultural extension in India, the coverage of such services is inadequate.

Review of Extension Approaches

Over the years, the country has tried many extension and development strategies that helped increase and enhance farm production, productivity and overall quality of human life. After the independence, first extension direction to agricultural development was Community Development approach (1952). Besides land reforms, the programme tried to assure cooperative credit, improved marketing facilities, large scale labour- intensive irrigation projects and the efficient use of resources with thc help of National Extension Service agency (1953). Though the basic idea of CD programme was to bring overall development of the rural community through community participation but not many positive results were seen due to lack of conviction, commitment and expertise and the compelling situation of food shortages.

During sixties, the agricultural production situation was very critical and intensification of agriculture with the use of high yielding varieties became must and agricultural development became the sole indicator and measure of rural development. The programmes such as Intensive Agriculture District Programme (1960), Intensive Agriculture Area Programme (1964), National Demonstration (1966) and High Yielding Varieties Programme gained momentum. At this point, the sole purpose was of increasing crop yields by using modern means of production like fertilizers, irrigation and high yielding seeds in particular. The strategy was found generally to be unsuitable for the conditions and needs of complex, diverse and risk prone agriculture. These approaches, though paid good dividend, generally failed to help especially the poor farm households and reduce inequity.

During seventies, the emphasis was broadened from agricultural development to rural development and various programmes like Small Farmers Development Agency, Marginal Farmers and Agricultural Labour Development Agency, Drought Prone Area Programme, Integrated Rural Development Programme etc. were launched.These programmes laid substantial emphasis on participatory approach for organizing and mobilizing poor farmers to enable to take advantages of the technology and services on a continual basis.

Superimposed on TOT approach was the T&V system (1974) with emphasis on the role of extension in technology transfer to encourage utilization of research results. T&V approach was found to be too narrow in its approach and not suitable for small farms and rain-fed areas, which are surrounded with so many uncertainties. Mean while, the ICAR launched its extension programmes namely,

Operational Research Project (1975), KrishiVigyan Kendra (1974) and Lab to Land (1979). During 7th plan, all these programmes including National Demonstration (ND) were later merged under the programme of KVK. The farming system research and extension programme was also started through KVKs. These programmes did receive some success, but could not make much impact, particularly in ensuring people's participation. It was realized that the modern technologies did not match with the farmers' need and resources, and therefore, farmers' participation was partial in the programmes. The situation demanded an emphasis on farmers' need-based and problem-oriented technology generation with active participation of the farmers. The need for technology assessment, refinement and transfer was felt and IVLP (Institute Village Linkages Programme) based on participatory methodology was launched (1995) in selected locations in the country. In addition, National Watershed Development Programme for Rain-fed Areas was also launched by Govt. of India in selected states with basic consideration of people's participation. National Agriculture Technology Project (NATP)integrated the activities of public and private agencies in some selected districts under Agriculture Technology and Management Agency (ATMA) using bottom up planning approach (1999). The project continued upto 2005 with refinement based on field experiences.This has resulted to launch Centrally Sponsored Scheme of "Support to State Extension Programmes for Extension Reforms" during 2005-06 and continuing till date in 630 districts of India.

Policy Guidelines for Extension Programmes

The extension programmes are implemented on the basis of National Agricultural Policies, National Policy for Farmers and Five-Year Plans of Planning Commission.

The following are the action points were highlighted under agricultural extension in National polices (Chandre Gowda,2013).

National Agricultural Policy, 2000

- Strengthen R-E-F linkages.
- Broad-base and revitalize the extension system.
- Encourage KVKs, NGOs, Farmers Organizations, Cooperatives, Corporate Sector and Para-technicians for organizing demand driven agricultural extension.
- Organise capacity building and skill upgradation programme for all extension functionaries.

- Move towards a regime of financial sustainability of extension services through a more realistic cost recovery of extension services .
- Mainstreaming.
- Gender concerns in agriculture.

National Policy for Farmers, 2007

- KVKs would take up training and lab-to-land demonstrations to provide skilled jobs in villages.
- Strengthening the State extension machinery and promote farm schools.
- Convergence of extension efforts at state level , district level and below.
- Linkage with Common Service Centres (CSCs) of the Department of Information Technology, Government of India for inclusive and broad-based development.
- Empowering farmers with the right information at the right time and place is essential for improving the efficiency and viability of small and marginal holding.

Five-Year Plans of Planning Commission, Government of India

Five Year Plans had many implications particularly for initiating new government programmes, policies and procedures.

The Sixth Five Year Plan period coincided with the operations of T&V system.

The Seventh Five Year Plan, the World Bank assisted National Agricultural Extension Projects I to III were in operation. A major effort had been to broad-base the extension activity by moving away from the narrow individual crop orientation to farming systems approach.

The Eighth Five Year Plan more emphasis were given on development of rainfed areas, using adoption of dry land farming system approach; involvement of NGO and Voluntary organizations in Extension, Women in Agriculture, and integration of six extension schemes of ICAR under KVK.

The Ninth Five Year Plan basically addressed the issue of food security. A regionally differentiated strategy based on agro-climatic zones to take into account agronomic, climatic, and environmental conditions to realise the full potential of every region was emphasised.

The Tenth Five Year Plan emphasised on reforms to make extension services demand driven, for which encourage the role of the non-government sector in agriculture extension and an innovative approach in the field of television/ radio broadcast,Communication networking, Strengthen the ACABC scheme to establish agri-clinics / agri-business centres / ventures by the agricultural graduates etc.

The Eleventh Five Year Plan reiterated the key role to be played by Public extension system in educating farmers and helping them to take right decisions. Infrastructure below district level is needed to support capacity building of farmers by up-scaling ATMA to all development districts through convergence. Promotions of research–extension–farmer–market linkages. Public-private partnership (PPP) in extension were also envisaged.

The Twelfth Five Year Plan envisaged National Mission on Agricultural Extension and Technology (NMAET) in which amalgamation of 17 different schemes of the Department of Agriculture & Cooperation will be done under the umbrella of Agriculture Technology Management Agency (ATMA).

NMAET consists of 4 Sub Missions:

i. Sub-Mission on Agricultural Extension (SMAE)

ii. Sub-Mission on Seed and Planting Material (SMSP)

iii. Sub-Mission on Agricultural Mechanization (SMAM)

iv. Sub-Mission on Plant Protection and Plant Quarantine (SMPP)

The common threads running across all 4 Sub-Missions are Extension & Technology.

The aim of the Mission is to restructure & strengthen agricultural extension to enable delivery of appropriate technology and improved agronomic practices to the farmers. This is envisaged to be achieved by a judicious mix of extensive physical outreach & interactive methods of information dissemination, use of ICT, popularisation of modern and appropriate technologies, capacity building and institution strengthening to promote availability of quality seeds, mechanization, plant protection etc. and encourage the aggregation of farmers into Farmers Interest Groups (FIGs) to form Farmer Producer Organizations (FPOs) etc.

Farmers' skill trainings and field extension as contained in all 4 Sub Missions of NMAET (Viz. SMSP, SMAE, SMAM and SMPP) will be converged with similar farmer-related activities going on through ATMA. Five-tiered modes (TV, Newspapers, Booklets, KCC, Internet, SMS) of extension will also be carried out in broadcast or interactive electronic modes in all the four Sub-Missions.

Technical, legal, administrative & regulatory functions and other components (not related to farmer centric extension) will continue to be discharged independently under the respective Sub-Missions.

Support to State Extension Programmes for Extension Reforms (ATMA Scheme)

This Scheme shall focus on the following key extension reforms:

- Encouraging multi-agency extension strategies involving Public/ Private Extension Service Providers.
- Ensuring an integrated, broad-based extension delivery mechanism consistent with farming system approach with a focus on bottom up planning process.
- Adopting group approach to extension in line with the identified needs and requirements of the farmers in the form of CIGs & FIGs and consolidate them as Farmers Producer Organisations.
- Facilitating convergence of farmer centric programmes in planning, execution and implementation.
- Addressing gender concerns by mobilizing farm women into groups.

The above objectives shall be met through strengthened institutional arrangements, dedicated manpower, Innovative Technology Dissemination component and revamped strategy.

The organizational structure at various levels has been depicted in the following diagram:

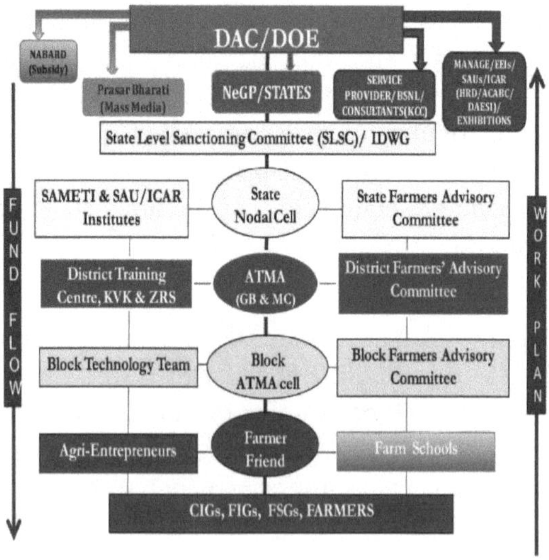

Extension through Krishi Vigyan Kendra

The Krishi Vigyan Kendra (KVK), or farm science center, is a multidisciplinary educational institution situated at the district level, with funding and technical supervision from ICAR. There are 631 KVKs, almost one for each district in India. Each center is under the administrative control of a state agricultural university, State Department of Agriculture, NGO, or central research institute. The performance of KVKs may vary depending on the administrative control. Each KVK is in one of 15 agro climatic zones and created a niche for themselves as the front line extension system performing the vital role of linking research-extension-farmers systems.

Extension through Private Sector

The private sector is increasingly playing a role in extension services in India.It develops context-specific models and using ICT tools to bring information directly to the farmer. Most of the agricultural input companies have an element of extension through various approaches. A few private companies have initiated 'one-stop farm solution centres' for their extension services. Some of them are:

Mahindra Krishi Vihar: Mahindra and Mahindra Limited has established one-stop solution centres with the establishment of Mahindra Shubhlabh Services subsidiary, since solution centres operate on a franchise basis and provide quality inputs, 2000. The one-stop solution centres operate on a franchise basis and provide quality inputs, rent farm equipment, credit in partnership with banks,farm advice by trained field supervisors who visit fields, and arrange contracts with processors for off-take of crop produce.

Tata Kisan Sansar by Tata Chemicals Ltd. Has a franchise-based 'hub and spokes' model of outlets. The extension services offered by these one-stop shops include soil testing, remote diagnostics and house brands for seeds, cattle feed, pesticides and sprayers.

Godrej Agrovet has chain or rural outlet and run in partnership with other companies to extend its products range. Its 'one-stop solutions' model offers agricultural equipment, consumer goods, technical services, soil and water testing, veterinary, financial and post office services, and pharmaceuticals.

Although few empirical studies have been carried out, the performance of private extension is said to vary widely; it tends to focus its services on areas with sufficient resources and is limited to a few crops and areas where profits can be assured (Sulaiman and van den Ban 2003). The private sector serves a corporate interest, working with individual farmers, so social capital is not built.

Moreover, private extension can only work well if farmers are willing and able to pay. One option suggested by Swanson (2008) is that the private sector could serve the needs of medium-size and commercial farmers, while the public sector could work in remote areas, which are currently not serviced well. This sort of system would require public-private partnerships that currently at initial stages in India.

Extension through NGOs

Considering the large number of marginal and small land holdings in India, Farmers organizations and SHGs could play important roles in articulating the needs of farmers to knowledge intermediaries. They can operate side by side with either NGOs or the public sector, but challenges exist in both sectors. Public capacity to build FBOs and SHGs is limited, while NGOs, which are not numerous, rely on donor funds and would need public support to develop the technical skills to facilitate the groups (Swanson 2006; Sulaiman and Hall 2004). Within FBOs or SHGs, problems related to social identity, including gender and caste, mean that these organizations may not be completely inclusive and are subject to elite capture problems. When farmer groups interact with other institutions, social identities and other social status perceptions mean that they may be too weak to articulate their concerns (Sulaiman and Hall 2004). Building the capacity of such groups, and promoting the development of leadership and management skills so that farmers can demand the information they need, is therefore an important component of agricultural extension approaches.

NGOs such as BAIF, Basix and PRADAN operate in numerous states and have been active in extension services for many years.

The Bharatia Agro-Industries Foundation (BAIF): BAIF is a Gandhian organization committed to imparting livelihood opportunities to the rural underprivileged families. BAIF currently works in round 50,000 villages in 12 states of India. BAIF's areas of work include water resources development, sericulture, agro forestry, post-harvest product management and marketing cattle feed ad forage production. BAIF facilitated the formation of farmers' cooperatives and federations of self-help groups.

Basix: Basix strategy is to provide a comprehensive set if livelihood promotion services which include Financial Inclusion Services (FINS) agricultural/Business Development Services (Ag/BDS) and institutional Development Services (IDS) to rural poor households under one umberall. Extension services for farmers is being provided. According to Basix its services reach around 800,000 farmers and involve productivity enhancement, risk mitigation, local value addition and alternative market linkage for synthetic bio-inputs and outputs.

PRADAN: A majority of the families that PRADAN works with belongs to the Schedule Tribes and Schedule Castes. PRADAN promotes the Integrated Natural Resources Management (INRM) of land, water, forest and biological resources to achieve and sustain potential agricultural productivity.

Extension through ICT applications

Kisan Call Centres (KCC) to harness the potential of ICT in agriculture, Ministry of Agriculture took the initiative of launching an innovative scheme "Kisan call Centres" on January 21,2004 aimed at offering solutions to farmers' queries on a telephone call.

aAqua ('Almost All Question Answered') is an internet based discussion portal initiated in 2003 by the Developmental informatics Lab of the Indian Institute of Technology in kiosks and cybercafés in Pune. A farmer can ask a question on aAqua from a kiosk or cybercafé other farmers or experts view the question and reply (in English, Hindi or Marathi).

Avaaj Otalo is a voice-based system for farmers to access and discuss relevant and timely agricultural information by phone. The system itself is a voice-XML based interactive Voice Response system. When a farmer calls iin, he/she hears audio prompts and is asked to enter a number from the keypad or say a word in order to navigate.

Digital Green is a non-profit organization with funding firm the Bill and Melinda Gates Foundation and the Deshpande Foundation. It disseminates agricultural information to small and marginal farmers through digital video.

e-Choupal is an intiative by the agri-division of ITC Ltd, the Indian Tabacco Company, Each e-Choupal is equipped with a computer connected to the internet. A local person acting as a sanchalak (coordinator) runs the village e-Choupal. Farmers can obtain daily updates on crop prices in local mandis, procure seed, fertilizer, and other products including consumer goods, and sell their crops for prices offered by ITC.

Extension through Farmers Organizations and Agri-preneurs

Agri-Clinics & Agri-Business Centres (ACABC) is a Central Sector Scheme under implementation since 2002. The scheme promotes the involvement of agri-preneurs to supplement the efforts of public extension system by way of setting up of agri-ventures in agriculture and allied areas. Agri-Clinics are envisaged to provide expert advice and services to farmers on various technologies, Agri-Business Centres are commercial units of agri-ventures established by trained professionals.

Mahagrapes: It is a Partnership firm of 15 co-operative societies, the main aim being to export grapes and other fresh produce to different parts of the world. Formation of co-operative societies has helped to reach every farmer.

Producers Companies: Producer Companies are the legal institutions, registered under GOI's Producer Company Act, or Companies (Amendment) Act, 2002. It is a hybrid of cooperative and private Company Act, 1956. Members have to be necessarily primary producers i.e., persons engaged in an activity concerned with or related to primary produce. Minimum 10 members must come together to form the producer company and there is no upper limit for number of members.

Vegetable and Fruit Promotion Council Keralam (VFPCK) is an ISO 9001-2000 certified company registered under section 25 on Indian Companies Act 1956 and has been established to bring about overall development for fruit and vegetable sector in Kerala. VFPCK is managed by a result oriented multidisciplinary team of professionals. VFPCK is a company with majority stake of farmers and the government an financial institutions as the other major shareholders.

Farm School to promote Farmer-to-Farmer Extension: Farm School under ATMA scheme is an attempt to institutionalize the concept of farmer-to-farmer-extension. These 'Farm Schools' are to be operationalized at Block/Gram Panchayat level and are set up in the field of achiever farmer who are the 'Teacher ' in the Farm School.

Conclusion

Due to changing agricultural conditions–including climate change, increasingly degraded and marginalized land coming into production, limited water availability, increasing use of inputs, rising fuel costs, and unknown market opportunities–farmers require access to timely, reliable, and relevant information that can support the complexity within which their farm enterprises operate. Although agricultural extension today has a broad mandate, this review shows that despite pluralistic extension approaches in India, the coverage and use of these services are limited. Considering the large number of marginal and smallholder farmers, particularly in rain fed regions, a major need is to build the capacity of farmers to demand and access information to increase their productivity, profitability, and incomes. The information must be reliable and timely. The public-sector extension system still receives significant investment from the central government and is increasingly pushed as the major source of knowledge through a presumed transfer of technology. ATMA is a key component of the system. The KVKs also face challenges that limit their ability to meet farmer's needs. Pluralistic extension system in India; with the public sector, private sector, and third sector all playing some roles. Public-private partnerships; for example, through the agri-clinic and agri-business center (ACABC) scheme; are one aspect that could be strengthened and encouraged. The public extension system dominates the provision of knowledge and information to smallholder farmers, especially in rainfed agricultural communities and largely concentrates on on-farm activities. The private-sector e-Choupal initiative and various small-scale models have tried to provide farmers with information not only regarding on-farm production but also regarding prices and accessing markets. However, these approaches work only for specific crops and regions where farmers have the incentive to take risks and are willing to pay for services. The impacts of the multiple ICT approaches are not documented and tend to work in small communities. Despite the variety of agricultural extension approaches that operate in parallel and sometimes duplicate one another, the majority of farmers in India do not have access to any source of information. This severely limits their ability to increase their productivity and income and thereby reduce poverty. Understanding the behavior of farmers in seeking information for their enterprises and communication through social networks will assist in the development of appropriate agricultural extension strategies. The existence of context-specific and relevant information for farmers also needs to be considered.

References

Chandre Gowda, M.J. (2013). Innovations in Extension Delivery in India.Paper presented at National Seminar on Futuristic Agricultural Extension for livelihood Improvement and sustainable Development, ANGRAU, Hyderabad. Jan 19-21, 2014.

Kokate, K.D. (2009). Developing Farmer –led and Market –Led Extension System:Paradign Shift and Experiences .Presented in National Seminar on Agriculture Extension 2009. Directorate of Extension, Department of Agriculture & Cooperation, Ministry of Agriculture, New Delhi.

Singh, B. and Sharma, P. (2004). Extension Approach For Agriculture Needs Re-Orientation,Paper presented at National Workshop on Communication Support for Sustaining Extension Services. 17-18 February 2004, Department of Extension Education, Institute of Agricultural Sciences, Banaras Hindu University, Varanasi, (India).

Sulaiman, R and A. Hall, (2004). India: The emergence of Extension-Plus: Future for extension beyond technology transfer? In Extension reform for rural development: Case studies of international initiatives, Vol. 1. Washington, D.C.: World Bank and the U.S. Agency for International Development.

Sulaiman, R., and A.W. van den Ban (2003). Funding and delivering agricultural extension in India, Journal of International Agricultural and Extension Education 10(1). [11].

Swanson, B. (2006). The changing role of agricultural extension in a global economy, Journal of International Agricultural and Extension Education 13(3). [12].

Extension Management Strategies for Sustainable Agriculture, pp 53-58

New India Publishing Agency, New Delhi, India

5

ICT enabled Knowledge Management in Agriculture: Status, Challenges and Opportunities

C. Kathiresan

1. Introduction

Democratisation of information makes it possible for ideas, opinions, knowledge and services to be accessible to everyone, anywhere, anytime. In this era of 'knowledge sharing', making access to the required information and knowledge is the key to innovation and empowerment of farming community. Rapid technological developments over the years, particularly internet and mobile, have made it possible to provide array of services on a single platform due to convergence. It is a unique opportunity to leverage upon the available ICT infrastructures and deliver the knowledge based services effectively to people. There are number of ICT initiatives being attempted in the field of Agriculture and other sectors. Some of the successful models (web-based and mobile based) have shown way for the future of e-extension in India.

2. Global Trend in ICT

The global trend in Information Communication Technology is changing rapidly, owing to new innovations and affordability of these technologies by common man. During the first decade of ICT revolution (1986-1995), it was regarded that *'Information is Wealth'* followed by *'Knowledge is Power'* during 1996-2005. But, in the ongoing era of ICT (Year 2006 onwards), *'Knowledge Sharing'* is considered as Wealth & Power. ICT facilitates Knowledge Sharing very effectively. The best example could be the use of Mobile phones and Social Media in India.

3. Evolution of Agricultural Extension System in India

Knowledge Management played a key role in Agricultural Extension System in India over the period. Though the traditional ICT resources like Radio and Television, are playing considerable role in this sector, it is fastly getting replaced with modern ICT resources like web and mobile technologies. Evolution of modern ICT over the period is schematically represented below.

Primary-Face to Face

- Friends / Relatives
- Progressive Farmers
- Extension workers / T&V
- Input dealers

Traditional ICT sources

- Television
- Radio
- Newspapers

Initial Modern ICT (India)

- Village Knowledge Centres (1996)
- Telecentres / CSCs (2004-2007)
- Kisan Call Centers
- ITC e-chaupal (1999)

New Modern ICT (India)

- Voice based services- IKSL (2007)
- SMS based services RML (2007)
- Video based Digital Green (2009)

Evolving technologies

- Spoken web
- Ubiquitous computing
- Augmented reality
- Social web

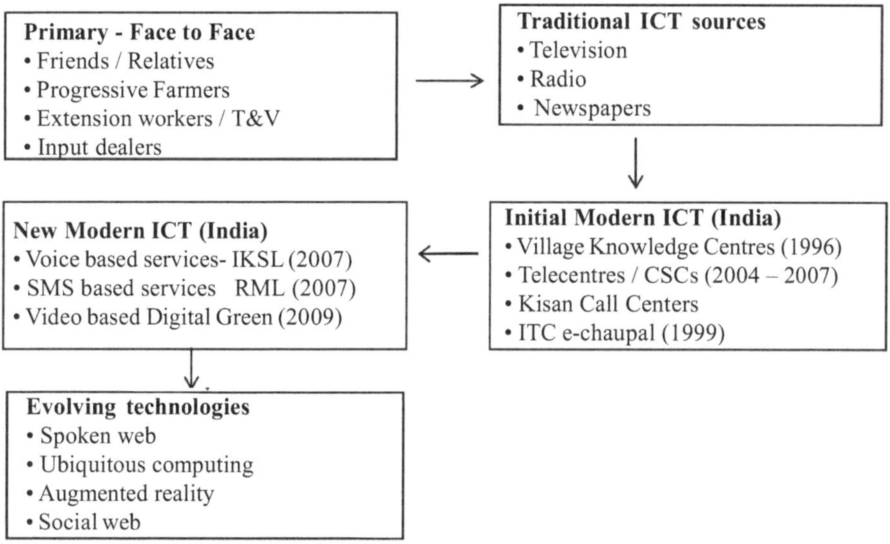

Fig. 1: Evolution of technology enabled Agricultural Extension system in India

Role of modern ICT

- Facilitates to bridge the information gap and asymmetry prevailing in extension system (Farmers: Extension Workers – 1700:1)
- Complement the existing extension services and information services (e.g. Radio, TV, Newspaper etc...)
- Reduces the time lag in adoption & affordability by farmers
- Influence the behaviour pattern of users (more personalised advisories possible)

4. Web-based Services in Agriculture & Allied sectors

Providing authentic content, which are region specific and in local languages, are key factors to be considered in designing any e-extension programmes. In India, most of the websites (80%), particularly Government websites, are available only in English and about 20% of the websites host bilingual content (Hindi/regional language). These websites are largely institute websites that have a greater focus on organization related aspects. The only predominantly available user centric information is that of policies and schemes of that particular institution. Limited scope for the users to share their experiences and knowledge with others and interact with experts or peers.

Some of the successful attempts in Agriculture like, Farmers' Portal, AGMARKNET, AGRISNET, Raita Mitra-Karnataka, Kissan Kerala, TNAU Agri-Tech Portal, Rice Knowledge Management Portal, Agropedia, Viakspedia, aAQUA etc. are attempting to address the issue of providing credible and timely knowledge to the farmers in regional languages.

Vikaspedia Portal

There was a felt need to develop an interactive multi-lingual knowledge platform to harness the strength of collective wisdom for development, particularly covering the key development oriented sectors including Agriculture. As an outcome, Department of Electronics & Information Technology, Ministry of Communications and Information Technology, Government of India started an initiative namely, India Development Gateway (InDG) in the year 2006 and is being executed by Hyderabad unit of Centre for Development of Advanced Computing (C-DAC), a scientific society under Ministry of Communications and IT, Government of India. As part of InDG activities, a multilingual knowledge portal 'Vikaspedia' has been developed and launched for the public use.

The multilingual portal www.vikaspedia.in is aimed at creating a versatile collective knowledge repository with a specific focus on domains relevant to social and economic development. It will serve as a collaborative content creation, sharing and utilization platform for a rainbow of stakeholders – NGOs, Government, Community based organizations, knowledge networks, CSRs, spread across the country. Vikaspedia is presently available in ten languages – Hindi, Assamese, Marathi, Bengali, Gujarati, Tamil, Telugu, Kannada, Malayalam and English. It is to be made available in 22 Indian official languages in phases in the near future. Information services related to key livelihood sectors (initially the six sectors of Agriculture, Education, Health, Social welfare, Energy and e-Governance) are currently available in the portal.

Vikaspedia portal currently hosts:

- Information/Content/Knowledge in regional languages (Success stories, technologies, best practices, Government schemes etc.).
- Interactive platforms (Discussion forums, polls, comments etc.).
- Multimedia products & e-Learning Courses in identified sectors.
- Value Added Services (Ask an Expert, e-Vyapar, Dynamic Market Information etc.).
- Mobile Apps (Krishi Vigyan Kendra Search, Ask An Expert, CSC finder etc.).

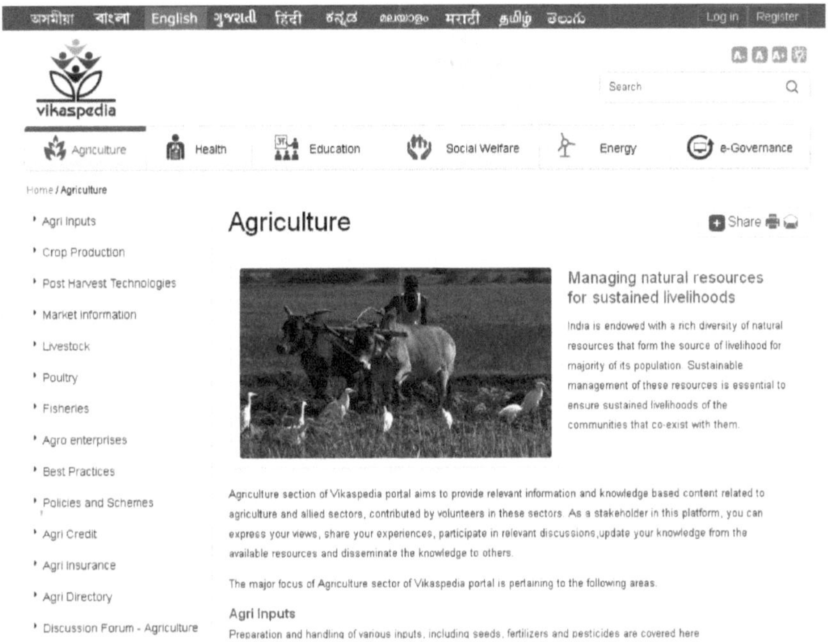

Fig. 1: Agricultural Home Page in Vikaspedia Portal

5. Mobile-based Services in Agriculture & Allied sectors

Mobile technology have brought revolutionary change in India for the past 5 years and also thrown open basket of options to integrate with our ongoing knowledge management process being adopted in Agriculture. As on 31st August 2014 (TRAI Report), there are 924 million mobile subscribers in India, with an overall teledensity of 74.2. About 46% of rural population has access to internet in their mobile phones. Competition among telecom service providers in India have led to cost reduction and affordability of this technology among the common man.

Many stakeholders in Agriculture have already made attempt to provide mobile based services to farmers. Initiatives like IKSL (IFFCO Kisan Sanchar Limited), RML (Reuters Market Light), mKRISHI, Fisher Friend, Annapurna, Kisan Call Centre etc have made considerable impact in the farming sector and being adopted by farmers in large scale.

6. Present Challenges

1. Lack of capacity building / building trust on ICT.
2. Local language support in low-end mobiles.

3. Lack of convergence among organisations.
4. Lack of appropriate content for mobile platform.
5. Lack of proper policy for implementation.
6. Service affordability by farmers.
7. Sustainability & scaling-up issues in Government organisations.

7. Opportunities

1. Support from government in the form of ICT policy.
2. Increased penetration of mobile networks.
3. Fast growing smart phones &Mobile based services.
4. Growing broadband connectivity.
5. Competition among Telecom Service Providers leading cost reduction.
6. Affordability in India.

8. Conclusion

The issues prevailing in the present Agricultural Extension system in India, i.e, reaching-out timely information to farmers at an affordable cost, could be addressed through effective use of ICT. Integrating the collective wisdom under one platform, overcoming the language barrier and providing demand based information and services for the farmers and customised knowledge products for the extension workers, is a major challenge. But, affordability of mobile phones, broadband connectivity, continued support from Government and availability of basket of services has opened-up many opportunities today. Convergence among stakeholders, capacity building and making availability of customised content, are the key factors for the success of any knowledge management initiative in Agriculture.

Extension Management Strategies for Sustainable Agriculture, pp 59-67

New India Publishing Agency, New Delhi, India

6

Gender Inclusive Strategies for Sustainable Development

K.R. Murugan

"Women's empowerment and gender equality are fundamental to FAO's vision of a world free of hunger and malnutrition"

– Jacques Diouf, Director-General, FAO

Agriculture is vital to the development of any nation, India being no exception. Development must include the youth and therefore the government seeks to encourage their participation in the agricultural sector. This effort seeks to change the negative perception the youth have of participation in agriculture as uneducated, unskilled, physical labourers with extremely low economic return. Modern agriculture is more than tilling the soil and animals. The sector today offers career opportunities in research, environment, financial management, engineering and other technical areas for the youth to explore. The introduction of the Youth in Agriculture is necessary and important to facilitate food and nutrition security for the sustainable development of the nation.

Young people are a major human resource for development and they are the key agents for social change and dynamic force for economic development and technological innovation. Particularly the rural youth are the future of the agricultural sector. With a growing world population and a decreasing agricultural productivity in combination with a rural migration, sustainable food production and supply are endangered. Young farmers play an important role in ensuring food security for future generations but they face many challenges.

Although agriculture now accounts for only 14 per cent of Gross Domestic Product (GDP), it is still the main source of livelihood for the majority of the rural population. As such rapid growth of agriculture is critical for inclusiveness. Important structural changes are taking place within the sector and there are definite signs of improved performance. Agricultural growth has accelerated compared to the Tenth Plan and diversification. The National Sample Survey Organisation (NSSO) data brings out that rural labourers are shifting to non-agricultural work, tightening the labour market in agriculture and putting pressure on farming. However, hope on agriculture remains unchanged among the rural self-employed whose average farm size continues to decline with population growth. But educated young members are less likely to want to stay in farming. The viability of farm enterprise, mostly small farms, must therefore be a special area of Plan focus by the Planning Commission in the Plan. The Twelfth Plan must also focus on other priorities such as resource-use efficiency and technology to ensure sustainability of natural resources, adaptation to climate change and improvements in total factor productivity.

Gender and sustainable development

Gender refers to the set of characteristics, roles and behaviours patterns that distinguish women from men which are not constructed biologically but socially and culturally. Gender is not determined biologically, because of sexual characteristics of either women or men, but is constructed socially. It is a central organizing principle of societies, often governing the processes of production and reproduction, and consumption and distribution.

Gender roles are the social definition of women and men, and vary among different societies and cultures, classes and ages, and during different periods in history. Gender-specific roles and responsibilities are often conditioned by the following:

- Household structure
- Accessible resources
- Specific impacts of the global economy
- Other locally relevant factors such as ecological conditions (FAO, 1997).

Almost everywhere, gender roles function in a way that discriminates against women, in their choices in life, their access to assets, and the voice they have in making decisions. This discrimination is not only reflected in individual relationships, but it also permeates social institutions.

Gender approach highlights the gender gaps between men and women in terms of employment, earnings, and access to control over and benefits from human and physical assets. It seeks to explain how these gaps affect power relations between women and men and thus their relative ability to influence decisions within their households and communities.

Gender inequalities result in lesser food and income, and higher levels of poverty and food insecurity. The following four sets of factors and theirinteractions with policies and institutions (at the global, national and local levels) affect rural livelihoods (FAO, IFAD and World Bank, 2008).

Assets

- Gender asymmetries in access to and control over assets (social, physical, financial, natural and human capitals).

Markets

- Gender asymmetries in participation and power in land, labour, financial and product markets.
- Gender differentiated distribution of risks and gains along value chains.

Information and Organization

- Gender asymmetries in market information, extension services, skills and training.
- Gender asymmetries in participation and leadership in rural organizations.
- Gender asymmetries in rights, empowerment and political voice.

Risk and Vulnerability

- Gender asymmetries in household compositions and labour availability (dependency ratio, migration and disability).
- Physical and agro-ecological risks and their gender differentiated impacts.
- Gender-responsive social protection measures.

Gender Mainstreaming and Gender Equality

Gender mainstreaming is the recognition of gender-based divisions in labour, rights, resources and voice. It is not about adding a woman's component or even a gender equality component into activities or projects. Mainstreaming a gender perspective "*is the process of assessing the implications for women and men of any planned action, including legislation, policies or programmes, in any area and at all levels. It is a strategy for making the concerns and experiences of women as well as of men an integral part of the design, implementation, monitoring and evaluation of policies and programmes in all political, economic and societal spheres, so that women and men benefit equally, and inequality is not perpetuated. The ultimate goal of mainstreaming is to achieve gender equality"* (ECOSOC, 1997).

Agriculture and food security are characterized by gendered dimensions in that women play a key role in agricultural production, food processing and marketing. They play a decisive role in dietary diversity and are responsible for nutrition in the home. In addition, women are involved in the production and domestication of plants and animals; they are knowledgeable in seed selection and vegetative propagation; they understand how plants and animals grow and reproduce; and they plant trees. Women comprise 20 to 50 percent of the agricultural labour force in developing countries (FAO 2011b) and 79 percent of women in least developed countries who are economically active report that agriculture is their primary economic activity (Doss 2011).

Gender is the word used especially for the female population of the society. Women's contribution to agriculture in the subsistence farming or commercial agriculture, when measured in number of responsibilities performed by women is greater than men. Their contribution in agriculture is aptly highlighted by a micro study conducted in Indian Himalayas which found that on a one hectare farm, a pair of bullocks work for 1060 hours, a man for 1212 hours and a woman for 3485 hours. Generally, operations performed by men are those that entail use of machinery and animals. Contrary to this, women always rely on manual labour using only their own energy. Not only are women overworked, their work is more arduous than that undertaken by men. Further, since women's work is largely based on human energy it is considered unskilled and hence less productive. On this basis, women are invariably paid less wage despite their working harder and for longer hours. This makes the woman's work invisible.

Gender Sensitization in Agricultural Extension Activities

The need for incorporating gender concerns in agricultural extension became apparent during the early 80s. This came about, as there has been an increasing realization that:

- Women play a major role in agriculture and agriculture continues to remain as an important source of livelihood for women;
- Women lack access to extension services as extension programmes rarely identify women as an integral part of its target audience;
- Extension needs of women and men vary considerably as often they are involved in different activities;
- Men and women have differential access to assets, information, markets, credits and other services; and

- Extension personnel are often unaware of these "gender issues". So, there is a need to sensitize them through training in "gender analysis" and "gender-sensitive agricultural planning" methods.

The most common approaches adopted to bring about greater gender sensitization in extension organizations are as follows:

- Organize training programmes for staff on gender difference in agriculture and gender analysis tools.
- Employ more women extension staff, preferably female para-extension agents to work closely with women groups.
- Target women as clients for extension and organize separate programmes (trainings, demonstrations, exposure visits) for them.
- Form groups of women farmers (common interest groups or thrift-and-credit groups); and provide them with the micro-capital assistance for enterprise development.

Trends in Gender Access to Agricultural Information and Technology

An agricultural innovation system is "a network of organizations, enterprises, and individuals focused on bringing new products, new processes, and new forms of organization into economic use, together with the institutions and policies that affect their behavior and performance. The innovation systems concept embraces not only the science suppliers but the totality and interaction of actors involved in innovation as well. It extends beyond the creation of knowledge to encompass the factors affecting demand for and use of knowledge in novel and useful ways."

Gender and Knowledge Systems

- Both women and men manage sectors of complex smallholder production systems.
- When gender is ignored, there is a cost to people's well-being and sustainable growth.
- Knowledge is not transferred; it is generated and exchanged in a continuous learning process.
- Farmers, agricultural educators, researchers, extensionists, and traders form parts of knowledge and information networks.

- Rules and mechanisms governing the way different actors, organizations, enterprises, and groups interact to supply and demand knowledge and technology are critical for equitable development.

Significant Role of Women in Agriculture

Society is the base for overall development of a nation. And the society's structure depends entirely upon the status and condition of its women. The condition of women was far better in ancient India. In early period Aryans treated men and women alike in all religious rituals and social functions. But condition of women has declined gradually.

Access to Resources

Despite their role as a backbone of food production and provision for family consumption in developing countries, women remain limited in their access to critical resources and services due to cultural, traditional and sociological factors.

• Access to land

Not even 2% of the land is owned by women, while the proportion of female heads continues to grow. Land reforms programme, together with break-up of communal land holding have led to the transfer of exclusive land rights to males as heads of the household which ignores both the existence of the female headed household and the rights of married women to a joint share.

• Access to agricultural inputs

Women access to technological inputs such as improved seeds, fertilizers and pesticides is limited as they are frequently not reached by extension services. In addition they often lack cash needed to purchase inputs even when they are subsidized.

• Access to education, training and extension services

Two- third of one billion illiterates in the world are women and girls. Available figures show that only 5% of extension services have been addressed to rural women, while no more than 15% of the world's extension agents are women.

• Access to research and appropriate technologies

Women farmers' roles and needs are often ignored when devising technology that may cause labour displacement or decrease workload.

Developmental Initiatives

Efforts have been made by department of agriculture to incorporate gender issues into development agenda and ensure full and equitable participation of women in all agricultural development programmes. In an effort to engender agricultural process, special programmes/ schemes should be implemented for women in agriculture sector, where a composite package of training, extension and input, managerial and entrepreneurial support was provided to selected farm women beneficiaries. Also different approaches should be followed to provide a complete support package to women farmers. The technical back-stopping was provided through a blend of skilled-based capacity building exercises coupled with extension support through on field Demonstration and Study Tours.

Women and Green Revolution

The green revolution of 1960 and 1970s with its package of improved seeds, farm technology, better irrigation and chemical fertilizers was highly successful in meeting its primary objective of increasing crop yields and aggregate food supply. Yet despite its success in increasing aggregate food supply, the green revolution, did not necessarily translate its benefits to the lower strata of rural poor including women. So a gender friendly second green revolution is to be ushered in.

Family Farming: Women and Youth

To cultivate the next generation of family farmers, we must invest in women and youth. The stereotype of the young male farmer has given way to the reality of an aging, female-fronted farming force. In the developing world, women make up 43 percent of the agricultural labor– and, in some countries, they make up 80 percent of agricultural labor. However, in some South American and European countries, rural young women are leaving the countryside in search of employment in urban centers leading to the masculinization of farming in these regions.

Women face many obstacles to productive farming. Compared to men, they have limited access to credit and lack control of family funds. In fact, the U.N. Food and Agriculture Organization (FAO) found that only 10 percent of credit in sub-Saharan Africa is available to women. Female farmers also face scant educational opportunities and gender discrimination at markets. Without available funds and proper training, women are unable to make improvements to their farming methods and continue to suffer from low food yields. Melanne Verveer, Ambassador-at-Large for Global Women's Issues, FAO, emphasizes, "In many developing countries, women are the backbone of the economy. Yet women farmers do not have equal access to resources and this significantly limits their potential in enhancing productivity." But encouraging experiences can be found, such in Brazil, where women have joint title to land, as part of innovative land reform policies, giving them equal ownership of land with their male partners, whether they are married or not. According to FAO, providing female farmers access to the same resources as men could reduce the number of hungry people in the world by 100-150 million people. And when women earn more and have better rights, they tend to invest more in the health and nutrition of their families.

Gender Inclusive Strategies for Sustainable Development

- Full and equal rights should be given to women to own land and other properties.
- Ensuring that, women's unpaid work and contribution to on-farm and off-farm production should be recorded in national account.
- Making appropriate technologies available to rural youth and women which focus on utilizing locally available materials.
- Revising legislation which guarantees equal pay for work of equal value, improving working conditions and enforcing legal standards.
- Promoting and increase in number of extension functionaries and agricultural researchers for women and youth.

Conclusion

Today's youth are tomorrow's family farmers, and maintaining interest in farming as a profession is vital to future food security. It is our duty and responsibility to cultivate the next generation of agricultural leaders not just farmers, but food entrepreneurs, scientists, agronomists, extension agents, union and government leaders. Governments, particularly in developing countries, but also elsewhere need to invest in gender specific policies and practices that provide access to land, credit and banking services, education and knowledge, and technical skills for young farmers. And these governments need to ensure that young people have access to markets, goods and service, employment opportunities and leisure to reside young people in the farm.

References

http://mofa.gov.gh/site/?page_id=1173

http://www.youthkiawaaz.com/2012/07/feminism-in-agriculture-a-study-on-gender-bias-in-indian-agriculture/

http://www.ifad.org/farmer/2012/youth/report.pdf

http://www.fao.org/fsnforum/forum/discussions/family-farming

ftp://ftp.fao.org/docrep/fao/011/aj288e/aj288e04.pdf
http://www.acdi-cida.gc.ca/acdi-cida/ACDI-CIDA.nsf/eng/ANN-612105312-KGB
http://www.reachingruralwomen.org/ifadandgender.htm

Extension Management Strategies for Sustainable Agriculture, pp 69-79

New India Publishing Agency, New Delhi, India

7

An Innovative Taxonomy of Farmers' Practices for Inclusive Development in Sugarcane Production System

C. Karpagam, P. Murali, T. Rajula Shanthy, D. Puthira Prathap and V. Venkatasubramaniam

Throughout the history of Indian agriculture, the contribution of farmers to agricultural technology development and dissemination has been very significant. Farmers' practices are useful not only for scientific technology development, but also for sustainable agricultural development. Rural economy will sustain only when it effectively utilise the available local resources. But recent agricultural development witnessed that those modern agricultural practices slowly replacing the farmers' practices. In this juncture the basic challenge for sustainable agriculture is to make better use of these internal resources. This can be done by minimizing the external inputs used, by regenerating internal resources more effectively or by combinations of both (Rolling and Pretty, 1997).

Shri. L. Narayan Reddy, a progressive farmer from Karnataka reported in LEISA, 2006 that "Documentation in agriculture, mostly in the form of research papers, is too elaborate with very little practical or result oriented information. Similarly, publications on package of practices published by the Department of Agricultural Extension and the Agricultural Universities are focused on high cost chemical external inputs. These are not only expensive, but also destructive to soil health, environment and crop yields. More documentation on low cost, organic and local available inputs, with the experience of successful farmers is necessary". It indicates the importance of documenting low and no cost farmers' practices in the current scenario.

Farmers practices cannot be merely considered as traditional or indigenous practices. Some practices may be very innovative or some may be entrepreneurial initiative or some may be unique practices. Review shows that there is always an overlapping issue while classifying farmers' practices under traditional practices, innovative practices etc. Hence, there is a need for classifying the farmers' practices under clear cut heads in extension research. In this regard, the present study has documented different farmers' practices in SPS and classified the practices into five different heads viz., Innovative, Traditional, Entrepreneurial, Unique and Success stories and discussed the criteria for each head.

Methodology

Among 32 districts of Tamil Nadu, 22 districts viz., Tiruneveli, Virudhunagar, Theni, Madurai, Ramanathapuram, Sivagangai, Pudukkottai, Dindigul, Coimbatore, Tiruppur, Trichy, Thanjavur, Ariyalur, Perambalur, Karur, Erode, Namakkal, Villupuram, Cuddalore, Tiruvannamali, Dharmapuri and Vellore were selected based on the predominance of sugarcane cultivation. Ten villages were selected from each district, thus, totally 230 villages have been surveyed throughout Tamil Nadu. But identifying the resourceful farmers is a difficult task felt by the research team. Hence, there were several Focused Group Discussions (FDGs) were arranged with sugar factories personnel and farmers to identify the resourceful farmers who practices any unique practices in sugarcane production system. By using the deliberations, individual farmers were identified and the farmers were interviewed in their respective farms with the help of an interview schedule.Documentation of farmer's practices was done by questionnaires, capturing photographs or by audio and video means and by different PRA tools. After collecting the information from the farmers, the each case was analysed with its scope and potential, further the practices were classified under five heads viz., Innovative, Traditional, Entrepreneurial, Unique and Success stories.

Focused Group Discussions with Sugar factory officials in the respective districts

↓

Village-wise farmers list for the concerned sugar factory area was prepared

↓

Focused Group Discussions were conducted with farmers in the respective villages

↓

List of farmers who practiced unique practices were identified

↓

Each identified farmers were interviewed at their field

↓

Details of the farmer's practices are documented by audio and video gadgets

↓

Each practices were anlysed and categorised under five different heads

Methodology adopted for documentation of the farmers' practices

Results and Discussion

Based on the extensive survey, there were 40 case studies were documented during the study. The documented cases were further categorised under five heads based on the suitability. The cases were have its own scope, potential, hence, one case under each head have been discussed with its' background, description and cost benefit analysis.

Head. I. Innovative practices in sugarcane production system

Innovative practice refers to the process through which individuals or groups within a given locality discover or develop and apply improved ways of managing the available resources – building on and expanding the boundaries of their knowledge in sugarcane production system. Farmers' innovation is something new to a particular locality, but not necessarily new to the world (Reij and Waters-Bayer, 2001).

Among the documented practices, which are all accomplish the above said criteria are classified under innovative practices. Thus, the study documented six innovative practices in SPS viz., 'innovative mechanisation of cattle waste management', 'modified off-barring practice','coir pith induced pre-sprouting in bud chip settling','innovative organic cultivation','innovative devices for wild boar management' and 'modified tractor mounted ridger'. Among the practices, one practice is discussed in this article.

Innovative mechanisation of cattle waste management

Cow dung is considered and used from time immemorial as the best natural fertilizer. Due to dwindling cattle population and labour scarcity, use of cow dung as a manure has become a very rare phenomena in modern day agriculture. But still cow dung and cow urine application is a predominant practice among the farmers who practice organic agriculture. Shri Sakthivel from Ganeshpuram village of Erode district practices organic sugarcane cultivation for more than a decade. He has also adopted the drip fertigation in his farm to reap the benefits of the modern and eco-friendly technology. He has enough number of cattle in his farm but cattle waste management was a big challenge for him. Collection and application of cow dung became more labour and time intensive. Mechanization of cattle waste management was thought of as one of the alternatives to overcome the problem. As he has already adopted the drip irrigation system, he mixes the cow dung and cow urine along with the drip irrigation water. But his efforts often resulted in clogged drip system due to the solid materials present in the cow dung. Over a period of time, he had evolved a technique to mix cow dung and cow urine, filtered it and fertigated through the drip system. Finally, he had designed a permanent four chamber cattle waste processing unit to prepare liquid manure for drip fertigation as well as bio-gas for house hold use.

This process is consists of four stages viz., mixing, filtering, purifying and fermentation. Cow dung and cow urine was collected into the first chamber by using a water blowing device. Once the required quantity is collected, a mixing device churns this mixture thoroughly. It helps to separate the solid and liquid portions of the mixture. The solid portion exiting from the first chamber is collected in the bio-gas plant and the liquid portion is passed through an outlet to the second chamber for further filtration.

The second chamber is a filtration bed in which layers of pebbles and sand are arranged one over the other separated by a net. After filtration in the second chamber, the liquid portion is passed through a pipe into the third chamber for purification and fermentation by using jaggery. After fermentation, the solid materials settle down and the clear liquid portion is passed through an outlet pipe into the fourth chamber, which is a 4000 litres capacity below ground sump. The drip fertigation unit is installed near this chamber and a venturi is placed in the chamber. The suction end of the venturi is placed at few inches above the bottom floor of the sump. The venture sucks the cow dung liquid and mix with irrigation water at fertigation inlet point. For every fertigation, the stored cow dung liquid manure is applied @ 300 litres per care.

The cost incurred for designing the structure was around Rs. 40,000 as one time investment. The savings in labourer cost and fertilizer cost was around Rs.10,000 per season. This innovative technology is being promoted as an effective technology for using drip fertigation in organic sugarcane production

Fig. 1: Full view of the mechanised cattle waste management system

system. It inspired the other organic sugarcane farmers to a greater extent. This innovative mechanised cattle waste management technology has been adopted by 10 farmers in the region and also farmers from Tirunelveli, Dindigul, Dharmapuri and Madurai are also approaching the innovator for the know-how of the technology.

Head. II. Indigenous practices in sugarcane production system

The term 'indigenous knowledge' is used synonymously with traditional, local and indigenous technical knowledge to differentiate the knowledge developed by a given community from the international knowledge system. According to International Institute for Rural Reconstruction (IIRR, 1996). Indigenous knowledge (IK) is defined as the knowledge that the people in a given community have developed overtime and contribute to development.Indigenous knowledge is not confined to tribal groups or the original inhabitants of an area. It is not even confined to rural people. Rather any community possessing indigenous knowledge is rural and urban, settled and nomadic, original inhabitants and migrants.

Based on the above said criteria the study documented nine indigenous practices in SPS viz., 'modified *panchagavya*', 'fish ensilage and lime egg solution',' mixed grain as green manure, 'locally followed planting methods','mud pot trap for white grub','plant based pest repellents', 'low cost practice for white grub','traditional practices for rat management' and 'traditional practices for wild animal menace'. Among the practices, one practice is discussed in this article.

Mud pot trap for white grub management

White grub (*Holotricha serrata*) is an important sub-terranean pest of sugarcane in tropical India. The best way to control white grub is when the adult comes above ground and congregate in Neem trees. Farmers have to collect the beetles on the very day of the receipt of summer showers and continue the collection for a week. This is the common recommendation for management of white grub in sugarcane. But, a farmer from Thalavadi village of Erode district has been following an indigenous practice to control white grub in sugarcane.

Castor seed based mud trap was used to control white grub. One kg of crushed castor seed (*Ricinus communis*) is mixed with four litres of water. The extract is kept in a five litre mud pot. The pots are buried in the soil to brim level earlier to the summer showers. Five such pots are buried per acre in the four corners of the field and one in the middle of the field and it serves as a trap to attract white grub beetle. The farmer reported that the odour emanating from the extract attracts white grub beetles and they are trapped into the pot. Total cost of the trap per acre is about Rs 325. (Five kg castor seeds cost around Rs. 125 and five pots cost around Rs. 200). It is evident from the field observation that this trap is attracting different insects in the sugarcane field, but the particular *Holotricha* spis a seasonal bound insect, so the extent of management of white grub by the trap has to be confirmed based on the field trails during the peak season when the adult activity is more.

Fig. 2: Mud pot trap at farmer's field

Head. III. Entrepreneurship initiatives in sugarcane production system

Entrepreneurship initiative refers to the act and art of being an entrepreneur or one who undertakes innovationsor introducing new things, finance and business acumen in an effort to transform innovations into economic goods. In this study, the most obvious form of entrepreneurship is that of starting businesses ventures related to sugarcane cultivation and allied sectors.

Among the documented practices, which are all accomplish the above said criteria are classified under entrepreneurship initiatives. Thus, the study documented seven entrepreneurial initiatives in SPS viz.,'Organic jaggery powder', 'bud chip settling production, 'trash based vermi-compost production, 'Agricultural Service Provider', 'Tricho card production', 'two wheeler mounted sprayer,'tractor mounted sprayer'. Among them one has been discussed in this article.

Organic jaggery powder – a value added business venture

Jaggery is a pure, traditional, unrefined form of sweeteners. Jaggery production in India is about 7-10 mt/year while its per capita consumption is about 5 kg.

Compared to sugar, jaggery gives good return to the farmers if the market price is higher. In this situation, taking up jaggery production as a small scale business will be a profitable. Further, profit may be many fold if it is produced organically or value added. By realizing the fact, a farmer from Puliangudi, Tirunelveli district is producing organic jaggery powder and selling in local markets as well as in export market and earning a good profit out of it.

Normally jaggery can be produced in different shapes. Shri Anthonisamy from Puliangudi of Sivagiri Taluk, Tirunelveli district thought of producing jaggery powder. He cultivates Co 86032 sugarcane variety from the time of its release and practicing organic farming. Organically grown sugarcane is crushed and juice is extracted. The extracted juice is filtered and boiled without adding any chemicals. After removing the scum, the juice is further condensed till it becomes semi-soild and removed from the furnace. The semi solid paste is cooled in a wide pan and scraped using wooden ladle to prepare jaggery powder. It has got greater demand in local area as well as in nearby states besides fetching higher prices. Further the farmer reported that the organic jaggery powder has good taste and more sweetness than others. He sells one kg jaggery @ Rs. 75, whereas, the normal rate of one kg jaggery in the commercial market is around Rs. 50. It was also know from him that the production cost of one kg jaggery powder including the cane cost is around Rs.30.

Fig. 3: Organic sugarcane field

Fig. 4: Organic jaggery powder

Head. IV. Unique practices in sugarcane production system

Unique practices refers to a practice/method/technique that has consistently shown results superior to those achieved with other means and the practice is being popular in particular locality by its special characteristics. Based on the above said criteria the study documented four such unique practices in SPS viz., betel vine followed by sugarcane planting with minimum tillage, paddy followed by sugarcane planting with minimum tillage, amla to improve the quality of bore well irrigation water and pheromone trap for early shoot borer in sugarcane. Among them one practice has been discussed in this article.

Pheromone trap for early shoot borer in sugarcane

Early shoot borer (ESB) is one of the major problems in sugarcane cultivation. Shoot borerlarva enters the plant laterally by making entry holes at the base of the stalk and bores downwards or upwards killing the growing point. This would sever the central leaf spindle which dries up forming a deadheart. Insecticide is the only option available with the farmers to manage early shoot borer. For Inter Node Borer, pheromone trap is the next option available with farmers. But for ESB, tying pheromone trap is not possible since crop is in too early stage (ESB damage the crop within 30-40 days). In this condition, a unique practice 'soil mounted pheromone trap' was documented during the field survey.

Shri Balamurugan from Periyakuchipalayam of Villupuram district practices soil mounted trap to manage Early Shoot Borer and got succeeded. In one acre sugarcane field, eight pits each with 1.5 feet depth and two feet width were dug and pit surfaces were lined with polythene cover. In each pit, water was added up to one foot height and kerosene/diesel/detergent was added along with water. To place the lure, a rod was kept across the pit (Fig. 5). The ESB moths were attracted by lure and got into kerosene in the pit. Thus the insect population can be controlled. For ESB, if the lure has been placed in a stick and kept in the field, the chances of falling down is more but the soil mounted trap is permanent one and it is low cost technology and all the farmers can easily design the trap.

Fig. 5: Soil mounted pheromone trap to manage ESB

Head. V. Success stories in sugarcane production system

Success stories refer to the account of achievement of success, fortune, fame or profit by adopting particular practices in sugarcane production system. The prime objective of documenting success stories is to draw key lessons and challenges. The successes stories also bring out the best practices adopted by the farmers. So, documenting the success stories will pave the way to upscale the best practices which are popular in one area to the other area where the practices are not familiar.

Among the documented practices, which are all accomplish the above said criteria are classified under success stories. Thus, the study documented seven success stories in SPS viz., 'wider row conversion', 'groundnut as intercrop,' tomato as intercrop', 'vegetable cowpea as intercrop', 'onion as intercrop' and ' raised bed nursery with single budded cuttings'. Among them one practice has been discussed in this article.

Groundnut as Intercrop in Sugarcane

Shri P. Manoharan from the Nacchipattu village of Thiruvannamalai district cultivates sugarcane variety CoSi 8 with intercrop groundnut in his half an acre land. After ten days of groundnut sowing, farmer takes up cane planting. Single bud cuttings are planted with 5 feet (150 cm) row spacing. In five feet row spacing, four rows of groundnut could be accommodated with row spacing of 30 cm. Groundnut matures within 3.5 months and it is harvested when sugarcane is in initial stage of development.

Initially land preparation is done by one disc plough and four normal ploughs. After land preparation, groundnut seeds are sown @ 20 kg per 0.5 acre. Land preparation and seed cost is around Rs.3000. For groundnut, 5 kg urea, 10 kg super phosphate and 15 kg potash with FYM are applied. Thus, the manuring expenditure is Rs. 1000 and for intercultural operations the farmer spent Rs.1500. After 3.5 months, groundnut has been harvested and harvesting cost is Rs. 1500. Thus, the total cost of cultivation for groundnut alone is Rs.7000/half an acre. Groundnut yield from half an acre is 10 bags (One bag = 60 kg). If the market price is Rs. 2000/bag, the gross income is Rs. 20,000 from groundnut cultivation. Thus, the net profit is around Rs. 13,000 from 0.5 acre for 3.5 months. It may be the additional profit, since, he utilizes the space effectively.

Scope and Utility

- As a leguminous crop, groundnut fixes atmospheric nitrogen and improves soil fertility.
- Due to intercrop, the main crop population and yield was not compromised.
- Farmer got additional profit of Rs. 13,000 from the same piece of land (0.5 acre) where the sugarcane was cultivated.
- Within 3.5 months, groundnut was harvested, so the other intercultural operations for sugarcane were not disturbed.

Fig. 6: Sugarcane intercrop with groundnut

Fig. 7: Interaction with farmer

Conclusion

This article gives insight about documentation and classification of farmers' practices in sugarcane production system. The farmers in one area may not know and / or are not familiar with the practices being followed in another area. The benefit of these best practices is localised and does not reach all those who need it. An authentic documentation of such practices by extensionists for dissemination to others is the need of the hour. By virtue of promoting modern technologies the farmer's practices are gradually disappearing. With the passage of time, unless a concise effort is made to document the valuable practices, it will be lost forever and cannot be regained in future at any cost. Hence compilation of farmers' practices could be a significant contribution to the history of agriculture.

This study is maiden attempt in sugarcane farming to classify the farmers' practices into different heads. The same procedure may adopt for other crop as well. If needed, further categories can be added based on the cropping system requirement. Different farmers' practices and initiatives will open up avenues for social scientists to promote community based ventures and to develop modern technology in sugarcane based cropping system. The study revealed that all practices are utilised the local resources and useful for other farmers and for the society as whole. Several societal problems in sugarcane cultivation viz. labour scarcity (mechanised cattle waste management), stakeholder conflict management (Agricultural Service Provider), human health deterioration (Organic jaggery powder), ill effects of pesticides (pheromone trap for ESB) were addressed by the farmers' practices. Therefore, Government and other institutional support are needed for further out-scaling of the initiatives for larger benefit of the society. The study also found that the grassroot level farmers' practices has lesser investment, greater utility, more social benefits (localised benefits are more) and higher sustainability compared to modern high investment ventures. When these initiatives are recognised by the institution and government then it lead for holistic rural development through sugarcane cultivation.

Acknowledgement

I express my sincere thanks to all seven respondents for their valuable inputs for this study and I thank my co-authors for their kind and moral support for writing this article. My sincere thanks to Director, Sugarcane Breeding Institute for given permission to take up the challenging task and to write this article.

References

Narayanreddy, L (2006). Doumenting farmers' knowledge. *LEISA India*. March 2006. p. 50.

Reij, C and A. Waters-Bayer (2001). Entering research and development in land husbandry through farmer innovation. In: Reij, C. and Waters-Bayer, A. (Eds.), Farmer Innovation in Southern Africa: A Source of Inspiration for Agricultural Development.London/Sterling, VA: Earthscan. p. 3-22.

Rolling, N and Pretty Jules, N (1997). Extension's role in sustainable agricultural development. In: Swanson, B. E., Bentz, R.P and Sofranko, A. J. (Eds.), *Improving Agricultural Extension – A reference manual,* Rome, FAO. p.181-191.

Extension Management Strategies for Sustainable Agriculture, pp 81-85
© 2016, Editors, T. Rathakrishnan, H. Philip, P.P. Murgan and Jagadeesan
New India Publishing Agency, New Delhi, India

8

Effective & Efficient Technology Delivery Mechanism – Experiences in TN IAMWARM Project

Senthilkumar M., Suganthi, A., Pandiyan M. and B.J. Pandian

Tamil Nadu Irrigated Agriculture Modernisation and Water Bodies Restoration and Management project (TN IAMWARM) is a multidisciplinary project funded by World Bank. It is being implemented by WRO, PWD as nodal agency in 63 sub basins along with line departments and TNAU as coordinating agency since 2007. TNAU is concentrating on up scaling water saving technologies including Sustainable Sugarcane Initiative (SSI) under Precision farming in sugarcane. Krishi Vigyan Kendra, Virinjipuram is implementing TN IAMWARM project in Cheyyar & Kiliyar sub basin since 2007. Cheyyar and Killiyar Sub - Basin are the tributaries of the Palar Basin that rises in KolarDistrict of Karnataka enters through Vaniyambadi and drains an area of about 13,470 Sq. K.M. of which 10,880 Sq. K.M. is in Tamil Nadu. The command area of this sub basin comes under the Cheyyar and Vandavasi Taluk in Thiruvannamalai and Uttermerur, Maduranthagam, and Seyyur Taluk in Kancheepuram Districts which covers 458 tanks in 12 blocks with a registered ayakut area of 36746 ha. The major crops grown in the sub basin area are paddy, sugarcane, groundnut and pulses.

The Precision farming in sugarcane intervention is implemented in select blocks namely Vandavasi, Maduranthakam and Acharapakkam and Cheyyar blocks of Tiruvannamalai district, wherein the sugarcane farmers are falling under the jurisdiction of the Cheyyar Co-op sugar mills, Cheyyar. The precision farming in sugarcane includes promotion of "Improved production technologies", SSI and Drip and fertigation in sugarcane. The project is initiated in 2011 when most of the tanks are dried and ground water potential is very low and the farmers are struggling to go for sugarcane and paddy.

Interventions carried out

The project envisaged implementation of Precision farming in sugarcane in 90 ha in the sub basin area over a period of three years from 2011. Research Fellows are posted at the project site to identify farmers and to implement the activities. The nearby sugar mill, bankers, farmers' organizations and PACBs were sensitized about the importance of the intervention and their support was utilized.

a. Awareness creation meetings / trainings / demonstrations / Exposure visits

Trainings followed by technical and financial support makes SSI and drip fertigation a reality and I recovered my investment
– Mr. Venkatesan, Farmer

The farmers were sensitized through series of awareness creation meetings at village levels and by distribution of leaflets and hand outs. The interested farmers are invited for Off and On campus trainings and demonstrations on SSI Techniques; Drip and fertigation techniques etc., The drip and fertigation companies were invited for participation in the programme. The cane Officers, Cane Inspectors, Administrators of the Cheyyar Coop sugarmills and progressive farmers and bankers are provided with a one day sensitization programme on SSI and need and importance of drip and fertigation in sugarcane. The interested farmers were also taken for an exposure visit to TNAU and other progressive farmers' fields in learning SSI techniques and drip fertigation technologies.

b. Technical and Financial assistance

The initial efforts of promoting SSI went on vein and farmers are reluctant to change

In order to compensate the risk and uncertainty of implementing any new intervention and to motivate farmers TN IAMWARM project provides 75 % subsidy to all farmers implementing drip fertigation under the programme. On successful installation of the drip fertigation system the farmers are provided with WSF fertilizer for the crop at no cost besides technical assistance in implementation of drip fertigation system and SSI as and when required. The cheyyar coop sugar mill provided financial support for establishing Sugarcane Nursery for installation of shade net for Nursery producing sugarcane seedlings for identified farmers.

Early Stages of Struggle and Break through

Majority of the farmers were concerned about SSI especially in terms of cost of cultivation and yield. Also, they are very much concerned on the high cost of contribution (25%) they have to invest for installation of drip fertigation system. Though the project was initiated in 2011, it took off only in 2012. By series of discussions, exposure visit arranged to select farmers to SSI farmers' fields and subsequently with intervention of sugar factory officials and PACB accepted for financing, few progressive farmers initially started drip and fertigation in sugarcane in conventional method. By seeing the success of the drip and fertigation system some more farmers joined the bandwagon in implementing SSI along with drip fertigation.

Progressive and big farmers were the early adoptors of SSI

Impact of the Intervention

- Between 2011 to 2014, a total of 90 ha were brought under drip and fertigation system and out of 90 ha, the adoption of SSI along with drip fertigation was only 10% in the first year; 32% in the second year and 80% in the third year.
- Over 50 farmers have shown interest for the implementation of SSI and drip fertigation in their farms. Sugar mills are requesting TNAU to continue the programme for some more years.
- A change from traditional method of cultivation with 3 ft x 2 ft spacing using 4 tonnes of setts / ac has been replaced by planting 4.5 ft x 1.5 ft spacing of 25-35 days old seedlings and adoption of sub surface drip irrigation is visibly noticed.
- Progressive farmers started adopting contract method of procuring sugarcane seedlings from nurseries.
- Farmers become entrepreneurs by establishing Sugarcane nurseries in the vicinity of the project area.
- Application of Water soluble fertilizers through drip fertigation is 100% in SSI farms.
- Farmers who have practised SSI were able to cultivate with less water and managed to raise the crop when the nearby farms with conventional method of cultivation failed during water stress conditions.
- More than 80% of the SSI farmers attained 30-40 % more yield than conventional method of cultivation and therefore harvest the crop in six months (About 30-35 tonnes/ ac) and sell to the factory as seed cane.

Action Photographs of SSI and Drip Fertigation

SSI Planted field in Perumbalai village of Anakavur block

SSI field in Keezhkodangalur village of Vandavasi block

Training on SSI at The Cheyyar Coop Sugarmill office at Keezhkodangalur village

Inputs received by farmers

Farmers in their sugarcane field

- About 20% of the farmers grow intercrops such as daincha, pulses and vegetables in SSI plots.
- The adoption rate of SSI and drip fertigation was high among the big farmers who have more than 4 ha.

Scaling up

- The experience of the current level of spread was more with progressive and big farmers or farmers who have access to credit facilities. To scale up, similar initiatives with training, financing and handholding support has to be provided.
- The staff capacity both at the project level and the hand holding organizations level has to be improved by rigorous trainings so as to handle the complexities in improved methods of Cultivation and drip fertigation.

Conclusion

The delivery of technology to a specific location, to a specific target audience, intensively on a project mode with the support of the local Government Institutions, Water User Associations, Panchayat and other stakeholders like Sugar factories, input dealers, suppliers and farmers when orgainsed trained, provided with sufficient knowledge, and inputs at subsidized cost on a participatory mode is highly successful method which is effective and efficient in technology delivery not only on a small scale but on a large scale.

Extension Management Strategies for Sustainable Agriculture, pp 87-92

New India Publishing Agency, New Delhi, India

9

An Effectiveness of Participatory Extension Approach for IPM Technologies

B. Narayanaswamy , M.R. Hegde, Saju George and R.K. Naika

Vegetables constitute about 55 per cent of horticultural crop production in the country with a total production of 85 million tonnes which is estimated to cross 100 million tonnes in near future. Successful cultivation of vegetables is hampered due to the incidence of several insect pests. Cultivation of hybrids, improved varieties, intensive agronomic practices, off season cultivation and indiscriminate use of insecticides has changed the pest complex in these crops.

Tomato is the second most important vegetable crop grown in India, next only to potato. It is grown in about 5.94 lakh hectare in India. It is mainly grown in Karnataka, Andhra Pradesh, Tamilnadu, Kerala, West Bengal, Northern Plains, Hills of Himachal Pradesh and UP, Assam, Maharashtra, Gujarat etc. However, the productivity is very low in India at 16.6 t/ha. This is mainly due to the loss caused by insect pests and diseases. The most serious pest of tomato is fruit borer. In addition, leaf miner, red spider mites are also other major pests. Among diseases, leaf curl, spotted wilt, early blight and late blight are serious. Our survey has shown that farmers are spraying more than 30 times for one crop of tomato in certain pockets near Bangalore and in spite of this borer incidence is always above 10% (George *et al*, 2012). Lack of Interactions , information sharing and effective communication with farmers is leading to less adoption of technologies. The study was conducted with the objective of finding the impact of communication of IPM technologies among tomato farmers through field demonstrations.

Methodology

A project for popularisation of IPM technology in tomato was implemented in two predominantly tomato growing districts of Kolar and Chikkaballapura in Karnataka with financial assistance from National Horticulture Board from September, 2011. The demonstration on IPM in tomato was implemented in 5 villages of the two districts with 34 tomato farmers. The selected farmers were briefed about, the IPM technology and its importance in initial group meetings with the selected farmers. Continuous monitoring of the crop was done by a multidisciplinary team of scientists. Regular feedback on progress of the crop, pest and disease incidence was collected from the field. Regular interaction meetings were conducted with farmers to integrate appropriate IPM technologies into their existing cultivation practices.

IPM technology for Tomato : The schedule for Integrated Pest Management in Tomato has been standardized at Indian Institute of Horticultural Research, Bangalore. The package includes use of neem cake, Nuclear Polyhedrosis Virus (NPV) sprays for fruit borer control. These IPM technologies along with growing leaf curl tolerant hybrids, vector management, and removal of basal diseased leaves and need based application of fungicides were extensively demonstrated in farmer's field near Bangalore between 2000-2005. By following the IPM, the disease and insect pest incidence were reduced to negligible. The percentage of borer was reduced to 2.07 % by the sprays of NPV and other majors. The mean cost of cultivation was Rs. 30,070 /ha. The mean tomato yield was 52.49 tons/ha and mean net returns was Rs. 97,574/ha. The number of sprays of pesticides was reduced to about 7 to 8 (botanical + chemical) as compared to 20 chemical sprays in non-IPM plots. The technology needs to be demonstrated ion varied other climatic situations so as to create awareness and also for safe and good quality production of tomato.

Results and Discussion

A cursory look at the Table 1 indicates that a majority of farmers (90 %) started using drip irrigation for the improvement of crop when compared to only 20 % before the implementation of the project. Also a considerable improvement in parameters like fertilization technology, management of pests at various stages, grading of fruits for marketing were improved 80 per cent over the 'before implementation' of the project. The use of foliar nutrition for improving the quality of fruits showed 100 per cent adoption. Important technology like mulching in tomato was not at all practiced by farmers before the implementation of the project as the farmers were not aware of the technology. But a drastic improvement (47.5 %) has been achieved in adoption of mulching technology through IPM technology project once the farmers realized its importance in water, weed, labor and pest management. This achievement could be possible because of the awareness and involvement through effective communication among farmers and project implementing scientists.

Table 1: Improvement in adoption of technological interventions with respect to IPM technologies due to project intervention (N= 40)

Sl. No.	Technology components	Change in Adoption of Technology			
		Before project implementation		After one year of project implementation	
		Freq (n)	Percent adoption	Freq (n)	Percent adoption
1.	Mulching in tomato	0	0	19	47.5
2.	Use of drip for irrigation	8	20	36	90.0
3.	Fertigation technology	6	15	32	80.0
4.	Use of NPV for control of fruit borer	0	0	17	42.5
5.	Management of pests at various stages	4	10	32	80.0
6.	Management of diseases at various stages	6	15	28	70.0
7.	Preparation on enriched FYM	4	10	27	67.5
8.	Grading of fruits for marketing	4	10	32	80.0
9.	Use of foliar nutrition for improving quality of fruits	2	5	40	100.0

It is implied from Table 2 that the contact of farmers with extension agents and scientists was much below (only 12.5 %), earlier to the implementation of the project. It has been improved *i.e* 70 % as 'more contact' and only 17.5 % were medium contact with extension agents and scientists. This is because extension agents and scientists showed the importance of the technologies, and its timely application in the field to achieve more economic returns. It has been realized by the farmers that regular contact with scientists will definitely paves the way for gaining a lot from them.

Table 2: Information seeking behaviour of farmers regarding IPM technology (N=40)

Sl. No.	Contact	Before implementation of project (Based on benchmark survey)		After implementation of project	
		Freq (n)	%	Freq (n)	%
1.	Less contact with extension agents and scientists (upto 7 score)	34	85	5	12.5
2.	Medium contact with extension agents and scientists (8 to 12 score)	6	15	7	17.5
3.	More contact with extension agents and scientists (Above 12 score)	0	0	28	70.0

From the Table 3, it is very clear that the season specific technologies like late blight management and its communication are to be needed in farmers plots to achieve maximum returns. Unfortunately a collective approach like integrated pest management practices by considering all the stages of the crop was never popularized among the farmers in the project area earlier. Through regular contacts and effective communication methods the farmers were convinced by the scientists in educating timely identification of insect pests as well as adopting the control measures. Accordingly a considerable improvement in crop recovery (65-72%) has been achieved in demonstration plots over it was only 40% in farmers practice.

Through the project the practice of spraying micronutrient through foliar application was also popularized for getting good quality fruits with increased shelf life and good color development for fetching more price for the farmers.

Table 3: Communication of season specific technology for the farmers

Particulars	Demonstration area (20 Nos)	Farmers practice plots (15 Nos)
Incidence of late blight in October end	45-60 %	75-90 %
Crop recovery by November end	65-72 %	40-50 %
Extent of yield reduction reported by farmers due to late blight	30-35 %	42-60 %
Foliar micronutrient sprays	3-4 sprays per crop	Nil
Advantage of Micronutrient sprays	Color development which fetches better price in the market	-

It is evident from the Table 4 the components of IPM technologies like reduction in chemical sprays played a pivotal role in achieving higher economic returns. Drastic reduction in usage of chemicals, frequency of sprays and unnecessary usages of chemicals have been thoroughly communicated to the farmers to achieve more economic returns. Accordingly demonstration farmers could get Rs 2.30 for every rupee invested compared to only Rs 1.7 in case of farmers earlier practices. A considerable improvement in economic returns as well as awareness regarding hazardous chemicals among the farmers have been achieved in this project. This in turn a lesson for fellow farmers of same villages as well as neighboring district tomato farmers.

Table 4: Economic improvement in tomato cultivation through IPM*

Sl. No	Particulars	Non IPM farmer	IPM farmer
1.	No. of plant protection chemical sprays	21	11
2.	Reduction in no. of chemical sprays	-	10
3.	No. of biological sprays	0	3
4.	No. of micronutrient sprays	0	3
5.	Yield (tones) per acre	16.1	18.2
6.	Cost of plant protection chemicals/acre	Rs 20,239	Rs 15,742
7.	Rate obtained per kg of tomato	Rs 7	Rs 8
8.	Total Revenue per acre	Rs 1,12,700	Rs 1,45,600
9.	Total cost per acre	Rs 66,325	Rs 63,829
10.	Net profit	Rs 46,374	Rs 81,770
11.	Benefit cost ratio	1.7	2.3

- Specific for the tomato grown during August to October, 2011.
- Economics have been worked out for the district of Chikkaballapur.

Conclusion

From the findings it can be concluded that an effective communication strategy is must for enhancement of adoption of IPM technologies among the farmers. The important components like management of pests in various stages, fertigation and irrigation technologies, mulching, foliar application of nutrients etc were effectively passed on to farmers by the scientists in convincing manner. The results showed that before implementation of the project there was no usage of NPV, for control of fruit borer, no mulching in tomato and very meager usage of foliar nutrition. Accordingly cent per cent adoption in foliar nutrition, 47.5 per cent adoption of mulching, 42.5 per cent practice of NPV have been achieved through effective communication methods. Communication of season specific technologies could fetch 65 to 72 per cent crop recovery in demonstration farmers, apart from improvement in more contact with extension agents and scientists (70 %). So more number of innovative technologies should be imparted to the farming community through the effective means like IPM demonstrations by the scientists in different locations. It will in turn definitely pave the way for achieving more economic returns by the farmers.

Acknowledgement

The authors greatly acknowledge the financial assistance rendered by National Horticulture Board (NHB) in implementing the project “Popularization of IPM in tomato” in the districts of Chikkaballapur and Kolar in Karnataka.

References

Bentley, J.W. (2009). Impact of IPM Extension for small holder farmers in the tropics. In *"Integrated Pest Management: dissemination and impact"*. Pub. Springer, Chapter 8, pp 333-346.

George, S., Hegde, M.R. and Doijode, S.D. (2012). Adoption of integrated Pest management practices in vegetable crops in Karnataka. *Pest management in horticultural ecosystems,* Vol. 18 (1):118-119.

Tiwary, K.B and Saxena, A. (2001). Economic analysis of FLD of oilseeds in Chindwara. *Bharathiya Krishi Anusandhan Patrika*, 16 (36/4): 185-188.

Singh, A and Singh, L. (2004). Yield gap assessment of Lentil under front line demonstration Uttar Pradesh. *Indian J. of Extn. Education.* 4 (1&2): 58-59.

Extension Management Strategies for Sustainable Agriculture, pp 93-100

New India Publishing Agency, New Delhi, India

10

A Study on Role Performance of Farmer Friend Under ATMA Programme

A. Anitha Pauline and C. Karthikeyan

The Revised ATMA Scheme provides for dedicated specialist and functionary support to take up training and extension initiatives at State, District and Block level. Besides this, improvement in extension outreach right down to the village level is expected to be achieved through Farmer Friend. The study mainly focused on farmer friend under ATMA scheme. The FF served as a vital link between extension system and farmers at village level (one for every two villages). The FF should be available in the village to advise on agriculture and allied activities. The FF mobilized farmers' groups and facilitate dissemination of information to such groups, individual farmers and farm women directly through one to one interaction individually or in groups and also by accessing information / services on behalf of farmers as per need through Common Service Centres (CSC) / Kisan Call Centres (KCC). (Government of India, 2010).

Support for Innovative Activity at Village Level (Farmer Friend)

1. It is necessary to identify and groom progressive farmers to act as focal points at village level. A progressive farmer identified as Farmer Friend (FF) will be placed @ one FF per two census inhabited villages. The FF will help in activating the much-needed village based, bottom-up planning process and serve as vital link, between extension system and farmers at village level. Farmer Friend will lead by example and is expected to have up-graded skills and would be available in the village to advise on agriculture and allied activities. The FF will be identified by Block Technology Manager on a resolution of Gram Panchayat which will, in turn, consult CIGs/ FIGs working in Panchayat area. It will be ensured, that Farmer Friend, so identified, is a progressive farmer, with requisite educational qualification

(Senior Secondary/ High School) and experience. In case of non-availability of progressive farmers with requisite qualifications, at some places, a progressive farmer, with required oral and written communication skills, may be considered.

2. It is proposed to provide, Farmer Friends, with special opportunity, for up gradation of their skills through trainings, study tours and visits to SAUs/ other institutes, by utilizing support available under ATMA. Government of India will provide '2000 per Farmer Friend, under the Scheme, which will be supplemented equally by State Govt. This total amount of' 4000/ - per Farmer Friend per year will be used to meet contingency expenditure which FF incurs for discharge of his duties, as Farmer Friend, including incidental expenses incurred on acquisition of knowledge. Payment of this amount of Rs. 4,000/- will be made based on Utilization Certification by Farmer Friend and counter signed by BTM.
3. Since male farmers are involved in varied agricultural activities which may require their frequent movements outside the village, more women farmers may be involved as 'Farmer Friends'. Women Farmer Friends will also be helpful in reaching out to women farmers / farm women.
4. The States may decide on a higher fixed amount of more than '4,000/- per annum for Farmer Friend' after taking into considerations, such as likely contingency/incidental expenditure or any other such factors. However, contribution of the Government of India shall not exceed '2000/ - per Farmer Friend' per year.

Objectives of the Study

1. To find out the factors motivated to become farmer friend.
2. To assess the awareness and role performance of farmer friend under ATMA scheme.
3. To study the association and contribution of independent variables to dependent variable.

Research Methodology

The study was carried out in Coimbatore district of Tamil Nadu with a sample size of 31 farmer friend. Coimbatore district constitutes 295 revenue villages and 12 blocks. Totally 144 farmer friend were available among the 295 villages. Three blocks were randomly selected viz. Thondamuthur (12), Madukkarai (10), and Periyanayakan Palayam (9). Saturated sampling procedure was followed to select all the respondents of the selected block.

Ex-Post facto research design was used for this study. Totally 12 independent variables and one dependent variable were considered for this study. The study was conducted with pre-tested and well-structured interview schedule and analyses using appropriate statistical tests i.e. frequency analysis, percentage analysis, mean and standard deviation. Ranks were assigned based on the frequencies; categorisation was done based on mean and standard deviation values. Correlation and regression analysis were used to study the association and contribution of independent variables to dependent variable.

Findings and Discussion

In the light of the objectives of the study, the findings together with discussion are presented under the following sub heads.

I. Factors motivated to become farmer friend

Motivation is an inner urge or goal directed behaviour. It plays a role in influencing the behavioural components. The data collected in this aspect are presented in Table below.

Table 1: Factors motivated to become Farmer Friend (n=31)

S. No.	Factors	Number	Percent	Rank
1.	Desired to transform the community through the improved livelihoods	21	67.74	V
2.	To access new knowledge	28	90.32	II
3.	To become famous	15	48.38	VIII
4.	To be empowered	19	61.29	VI
5.	Sprit of sharing knowledge	27	87.09	III
6.	Transform fellow farmers from subsistence to commercial farming	16	51.61	VII
7.	To receive training and gain knowledge	23	74.19	IV
8.	Selected by other farmers	2	6.45	X
9.	Opportunity for exchange visits	10	32.25	IX
10.	Interest in farming	30	96.77	I

*Multiple responses obtained

The above findings revealed that overwhelming of the respondents (96.77%) were interested in farming followed by access new knowledge (90.32%), sprit of sharing knowledge (87.09%), to receive training and gain knowledge (74.19%), desired to transform the community through the improved livelihoods (67.74%), to be empowered (61.29%), transform fellow farmers from subsistence to commercial farming (51.61%), to become famous (48.38%), Opportunity for exchange visits (32.25%) and Selected by other farmers (6.45%). According to ATMA guidelines, selection of farmer friend should be a progressive farmer. So the characteristics of progressive farmer was interested in farming, to access new knowledge, sprit in sharing information to fellow farmers, gain more knowledge through exposure visits. According to ATMA revised guidelines farmer friend should be a progressive farmer. From the findings it could be concluded that all the farmer friend were qualified the guidelines of ATMA revised scheme, 2010.

II. Awareness about roles and responsibilities of farmer friend under ATMA scheme

The level of awareness of farmer friend indicates how much they know about their roles and responsibilities under ATMA scheme. The overall awareness of farmer friend's roles and responsibilities are studied and results are given below Table 2.

Table 2: Distribution of respondents based on their overall awareness about their roles (n=31)

Sl. No	Category	Frequency (n=31)	Percent
1.	Low	-	-
2.	Medium	3	9.67
3.	High	28	90.33
Total		**31**	**100.00**

The above findings revealed that more than 90.00% of the respondents had high awareness about their roles and responsibilities. Nearly one-tenth of respondents had medium level of awareness about their roles. The reason might be due to close contact with extension agency contact and high level of educational qualification.

Activity wise awareness about their roles

Further an attempt was made to assess the awareness of farmer friend about their roles and responsibilities were studied. In this regard, eight roles identified according to revised guidelines of ATMA scheme, 2010. The ATMA activity wise awareness level was studied and results are given in Table 3.

Table 3: Distribution of respondents based on their activity wise awareness level (n=31)

S. No.	Responsibilities / Duties	Number	Percent
1.	Mobilization of farmers	31	100.00
2.	Constitution of Farmer Interest Groups	28	93.33
3.	Facilitate in Conducting field demonstrations, Kisan Goshties	27	87.06
4.	Preparation of Village Research Extension Action Plan	29	93.54
5.	Liaison with SMS at Block level for exchange of information	31	100.00
6.	Attending meetings of Gram Sabha	30	96.77
7.	Maintaining a daily diary of activities	25	80.64
8.	Ensuring dissemination of information through multi media	31	100.00

From the Table 3, it is evident that, cent per cent of the respondents had awareness about mobilization of farmers, Liaison with SMS at block level for exchange of information and ensuring dissemination of information through multi-media. More than ninety per cent of the respondents had awareness about constitution of farmer interest group, attending meetings of Gram Sabha and preparation of village resource extension plan. From the above findings it could be concluded that Farmer Friend were well aware about their roles and responsibilities.

Role performance level of farmer friend under ATMA scheme

The data with respect to the role performance of farmer friend's are presented in Table 4.

Table 4: Distribution of respondents based on their overall role performance level (n=31)

S.No	Category	Number (n=31)	Per cent
1.	Low	7	22.58
2.	Medium	21	67.74
3.	High	3	9.68
Total		**31**	**100**

The results indicated that more than two-third (67.74%) of the respondents belonged to medium level of role performance followed by low level of role performance (22.58%) and nearly one-tenth of respondents (9.68%) had high level of role performance. It could be inferred from the Table 5 that most of the respondents had medium to high level of role performance in ATMA activities. The reason for medium level of role performance might be due to high level of awareness and close contact with extension agency.

Table 5: Distribution of respondents based on their role performance level (n=31)

S.No	Particulars	Good	Average	Poor
1.	Contacting with officials	27 (87.09%)	3(9.67%)	1(3.22%)
2.	Contacting with fellow farmers	24(77.41%)	5 (16.12%)	2(6.45%)
3.	Conducting extension activities in your villages	12(38.70%)	9 (29.03%)	(32.25%)
4.	Maintenance of record	2(6.45%)	1(3.22%)	28(90.32%)
5.	Giving recommendation to fellow farmers	25(80.64%)	4(12.90%)	2(6.45%)
6.	Availability in village	26(83.87%)	3(9.67%)	2(6.45%)
7.	Preparation of Village Resource plan	19(61.29%)	8(25.80%)	4(12.90%)

The above findings revealed that most of the respondents (87.09%) had good performance in contacting with officials followed by availability in village (83.87%) and giving recommendation to fellow farmers (80.64%), contacting with fellow farmers (77.41%), preparation of village resource plan (61.29%) and conducting extension activities in their villages (38.70). Meagre percentage of respondents (6.45%) had good performance of maintenance of record. More than 90.00% per cent of respondents had poor performance in maintenance of record. The results showed that farmer friend had average performance in conducting extension activities like field days, demonstration etc. The reason might be difficult to mobilize farmers in their villages, less response among farmers and lack of physical arrangements in their village.

III. Contribution and association of independent variables with dependent variable role performance

Correlation analysis

This section gives the association and contribution between the characteristic variables role performance of farmer friend. For studying the association and contribution of independent variables towards dependent variables, the statistical tools namely simple correlation, multiple regression was employed respectively and the results have been presented in table 6 and discussed as follows.

Correlation analysis was performed to find out the association of independent variables namely Age (X_1), Educational status (X_2), Occupational status (X_3), Farming experience (X_4), Farm size (X_5), Annual Income (X_6), Social participation (X_7), Extension agency contact (X_8), Information dissemination behaviour (X_9), Participation in extension activities (X_{10}), Role perception (X_{11}) and Awareness about their roles (X_{12}), with the dependent variable role performance.

Table 6: Association and contribution of characteristics with role performance **(n=31)**

Sl.No	Name of the variables	"r" value	Regression co-efficient	Standard " error	t" value
1.	Age	0.352^{NS}	0.139	0.196	0.707
2.	Educational status	0.129^{NS}	0.150	0.113	1.324**
3.	Occupational status	0.105^{NS}	0.124	0.098	1.266
4.	Farming experience	-0.145^{NS}	-0.374	0.205	-1.823
5.	Farm size	0.368*	0.150	0.138	1.084
6.	Annual income	-0.086^{NS}	-0.143	0.135	-1.058
7.	Social participation	0.104^{NS}	0.226	0.102	2.211**
8.	Extension agency contact	0.580**	0.135	0.124	1.084**
9.	Information dissemination behavior	0.516**	0.119	0.170	0.698*
10.	Participation in extension activities	0.439*	0.083	0.167	0.498*
11.	Role perception	0.663**	0.458	0.162	2.233**
12.	Awareness about their roles	0.504**	0.15	0.210	0.071**

** Significant at one percent level *Significant at five percent level R Square = 0.703 F=8.816**

The results in Table 6 exhibited that out of twelve variables studied, ten variables had shown positive significant relationship, out of 12 variables Extension agency contact (X_8), Information dissemination behaviour (X_9), Role perception (X_{11}) and Awareness about their roles (X_{12}) had shown positive and significant association with role performance at one per cent level of probability. Two variables viz., farm size (X_5) and participation in extension activities (X_{10}) had shown positive and significant association with role performance at one per cent level of probability.

The rest of the variables viz., Age (X_1), Educational activity (X_2), Occupational activity (X_3) and Social participation (X_7) had shown non-significant relationship with participation. Farming experience (X_4) and Annual income (X_6) had shown negative and non-significant relationship with role performance.

Multiple regression analysis

In order to find out the relative influence of each independent variable towards role performance, multiple regression analysis was performed and the results are presented in Table 6.

It could be discerned from the Table 6 that R^2 value was 0.703 which indicated that 70.30 percent variation in the role performance was explained by the given 12 independent variables and remaining 29.70 percent of the variation in dependent variables unexplained. The 'F' value was significant at one per cent level of probability. Thus the cause and effect relationship between the dependent and independent variables are fitted at one percent level of significance. From the equation it is clear that intercept Educational qualification (X_2), Social participation(X_7), Extension agency contact(X_8), Role perception (X_{11}) and Awareness about their roles (X_{12}) are significant at one per cent level. The Table 6 indicated that Information dissemination behaviour (X_9), and participation in extension activities (X_{10}) were significant at five per cent level.

From the results, it could be revealed that one unit increase in the following independent variables Educational qualification(X_2), Social participation (X_7), Extension agency contact (X_8), Information dissemination behaviour (X_9), participation in extension activities (X_{10}), Role Perception (X_{11}), Awareness about their roles (X_{12}) would cause to increase the level of participation 0.15, 0.22, 0.13, 0.11, 0.08, 0.45 and 0.15 units respectively.

Conclusion

The present study revealed that most of the respondents had medium level of role performance followed by high level of role performance. On the basis of findings of this study, certain implications have been brought which might be useful for organizing programmes for enhancing participation in ATMA.

Implication of the study

- Record maintenance should be made compulsory among the respondents.
- Monthly meeting shall be conducted to evaluate Farmer Friend Performance.
- Motivate the farmer to participate more extension related activities.
- Honorarium may be split in to quarterly or monthly basis so that response feel more satisfaction.
- In revised ATMA guidelines there is no specific period was not mentioned to work as farmer friend in village. Specific period should be mentioned so that other farmers also can become farmer friend.

References

'Programmes for Extension Reforms' Scheme (2010). Department of Agriculture & Cooperation, Ministry of Agriculture, GOI: PP 72-74.

Chandra Sekar, (2013). Impact Analysis of Agricultural Technology Management Agency (ATMA) Programme amongst end-users in the process of Agriculture & Rural Development. Unpub. M.Sc. (Ag.) Thesis, AC & RI, TNAU, Coimbatore.

Extension Management Strategies for Sustainable Agriculture, pp 101-106

New India Publishing Agency, New Delhi, India

11

Krishi Vigyan Kendra: A Boon for Farmers

Ramakrishnan, K., T. Rathakrishnan, Paulpandi, V.K. and R. Karthik

Krishi Vigyan Kendra is a Government of India funded Indian Council of Agricultural Research Project implemented at district level for enhancing the productivity and livelihood through scientific extension works. Realizing the importance of technology dissemination in the changing scenario of food and nutritional security, ICAR intervened in a big way. All the first line extension projects were merged and brought under single umbrella of KVK. The major emphasis was to enhance the production and productivity as well as to generate household income and employment of farming community. It was also envisaged that these extension works will be act as model for extension agencies and also catalyzed to improve the existing systems for better delivery mechanism.

Krishi Vigyan Kendras have effective institutional link between agricultural research system and extension education network in the country. KVKs have grown into a large network in the country and working under the administrative control of various organizations. KVK conducts on-farm testing to identify the location specificity of agricultural technologies under various farming systems and frontline demonstrations to establish its production potentials on the farmers' fields. It trains the farmers to update their knowledge and skills in modern agricultural technologies and trains extension personnel to orient them in the frontier areas of technology development.

KVK works as resource and knowledge centre of agricultural technology for supporting initiatives of public, private and voluntary sector for improving the agricultural economy of the district. In addition, in order to speed up the process of dissemination of technology, a large number of various extension activities and production of seeds and planting materials are taken up by the KVKs.

Activities of KVK

KVK is the grassroot level vocational training Centre in the State of Nagaland. Krishi Vigyan Kendra (KVK) is a project of ICAR for testing and transfer of Agricultural technologies to bridge the gap between production and productivity and to increase self employment opportunities among the farming communities. The trainings offered here follow the principles of "Learning by doing" and "seeing is believing". It offers skill and knowledge oriented trainings in multidisciplinary areas like crop production, plant protection, horticulture, Animal Sciences, soil conservation and Genetics and Plant Breeding. The KVK is the light house of knowledge to the farming community of the State.

The Krishi Vigyan Kendra has several activities like conducting ON as well as OFF campus trainings in various disciplines, conducting frontline demonstration of proven technologies on farmers fields in cereals like rice, oilseeds like groundnut and horticultural crop like vegetables. It has a programme of On Farm Testing (OFT) of technologies on farmers fields and collecting feedback for refinement.

KVK also established functional linkages with other related organizations. It actively participates in extension activities like Exhibitions, field days, seminars, Kisan Melas and also contributes in mass Media like All India Radio and Doordarshan programmes, publishing popular articles and Extension literature like folders, booklets, technical reports etc. It also conducts village surveys for assessing the technological gaps and training needs of farmers, rural youth and farm women and offers technical guidance to needy farmers.

Objectives of KVK

1. Planning and conducting survey of the operational area in order to prepare the resource inventory with special reference to identifying the training needs of the farming community.
2. Planning and conducting production- oriented, need-based short and long duration training courses both on campus as well as in the villages for various target groups with priority on the weaker and the poor.
3. Developing and organizing non-formal educational programmes by way of field days, farm visits, farmers fair, radio talk, Farm Science clubs etc. as the follow up information support to training courses.
4. Organizing farm science clubs, both in rural schools and in villages in order to induce in younger generation a liking for and an interest for agricultural and allied sciences and scientific farming through supervised projects.

5. Developing and maintaining the campus farms and demonstration units on scientific lines as the facilities for providing work experience to the trainees as also disseminating the latest technical know how.
6. Providing practical facilities of the Kendra to the teachers and the students of the vocational agriculture of the higher secondary schools.
7. Imparting some general education to rural illiterates and school drop-outs in order to make them not only good farmers but also better citizens.
8. Providing added training facilities in the areas for home making and nutrition education for rural community.
9. Gradually enlarging the training facilities to encompass other important areas such as home crafts, cottage industries etc. consistent to the requirements of the integrated rural Development in collaboration with concerned organization.
10. Implementing all such schemes of the ICAR and other related organizations which intend to strengthen the training programmes of the Kendra.

Mandates of KVK

- Conducting "On-Farm Testing" for identifying technologies in terms of location specific sustainable land use systems.
- Organising training to update the extension personnel with emerging advances in agricultural research on regular basis.
- Organising short and long term training courses in agriculture and allied vocations for the farmers and rural youths with emphasis on "Learning by doing" for higher production on farms and generating self-employment.
- Organising Front Line Demonstrations (FLDs) on various crops to generate production data and feed back information.

Uniqueness of KVK

- Creation of valuable resources in terms of man power and assets.
- Confirmation of technology to suit local specificity.
- Showcasing the frontier technologies.
- Capacity building among stakeholders.
- Front runner in technological application, information and inputs.
- Participatory approach in planning, implementing, executing and evaluation.

On Farm Trial

Process for bringing together the knowledge and research capacities of the local farming communities with that of the commercial scientific institutions in an interactive way. The process where farmer investigates, measures, and studies in collaboration with researchers.

Importance of OFT

Agricultural technologies may be highly appropriate in one region but completely useless in another. Thus adaptation is a must. Refinement is a continuous process in the wake of new options and changing problems. Scientific recommendation and ITK on their own fail to provide long lasting solutions, thus necessitating their integration. Farmers have been experimenting on their own and hence OFT standardizes the existing process.

Experimentation

On-farm experiments are conducted in the fields of representative farmers and to examine a small number of experimental variables. The experimental variables in an on-farm experiment may represent exploratory factors possible solutions, or some combination of the two. Non experimental variables are usually set at the level of representative farmer management.

Front Line Demonstration

Systematic creation of local proof of both the applicability and profitability of the recommended technology with the cooperation and participation of the farmers and under the personal guidance of the scientists and/or extension personnel.

Principle: Seeing is Believing.

Objectives of Frontline Demonstrations

- To demonstrate the productive potential of newly released technologies to farmers' conditions.
- To get first hand feedback on the contributory or limiting factors for achieving the productive potential of the new technology.
- To prepare technical leadership in the village, and
- To organise need based training programmes for Subject Matter Specialists, Extension Workers, Farmers etc.

Training

The main objective to change the participants knowledge, attitude or skills.

Knowledge

The basic requirement for acquiring or altering knowledge (concepts, ideas, facts etc.) is that information be presented clearly lo the learner. He/She should receive feedback on attempts to communicate an understanding of the new information.

Appropriate techniques for communicating information include interactive lectures, (including two-way question/ answer); open discussion; symposia where members of the audience are given the full opportunity to question the speaker; television; films; tape recordings or readings which are analyzed by the group of participants with the direction of the trainer.

Skills

Change in the actual ability to do something usually requires guided practice with feedback about the success or failure of the practice. Many people believe that group- relevant skills, such as keeping a discussion on the topic, can be learned through practice. It is possible to learn all about driving a car by reading a book, but the actual driving can only be learned by doing, by seeing the results, and doing it again. Intermittent process analysis use of group observers coaching sessions critiques and role plays.

Attitude

If attitude change is desired, it is appropriate to utilize small, informal, open-ended discussion groups where the individual will not feel threatened. To the degree that these discussion groups are important or valuable to the participants, attitude changes are more likely to be lasting.

Conclusion

Krishi Vigyan Kendras have effective institutional link between agricultural research system and extension education network in the country. Krishi Vigyan Kendra strives for testing of technologies through various trials, vocational trainings, interventions for technology dissemination and application, entrepreneurship development and farmers' trainings. Extension interventions are made through on-farm demonstrations, on/off campus trainings, farmer-scientist interface and field days for wider dissemination of improved varieties and practices.

References

Prabhu Kumar, S. and B.T.Rayudu. (2011). Innovative experiences of KVK. Souvenir of 6th National Conference on KVK-2011, JNKVV, Jabalpur (MP). Pp: 40-42.

Kirar, B.S, R. Nashine, A.K. Gupta and S.C. Mukherjee. (2006). Demonstration: An effective tool for increasing the productivity. Ind. Res. J. of Ext. Edu. **6** (3): 89-97.

http://agritech.tnau.ac.in

http://www.icar.org.in

Extension Management Strategies for Sustainable Agriculture, pp 107-113

New India Publishing Agency, New Delhi, India

12

Impact of Paddy Drum Seeder Front Line Demonstrations on Farmers Knowledge and Adoption Levels

O. Sarada, G.V. Suneel Kumar and P. Sandya Rani

Rice is a staple food for millions of people in the world, particularly in developing countries like India. The demand for rice is growing with ever increasing population. In India more than 70 per cent of the ground and surface water is being used for Agriculture and out of this 70 per cent is allocated to rice cultivation. Each kg of rice produced with irrigation requires 3000- 5000 litres of water (Anonymous, 2011). The demand of rice in India is increasing with increase in population and is expected to be 140 m.t. by 2025 (Pandey *et al.*, 2008).

Andhra Pradesh is the fifth largest state in India accounting for 9 and 8% of the country area and population respectively. The state has two main regions viz., Coastal (9 districts) and Rayalaseema (4 districts) and a fairly extended tribal belt, along the Northern and North-Eastern regions. Andhra Pradesh has three major river basins (Krishna, Godavari and Penna) and five other small rivers which drain into bay of Bengal. Rice is the principle food crop cultivated throughout the Andhra Pradesh state providing food for the growing population, fodder to the cattle and employment to the rural masses. It is cultivated in an area of 28.03 lakh ha in *Kharif* and 15.84 lakh ha in *rabi*. The main source of irrigation is canals (52.0%), tube wells (19.3%), tanks (16.25), other wells (8.8%) and other sources (3.7%). Any decline in its area and production will have a perceivable impact on the state economy and food security.

Due to rise in input costs, high competition in international market for rice and problems in managing food grain stocks in India, rapid degradation of rice ecologies due to imbalanced use of fertilizers and improper water management practices adopted, it has put tremendous pressure on the rice growers to make rice farming economically viable and ecologically sustainable. Of late, farmers involved in paddy cultivation are facing several problems *viz*., uncertainty in availability of canal water, paucity of labour availability coupled with hike in labour wages leads to rice cropping becomes lack luster and less profitable, which is finally forcing farmers to give up rice cultivation citing low productivity and high labour costs as reasons. By considering the above prevailing problems Krishi Vigyan Kendra (KVK), Darsi popularized paddy drum seeder technology in adopted villages by using Front Line Demonstrations (FLDs) and trainings as a means of transfer of technology during 2011-2013. FLDs are the important interventions of Krishi Vigyan Kendra to demonstrate and popularize the cost effective, sustainable, feasible and profitable crop production and protection technologies under different agro-climatic regions and farming situation. The transplanting of rice seedlings which is a highly labour-intensive and expensive operation can be replaced by direct seeding that can reduce labour needs by more than 20 per cent in terms of workinghours required (Santhi *et al*., 1998.). To step up the profitability in rice cultivation cost cutting technology is the only way despite of the productivity aspects. Hence several on-campus and off-campus training and demonstration programmes were organized by the KVK to make horizontal spread of the same to the end users. Due to the concerted effortby different ways and means drum seeded rice cultivation is being in more than 1400 acres in the district as the paddy farmers witnessed the savings on labour costs involved in nursery management, pulling of seedlings and further transplanting in the main field, Moreover, farmer's can take up sowing with family labour/ limited labour in event of peak. An attempt wasmade to assess the impact of paddy drum seeder FLDs organized with the following objectives.

1. To study the knowledge and adoption levels of farmers on recommended paddy drum seeder technology.
2. To analyze the constraints in adoption.
3. To study the perceptionon paddy drum seeder technology.

Material and Methods

Ex-post facto research design was adopted for the study purpose. Study was conducted in fivevillages *viz*; Lakkavaram, Pulipadu, Nagambotlapalem, Lambadithanda and Allurufrom Thalluru, Mundlamuru and Kottapatnam mandals of Prakasam district where Front Line Demonstrations on paddy drum seeder were organized by Krishi Vigyan Kendra, Darsi during the years 2011-12 and 2012-13. Data was collected from 90 farmers where demonstrations have been carried out. Data was collected from the sample of farmers by personal interview method using structured interview schedule. Knowledge was operationalized as the amount of information and understanding possessed by the paddy farmers about the practices demonstrated during the FLDs and trainings. Adoption was operationalized for the purpose of investigation as practicing the recommended and demonstrated technologies by the respondents.Knowledge and adoption of the farmers was tested against ten items related to paddy drum seeder technology demonstrated. The package of practices recommended by Acharya N. G. Ranga Agricultural University and demonstrated by KVK, Darsi were included in the study to assess the knowledge and to measure the extent of adoption. Response for each item was analyzed using frequencies and percentages. Perception of the farmers was studied using structured schedule containing 11 statements on 3 point continuum i.e., Agree, Undecided and Disagree. To analyze constraints open ended questions were posed to the respondents.

Results and Discussions

Knowledge of farmers on recommended paddy drum seeder technology

It is evident from table 1 that great majority of the farmers had correct knowledge on main field preparation (90.00%), mechanical weeding as well suitable soils (86.67%), season (83.33%), variety (80.00%), seed rate (78.89%), seed soaking (77.78%), water management as well potash fertilizer management (76.67%), pre emergence herbicides (75.56%), sowing (74.44%), nitrogen fertilizer management (72.22%) and phosphorus fertilizer management (60.00%). But considerable percent of paddy farmers had incorrect knowledge about post emergence herbicides (46.67%) and phosphorus fertilizer management (40.00%). The reason behind this scenario was they had strong apprehension that without application of complex fertilizers throughout the crop season attaining good yields was literally impossible.

Adoptions of recommended paddy drum seeder technology by the farmers

It could be inferred from table 2 that great majority of the paddy farmers fully adopted variety (91.11%), season (78.89%), seed soaking (71.11%), suitable soils (68.89%), mechanical weeding (67.78%), sowing (66.67%), main field preparation (65.56%), seed rate (63.33%) and potash fertilizer management (57.78%). Where as considerable per cent of farmers were partially adopting phosphorus fertilizers (53.33%) and nitrogenous fertilizers (43.33%). The reason for this partial adoption was framers were not practicing the balanced use of fertilizers and Integrated Nutrient Management practices.

Perception of the farmers on Paddy Drum seeder technology

Majority of the paddy farmers perceived that 7-10 days time is saved in paddy drum seeder technology (86.67%), perfect leveling of the field is very difficult (85.56%), it could be practiced even in smaller area (84.44%), the incidence of pest and diseases was low as well there is no risk in taking up the technology (68.88%), weed management is very difficult (64.44%), the yields in drum seeder paddy are sustainable (63.33%), over all yield potential was increased (61.11%) and more than half of the respondents have disagreed for the statement that drum seeders are not available (52.22%). These results are in agreement with Malleswara Rao *et al.* (2014), who reported that the cost of cultivation was reduced by 26% because of drum seeder paddy cultivation and net returns per acre was also increased by 34%, the average increase in yield drum seeder paddy(Kharif 2011, 2012 and 2013)was 14% when compared withfarmers practices and the crop duration was also reduced by 7-10 daysby using paddy drum seeder technology. Wang and Sun (1990) noticed that duration can beshortened by 7-15 days in direct seeded rice compared to transplanted rice. Chandrasekhararao *et al.* (2013) confirmed that crop duration was reduced by 8-10 days and yields were increased to an extent of 8.3% in Kharif and 11% in Rabi by adopting drum seeder paddy technology.

Constraints expressed by farmers in adopting drum seeder paddy technology

Perfect leveling of the field was one of the major constraint expressed by the most of the respondents (87.79%), followed by weed management (82.22%), non availability of weeders suitable to drum seeder paddy ((75.55%), germination used to be affected by heavy rains (73.33%) and water management (62.22%). Pathak *et al* (2011) concluded that direct sown paddy faces a potential threat from high weed infestation and weed species are difficult to control.

Knowledge of farmers on recommended drum seeder paddy technology **N=90**

S.No	Recommended drum seeder package of practices	Knowledge			
		Correct knowledge		Incorrect knowledge	
		Freq	%	Freq	%
1.	Suitable soils	78	86.67	12	13.33
2.	Main field preparation	81	90.00	9	10.00
3.	Season	75	83.33	15	16.67
4.	Variety	72	80.00	18	20.00
5.	Seed rate	71	78.89	19	21.11
6.	Seed soaking	70	77.78	20	22.22

S.No	Recommended drum seeder package of practices	Knowledge			
		Correct knowledge		Incorrect knowledge	
7.	Sowing	67	74.44	23	25.56
8.	Fertilizer management				
a)	Nitrogen fertilizer management	65	72.22	25	27.78
b)	Phosphorus fertilizer management	54	60.00	36	40.00
c)	Potash fertilizer management	69	76.67	21	23.33
9.	Weed management				
a)	Pre emergence	68	75.56	22	24.44
b)	Mechanical weeding	78	86.67	12	13.33
c)	Post emergence	48	53.33	42	46.67
10.	Water management	69	76.67	21	23.33

Adoption of recommended drum seeder paddy technology by the FLD farmers **N=90**

S.No	Recommended drum seeder package of practices	Adoption					
		Full adoption		Partial adoption		No adoption	
		Freq	%	Freq	%	Freq	%
1.	Suitable soils	62	68.89	20	22.22	8	8.89
2.	Main field preparation	59	65.56	18	20.00	13	14.44
3.	Season	71	78.89	19	21.11	0	0.00
4.	Variety	82	91.11	6	6.67	2	2.22
5.	Seed rate	57	63.33	21	23.33	12	13.33
6.	Seed soaking	64	71.11	24	26.67	2	2.22
7.	Sowing	60	66.67	18	20.00	12	13.33
8.	Fertilizer management						
	a) Nitrogen fertilizer management	42	46.67	39	43.33	9	10.00
	b) Phosphorus fertilizer management	36	40.00	48	53.33	6	6.67
	c) Potash fertilizer management	52	57.78	26	28.89	12	13.33
9.	Weed management						
	a) Pre emergence	50	55.56	22	24.44	18	20.00
	b) Mechanical weeding	61	67.78	26	28.89	3	3.33
	c) Post emergence	39	43.33	27	30.00	24	26.67
10.	Water management	35	38.89	48	53.33	7	7.78

Perception of the farmers on Paddy Drum seeder technology **N=90**

S. No	Statement	Agree		Undecided		Disagree	
		Freq	%	Freq	%	Freq	%
1.	Cost of cultivation is less in direct sown paddy due to reduced cost on nursery raising and transplanting	71	78.89	11	12.22	8	8.89
2.	Overall yield potential is increased	55	61.11	13	14.44	22	24.44
3.	There is a risk in taking up drum seeder paddy	18	20.00	10	11.11	62	68.88
4.	7-10 days time is saved in drum seeder paddy	78	86.67	7	7.78	5	5.55
5.	Perfect leveling of the field is very difficult	77	85.56	6	6.67	7	7.77
6.	Weed management is very difficult	58	64.44	22	24.44	10	11.11
7.	Non-availability of drum seeder	28	31.11	15	16.67	47	52.22
8.	Drum seeder paddy cultivation is easy to practice	46	51.11	34	37.78	10	11.11
9.	It could be practiced even in smaller area	76	84.44	6	6.67	8	8.89
10.	The incidence of pest and disease is low	62	68.88	22	24.44	6	6.68
11.	The yields in drum seeder paddy are sustainable	57	63.33	16	17.78	17	18.89

Constraints expressed by paddy farmers in adopting Drum seeder Paddy technology

Constraint	Ferq	%
Perfect leveling of field	78	87.78
Weed management	74	82.22
Non-availability of weeders	68	75.55
Germination used to be affected by heavy rains	66	73.33
Water management	56	62.22
Insufficient time for field preparation	48	53.33

References

Anonymous, (2011). http://www.rkmp.co.in

Chandrasekhararao. C., Jitendranath. S and Murthy, T.G.K. (2013). Resource Optimisation in Rice through Direct Seeding by Drum Seeder, International Journal of Agriculture and Food Science Technology, 4 (3):239-246.

Malleswara Rao, S.S.N., Patil D.V., Srihari Rao, B. and Rajender Reddy, G. (2014). Performance evaluation of a manually operated paddy drumseeder- a cost saving technology for paddy cultivation,Agric Eng Int: CIGR Journal, 16 (1):75.

Pandey, N., Vema, A.K. and Tripathi, R.S. (2008). Effect of planting dates and N levels on N concentration in the leaf, grain yield and N uptake by hybrid rice. Oryza, 45 (1):18-22.

Pathak.H , Antewari, S. Sankhyan, Dubey D.S, Mina.U, Virender. K Singh, N. Jain and Bhatia.A, (2011). Direct-seeded rice: Potential, performance and problems–A review, Current Advances in Agricultural Sciences 3 (2):77-88 (December 2011).

Santhi, P., K. Ponnuswamy and N.K. Chetty. (1998). Effect of seeding methods and efficient nitrogen management practices on the growth of lowland rice. Journal of Ecobiology. 10 (2):123-132.

Wang H. Y. and Sun T. S. (1990). The characteristics of machine direct-sown rice following wheat and the corresponding techniques. Acta Agricultural Universities, Jiangxiensis 12:34-39.

Extension Management Strategies for Sustainable Agriculture, pp 115-128

New India Publishing Agency, New Delhi, India

13

Adoption of Composite Fish Culture Technology Among the Fish Farmers of Thanjavur District

K. Veerabhadran, G. Arul Oli and K. Senthil Kumar

Theworld fashioned and occupied by three fourth of the water. Fisheries activities are an indispensable part in the life and livelihood of the peoples. Aquaculture is the most important role for the upward economic strengthening and social mobility. Most of the inland fish farmers are poor due to lack of financial resource. Because of the poor production through the inland fish culture, the most of the farmers in Tamilnadu are undertaking composite fish culture technology. Demand for fish is rapidly increasing due to the growing population and awareness about the superiority of fish to other meats. The marine and freshwater sectors growth were 2.5 and 8 per cent respectively (Ayyapan and Diwan, 2004). The stagnation in marine production over year and its slow growth rate necessitates looking at inland fishery resources to meet the rapidly increasing demand. During 2004, the World Fish Center and National Centre for Agricultural Economics and Policy Research reported that the annual production of inland fish was in the range of 3.6-3.7 mmt in 2005 and going to reach upto 4.6-5.5 mmt by 2015. Aquaculture is considered as an important means for increasing inland fish production of the growing human population inland aquaculture. As fish production can be increased to a considerable extent through aquaculture. In India, aquaculture can contribute considerably to increase food supplies, improving the quality life of the rural poor, improving productivity and the national economy. Fresh water aquaculture in India is mainly carp based and carp contribute as much as 90 per cent of the total freshwater aquaculture production of the country (Tripathi, 2003). The freshwater fish culture technology namely composite fish culture aims at achieving the highest possible fish production from ponds (Sinha and Ramchandran, 2002). About 1.6 million ha of cultivable water spread area are available in the forms of ponds and lakes of which, only 38 per cent alone are utilized at present (Stephen *et al.,* 2007).

Polyculture of indigenous and exotic carps is popularly called as "Composite fish culture". The aims of the composite fish culture are to produce maximum quantity of fish per unit area from a scientifically managed water body by stocking fast growing, economically important and compatible species having shortest food chain utilizing the all ecological niches of the water body (Talukdar and Sontaki, 2005). The composite fish culture technology has been demonstrated to yield a production of 5-10 tonnes of fish under scientific management in different agro-climate regions of this country. Tamilnadu ranked 8th position in Inland fish production during 2003 (Handbook of Fisheries Statistics, 2004). The fish farmers of Andhra Pradesh produce to the tune of 5-8 tonnes/ha/year, where as their counter parts in Tamilnadu produce only 2 tonnes/ha/year (Ayyappan and Jena, 1998). The Department of Fisheries, Tamilnadu promotes freshwater aquaculture as a livelihood alternative for poor farmers. At present, 12 Fish farmers development Agencies are functioning in the state under the centrally sponsored scheme and so far 6,000 fish farmers have been benefited. During 2005-06, Tamilnadu Fisheries Department have taken about 32,000 ha of major and minor irrigation tanks for stocking with Indian major carp seeds and exploiting with the help of local farmers and about 5,000 ha of village ponds and tanks are being developed to adopt composite fish culture technology through Fish Farmers Development Agencies (www.tn.gov.in).

Thanjavur district is also richest in inland fishing due to the presence of Cauvery river system. The irrigation channels, canals, major and minor tanks are richest in high varieties. The inland fishing consists mostly of local carp, major carp and other varieties such as catfish, freshwater prawn, murrels etc. Thanjavur district has 232 inland fishing villages, about 7,631 fishermen were engaged in fishing and the production of fish from inland water sources. The total water resource available was 2,400 ha in which, farmers own ponds were 400 ha. The seeds of catla, rohu, mrigal and their early fries were brought from the Fish Seed Production Centre of Thanjavur district and the silver carp, common carp and grass carps were brought from West Bengal and reared by private fish seed producers. This study was undertaken mainly to identify and analyse the adoption of the composite fish culture technology among the fish farmers of Thanjavur district with the following objectives:

1. To study the socio-personal characteristics of fish farmers.
2. To asses the knowledge level of fish farmers.
3. To study the extent of adoption of composite fish culture technology among fish farmers.
4. To identify the constraints among fish farmers and to overcome the same.
5. To develop the suitable strategies for the adoption of composite fish culture technology.

Approach Methodology

For this study, fish farmers of Thanjavur district was selected as the population for sampling. Among the 213 fish farmers, the numbers of respondents were selected proportionally to the number of fish farmers of each taluk (Thiruvarur Collectrate, 2008). A total of 120 fish farmers were selected by using simple random sampling method. The entire district was selected for the study and taluk-wise distribution of fish farmers and total number of respondents were 120 which were selected for the study. A total of 18 independent variables were selected for analyzing the adoption of composite fish culture practiced by fish farmers, each independent variable was individually measured to know its influence on adoption. The interview schedule was formulated in the appropriate manner and which is used as device to collect the data from the fish culture farmers. Before completing the interview schedule, it was pre-tested with fish the farmers of Thirunelveli district. The data collection was done by personally contacting the individual respondents. The statistical tools such as percentage analysis, mean, deviation and correlation co efficient were used for the study.

Results and Discussion

Socio-personal characteristics of the composite fish culture farmers

- Age was resulted that more than half (53.34 per cent) of the respondents of the study belong to middle aged category followed by old (34.16 per cent) and young (12.50 per cent) age categories.
- Among the various levels of formal education undergone by the respondents, primary level education had the maximum percentage of respondent (26.67 per cent) followed by middle school level (20 per cent). It could be concluded that majority of the respondent were literate.
- Morethan two-thirds (69.17 per cent) of the respondents had farming as their only occupation, followed by farming and allied activities (30.83 per cent).
- Majority (60.55 per cent) of the respondents had undergone farming experience ranging from upto 2 to 6 years. Among the various levels of farming experience undergone by the respondents, upto 4 years (18.33 per cent) followed by upto 6 years (12.50 per cent). It could be concluded that majority of the respondent had upto 6 years farming experience.
- More than one-third (35.84 per cent) of the respondents were found to be marginal farmers and more than one-fourth (28.33 per cent) were small

farmers. Distribution among the respondent was 23.33 per cent and 12.50 per cent for medium and big farmers respectively. It could be concluded that majority of the composite fish farmers in study area were marginal and small farmers.

- Majority (65.83 per cent) of the respondents had undergone the annual income ranging from upto Rs.20,000 to Rs.50,000. Among the respondents were 26.67 per cent in the annual income category of Rs.20,001 – Rs.30,000 followed by Rs.40,001 – Rs.50,000 (22.50 per cent). It was concluded that majority of the respondents had Rs.20,001 – Rs.50,000 as annual income.

- Majority (60.84 per cent) of the respondents had undergone the annual expenditure ranging from upto Rs.20,001 to Rs.50,000. Among the various respondents 26.67 per cent were in the category of Rs.40,001 – Rs.50,000 followed by Rs.50,001 – Rs.60,000 (24.17 per cent).

- Majority (58.34 per cent) of the respondents had medium level of fish farm implements followed by low (30.83 per cent) and high (10.83 per cent) levels. It was that most of the fish farmers in the study area possessed fish farm implements at low to medium level.

- More than three-fourth (78.34 per cent) of the respondents had medium level of extension agency contact followed by low (15.83 per cent) and high (5.83 per cent) levels. It was concluded that most of the fish farmers in the study area had extension agency contact at medium to low levels.

- It was foundthat 40 per cent of the respondents had medium level of economic motivation followed by high (35.83 per cent) and low (24.17 per cent) levels. This implies that majority of fish farmers had medium level of economics motivation.

- More than four-fifth (82.50 per cent) of the respondents possessed medium level of credit orientation followed by high (9.17 per cent) and low (8.33 per cent) levels. Most of the fish farmers in the study area had credit orientation at medium to high levels.

- Majority (73.33 per cent) of the respondents had medium level of social participation followed by low (17.50 per cent) and high (9.17 per cent) levels. Most of the fish farmers in the study area had social participation at medium to low levels.

- Majority (82.50 per cent) of the respondents had medium level of risk orientation followed by low (11.67 per cent) and high (5.83 per cent) levels. Most of the fish farmers in the study area had risk orientation at medium to low levels.

- It was found that 75.83 per cent of the respondents had medium level of mass media exposure followed by low (16.67 per cent) and high (7.50 per cent) levels. Most of the fish farmers in the study area had mass media exposure at medium to low levels.
- More than two-fifth (46.66 per cent) of the respondents had medium level of information seeking behaviour followed by low (31.67 per cent) and high (21.67 per cent) levels.
- It was inferred that majority (80.84 per cent) of the respondents had medium level of innovativeness followed by low (13.33 per cent) and high (5.83 per cent) levels. The major group (47.22 per cent) belonged to medium level of innovativeness followed by low level (29.45 per cent).
- It was found that three-fourth (74.17 per cent) of the respondents had medium level of decision making behaviour followed by high (20 per cent) and low (5.83 per cent) levels. Most of the fish farmers in the study area had decision making behaviour at medium to high levels.

Knowledge Level of the Fish Farmers

Knowledge refers to the body of information possessed by an individual which is in accordance with the established fact that influences them to behave in a particular manner.

Table 1: Distribution of the respondents according to their knowledge level

Sl.No.	Category	Number	Percentage	Mean	Standard deviation
1.	Low (below 26.61)	17	14.17	30.39	3.78
2.	Medium (26.62 – 34.16)	91	75.83		
3.	High (above 34.17)	12	10.00		
	Total	**120**	**100.00**		

It was concluded that the majority (75.83 per cent) of the respondents had medium level of knowledge followed by low (14.17 per cent) and high (10.00 per cent) levels. Most of the fish farmers in the study area had knowledge at medium to low levels.

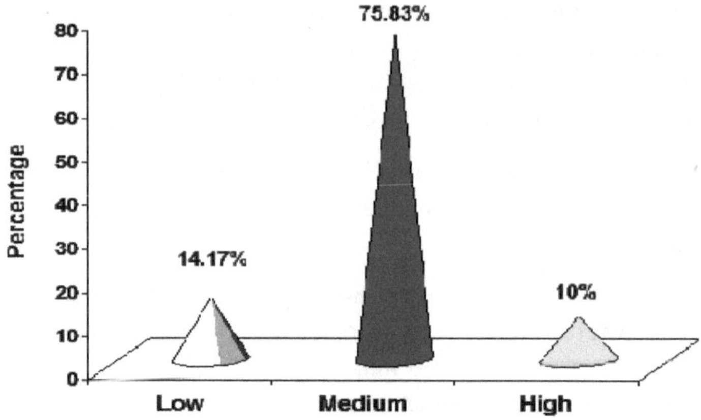

Fig. 1: Knowledge level of the respondents

Extent of Adoption of different culture practices under composite fish culture technology

Among the 20 practices taken up for the study, wide variations were observed in respect to adoption of the individual practices. Majority of the respondents were found to have fully adopted six practices viz., construction of ponds with rectangular shape (85.30 percent), recommended stocking density in the range of 5,000-10,000 nos/ha (72.50 per cent), covering the inlets with screens to prevent the entry of unwanted fishes (64.17 per cent), monthly periodical fish sampling (63.33 per cent) disease control (60.00 per cent) and supplementary feeding (52.50 per cent). In the case of a few practices *viz.*, application of potash (75.83 per cent), suitable species combination (75.83 per cent), application of liming (74.20 per cent), application of super phosphate (72.50 per cent), number of species stocked (72.50 per cent), application of organic manure (70 per cent) and maintenance of ideal water depth of 4-5 feet (69.17 per cent), a majority of the respondent reported as partial adoption.

Further, it was found that a majority of the respondent have not adopted some of the practices *viz.*, maintaining the ideal water pH value (77.50 per cent), control of aquatic weeds during pond preparation (76.16 per cent), maintaining of oxygen level (72.50 per cent), application of urea (65.17 per cent) and control of weed fishes (60 per cent). Moreover, in the case of two practices *viz.*, desilting of pond (Figure 9) and measurement of water transparency using Secchi disc, most of the respondents belonged to partial and non-adoptions. The percent of respondents under partial and non-adoption categories for these two practices ranged from 77.50 to 78.33 per cent respectively.

Table 2: Distribution of the respondents according to their adoption level of the different practices under composite fish culture technology

Sl. No.	Practices	Adoption		Partial Adoption		Non Adoption		Total	
		No.	%	No.	%	No.	%	No.	%
1.	Shape of the pond	103	85.30	3	2.50	14	11.67	120	100
2.	Inlet	77	64.17	14	11.67	29	24.16	120	100
3.	Desilting	26	21.67	38	31.67	56	46.66	120	100
4.	Application of liming	7	5.80	89	74.20	24	20.00	120	100
5.	Application of organic manure	33	27.50	84	70.00	3	2.50	120	100
6.	Application of urea	0	0.00	43	35.83	77	64.17	120	100
7.	Application of super phosphate	0	0.00	87	72.50	33	27.50	120	100
8.	Application of potash	0	0.00	91	75.83	29	24.17	120	100
9.	Stocking rate	87	72.50	29	24.17	4	3.33	120	100
10.	Species combination	18	15.00	91	75.83	11	9.17	120	100
11.	Stocking ratio	19	15.83	87	72.50	14	11.67	120	100
12.	Supplementary feeding	63	52.50	44	36.67	13	10.83	120	100
13.	Periodic fish sampling	76	63.33	23	19.67	21	17.50	120	100
14.	Control of aquatic weeds	11	9.17	17	14.67	92	76.16	120	100
15.	Control of weed fishes	17	14.17	31	25.83	72	60.00	120	100
16.	Disease control	72	60.00	17	14.17	31	25.83	120	100
17.	Ideal depth of the pond	34	28.33	83	69.17	3	2.50	120	100
18.	Maintaining of oxygen level	12	10.00	21	17.50	87	72.50	120	100
19.	Water pH value	14	11.67	13	10.83	93	77.50	120	100
20.	Water transparency	27	22.50	39	32.50	54	45.00	120	100

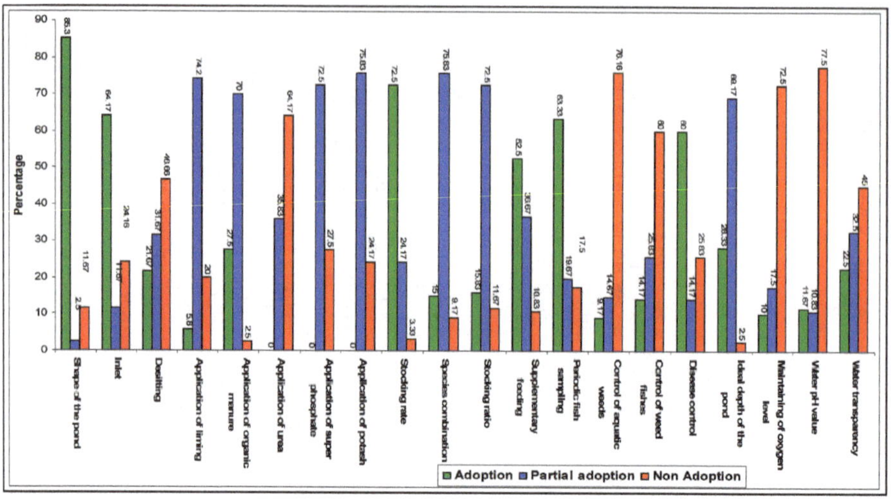

Fig. 2: Adoption level of the individual practice of composite fish culture technology

Constraints of the composite fish culture technology

It was found that a vast majority (95.83 per cent) of the fish farmers felt high tariff for electricity being charged now for fish culture as a major constraint in the adoption of recommended composite fish culture practices. Nearly 85 per cent of the fish farmers highlighted the high cost of the feed in composite fish culture because of lesser number of aqua feed shops. About 70.83 percent of the fish farmers had opined that non-availability of institutional credit is the third major constraint in composite fish culture. The fourth constraint such as non-availability of exotic carp seeds were faced by nearly 62.50 per cent of the farmers adopting composite fish culture technology.

The fifth, sixth and seventh constraints faced by the farmers adopting composite fish culture practices are inadequate water body (59.17 per cent), non-availability of quality seeds (55 per cent), high marketing charges (51.67 per cent) respectively. The other minor constraints faced by the farmers adopting composite fish culture practices *viz.*, requirement of high investment (40 per cent), high cost of fertilizer (33.34 per cent), inadequate power supply (32.50 per cent), very low price during some seasons (26.67 per cent), disease (25 per cent), costly labour for harvesting (20.83 per cent) and less consumer preference for carps (5.84 per cent). The level of adoption of composite fish culture practices was medium level and the carp production was also low when compared to other states such as Andhra Pradesh and West Bengal. Based on the above constraints faced by the farmers of Thanjavur district adopting composite fish culture practices, solving oriented problems will be analyzed and suitable strategies will also be developed in order to increase the adopting level and production level among the farmers.

Table 3: Constraints of the composite fish culture technology

No.	Constraints	Percentage
1.	High electricity tariff	96.83
2.	High cost of feed	85.00
3.	Non-availability of institutional credit	70.83
4.	Non-availability of exotic carp seeds	62.50
5.	Inadequate water body	59.17
6.	Non-availability of quality seeds	55.00
7.	High marketing charges	51.67
8.	Requirement of high investment	40.00
9.	High cost of fertilizer	33.34
10.	Inadequate power supply	32.50
11.	Very low price during some seasons	26.67
12.	Disease	25.00
13.	Costly labour for harvesting	20.83
14.	Less consumer preference for carps	5.84

Strategies for the adoption of composite fish culture technology

Normal electricity tariff pattern may be treated and given to small scale farmers to lower the cost of production, for which government has to initiate to revise the existing commercial tariff pattern.

Strategies for improving the adoption of composite fish culture technology

1. Number of aqua feed shops will be opened through society so as to lower the cost of the feed. Further, farmers are also advised to prepare pelleted or compounded feed in their own farm by utilizing locally available ingredients.
2. Credit should be increased for the sustainable adoption of the farmers. Further, new projects or scheme could be arranged on subsidy basis.
3. Exotic carp seed production centers should be created so as to supply quality with lesser price of seed to the farmers.
4. Most of the ponds are seasonal based culture system. With the available seasonal pond water, production could be enhanced further by increasing stocking size i.e. fingerlings.

5. Inbreeding should invariably be avoided. Further, brooders used for breeding purpose should be limited to maximum of five times. Moreover, brooders should be collected from different regions of the state. By following the above strategies, quality carp seed should be produced and supplied to the farmers.

6. Transport facilities should be increased. Further, middle man involved in marketing of carps should be eradicated so as to reduce the marketing charges.

7. Farmers should be encouraged to start the farm at small scale level (0.5 to maximum of 1 acre) towards investment. Moreover, schemc will also be available in Agricultural Engineering Department of Tamilnadu under Rainwater Harvest Structure by paying only Rs. 4,000 for pit size of 120 X 120 X 5 ft.

8. Stand by power system should be installed in fish farm. Further, farmers are encouraged to dig excavated type of ponds for easy gravitational flow to supply water.

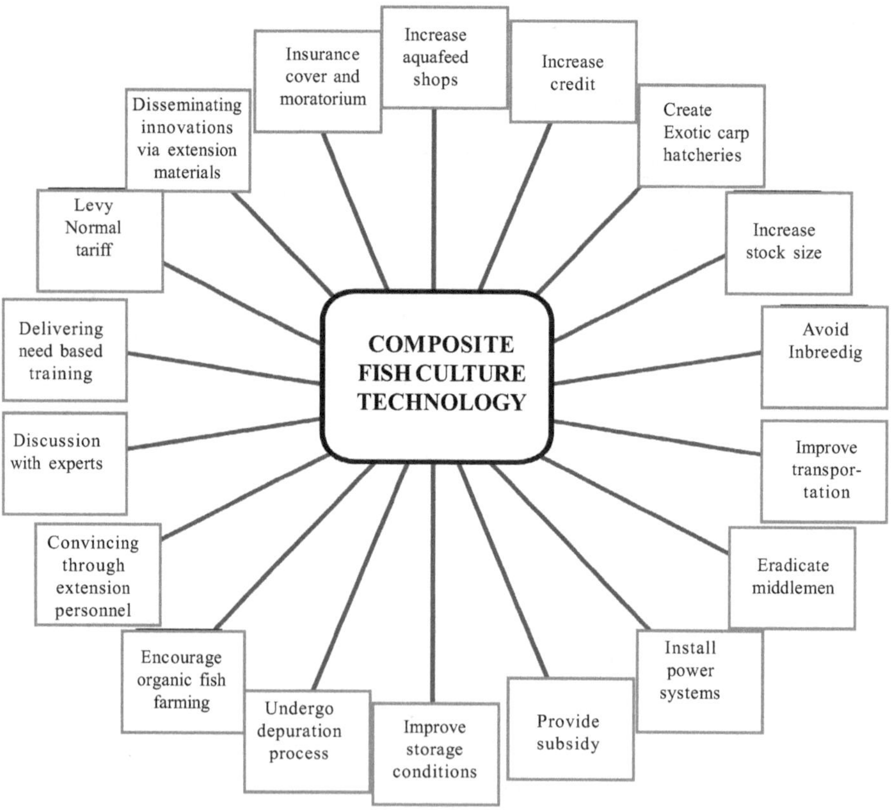

9. Subsidy will be given for buying in organic fertilizers by the governments to reduce the cost of inorganic fertilizers.
10. Facilities such as cold storage and pre-processing unit should be created to have quality carps in order to fetch high price in the market.
11. Hybridization carp of potla should be removed in the system because of fish infection of *Argulus* faced by the farmers.
12. Harvesting pit should be provided in the ponds (both inside & out side) to have easy and quicker method for harvesting.
13. All the farmers should invariably be advised to undertake depuration process to reduce musky or earthy or muddy odour from the harvested carps. Further, this process will fetch more prices.
14. Encourage organic fish farming practice among farmers.
15. The primary need is to make them and others conscious about the role play person.
16. Steps should be taken by the extension personnel to convince the farmers for adopted technology.
17. Front line demonstration unit should be increased.
18. Discussion with scientists and subject matter specialized persons.
19. Training programmes should be conducted for the fish farmers in off campus in relevant areas based on the actual needs of the farmers.
20. Bulletins and newsletter should be communicated to the farmers for every months.
21. Transfer of technology through trained extension personnel would further improve the adoption of innovations in the composite fish culture technology.
22. Corpus should be created for assisting the farmers by crop losses.
23. Insurance cover needs to be immediately expanded to cover the entire country.
24. A moratorium on debt recovery method should be introduced.
25. Farmers are advised to undertake harvest on need basis in and around region so as to fetch more price and constant steady income.

Relationship between socio-personal characteristics and adoption of composite fish culture technology.

Variable No.	Constraints	Correlation co-efficient
X_1	Age	0.040 NS
X_2	Educational status	- 0.179*
X_3	Occupational status	- 0.248**
X_4	Farming Experience	0.325**
X_5	Fish farm size	- 0.048 NS
X_6	Annual income	0.176 NS
X_7	Annual expenditure	0.219**
X_8	Fish farm implements	0.200**
X_9	Extension agency contact	0.098 NS
X_{10}	Economics motivation	0.089 NS
X_{11}	Credit orientation	- 0.370 NS
X_{12}	Social participation	0.112 NS
X_{13}	Risk orientation	0.310**
X_{14}	Mass media exposure	0.390**
X_{15}	Information seeking behaviour	0.409**
X_{16}	Innovativeness	0.401**
X_{17}	Decision making behaviour	0.396**
X_{18}	Knowledge	0.222*

** Significant at 1 percent level   * Significant at 5 percent level   NS – Non-significant

A total of 14 of the independent variables had positive relationship with the dependent variable and the remaining four variables exhibited negative relationship. Independent variable *viz.*, age (X_1), farming experience (X_4), annual income (X_5), annual expenditure (X_6), fish farm implements (X_8), extension agency contact (X_9), economic motivation (X_{10}), social participation (X_{12}), risk orientation (X_{13}), mass media exposure (X_{14}), information seeking behaviour (X_{15}),innovativeness (X_{16}), decision making behaviour (X_{17}) and knowledge (X_{18}) were found to have positive relationship with dependent variable, extent of adoption of composite fish culture.

Among 14 variables that exhibited positive relationship to the adoption seven independent variables *viz.*, farming experience (X4), fish farm implements (X_8), risk orientation (X_{13}), mass media exposure (X_{14}), information seeking behaviour (X_{15}), innovativeness (X_{16}) and decision making behaviour (X_{17}) had significant relationship at one per cent level. While another independent variable namely knowledge had significant relation with adoption at 5 per cent level. Among the 4 variables that exhibited negative relationship to the adoption two independent variables *viz.*, fish farm size (X_5) and credit orientation (X_{11}) had non-signification relationship with the dependent variable. While independent variable namely educational status (X_2) and occupational status (X_3) had negative relationship with significance at 5 per cent level and 1 per cent level respectively.

Summary and Conclusion

Indian Government realized the needs for developing fish culture and set up different agency on a pilot basis during the 11th five year plan to develop the inland fisheries. Training programmes in composite fish culture are organized to help the people to start the fish farm in their villages. The people trained by the organized institute have started fish culture in different part of the Tamilnadu. However, so far, no scientific and systematic study has been made to infer to what extent the practices recommended for the adoption of composite fish culture technology were adopted in Thanjavur district. If any improvement is to be made on the activities of inland fisheries, all the recommended practices should have been adopted by the fish farmers with out any gap. Then only, the inland fisheries production can be increased and their socio-economic status can be raised. Therefore, it was felt essential to conduct a scientific study to know the extent of adoption of composite fish culture practices by the fish farmers.

Providing normal electricity tariff , increasing the number of aqua feed shops, preparing pelleted or compounded feed in their own farm, increasing credit, low price and quality exotic carp seed production, avoiding inbreeding, sufficient transport facilities for marketing of carps, eradicating the middle man, giving subsidy for buying organic fertilizers, creating cold storage for storage and marketing of carps, encouraging organic fish farming practice, convincing the farmers through extension personnel for adopted composite fish culture technology, conducting training programmes for the fish farmers, communicating bulletins and newsletter to the farmers and insurance were required for improving the adoption level of the composite fish culture technology.

References

Arpita Sharma and Radha C. Das (2001). Functional training for women in fisheries. Aquacult, 2(2):159-162.

Ayyapan. S. and A. D. Diwan (2004). Road map to increase fish production and productivity India. Fishing Chimes. 24(4): 24-26.

Ayyapan, S. (2000). Technological potential for the development of freshwater Aquaculture. In. M. Krishan and Pratap Birthal (Ed.) Aquaculture Development in India: Problems and Prospects workshop Proceedings 7, National Center for Agriculture Economics and Policy Research, New Delhi. 76-78.

Balasubramaniam, S., M.P. Ramesan and Nikita Gopal (2003). Fish catch variations and associated variables among fishermen operating plank-bulit crafts. Fishery Technology, 40(2): 139-144.

Balasubramaniam, S. and G. Perumal (1990). Adoption Behaviour and impact of Technology Transfer among Fish Farmers. Fishery Technology. 27: 75-79.

Chandran, K.K. (1986). Productivity and conservation of prawns, Sea Food Export Journal, 16(3): 21-25.

FAO (1998). Aquaculture production statistics. Food and Agriculture Organization of the United National, Rome, pp 197.

Goswami, A., Chand., B.K., Ukanta Biswas., Ghosk, R.K. and Roy, K.C. (2004). Impact of training on ornamental fish farming of fishery extension officers. Aquacult, 5(1):109-113.

Govind Gowda, S., M.T. Lakshminarayan and B. Krishnamurthy (2000). Mass media exposure of women leaders a study in Karnataka. Rural India, 63 (10):190-194.

Gupta, M.V., M. A. Mazid., M.S. Islam., M. Rahman and M.G. Hussain (1999). Integration of Aquaculture into the farming system of the floodprone ecosystem of Bangladesh: An evaluation of adoption and impact. ICLARM, Metro manila.

Handbook of Fisheries Statistics (2004). Ministry of Agriculture. Govt. of India.

Jayaraman, R. (1998). An economic analysis of carp culture in Thanjavur District, Tamilnadu state, India. In: Proceeding of the ninth conference of the international institute of fisheries economics and trade. 8-11 July, at the Norwegian College of Fisheries Science, Tromso, Norway. 71-82.

Kumaran, R. (2002). Shrimp farm enterprise budgeting in Thanjavur district of Tamilnadu. Unpub. M.F.Sc.Thesis. FC&RI. TANUVAS. Thoothukudi.

Kumaran, M., K. Ponnusamy and N. Kalaimani (2003). Diffusion and adoption of shrimp farming technologies. Aquaculture Asia, VIII(2): 20-23.

Nagabhushanam, K. and K. Nanjaiyan (2002). Factors of sustainable agriculture in a watershed. Indian Journal of Extension Education, XXXVIII (384): 159-161.

Ponnusamy, K. (2004). Socio–Economic profile of shrimp Entrepreneurs. Fishing Chimes, 23(10&11): 115-117.

Ponnusamy, K., K. Gopinathan., M. Kumaran and M. Krishnan (2001). Constraints in the adoption of shrimp farming in Tamilnadu. Applied Fisheries & Aquaculture. 1(1): 103-105.

Sharma, B.K. (2005). Strategy for Transfer of Composite Fish Culture Technology. In: Lecture notes on composite fish culture and its extension in India. KVK. Central Inland Fisheries Research Institute (ICAR), Kausalyaganga, Bhubaneswar. Orissa.

Sinha, V. R. P. and V. Ramchandran (2002). Freshwater fish culture. Indian Council of Agriculture Research, New Delhi. pp: 4-8.

Stephen Sampath Kumar, J., C. Sudha and K. Anantharaja (2007). Business Opportunities in Aquaculture. Aqua International, 15(10): 23-25.

Talukdar, P.K. and B.S. Sontaki (2005). Correlates Adoption of composite fish culture practices by fish farmers of Assam, India. The Journal of Agriculture Sciences, 1(1) :12-18.

Tripathi, S.D. (2003). Inland fisheries in India. In: Fish for all, National Launch, Kolkata. India: 27-35.

Vasantha Kumar, J. (1986). Report of Research Project (Agri. Eco. 102), Fisheries College and Research Institute. Thoothukudi-8.

Extension Management Strategies for Sustainable Agriculture, pp 129-135

New India Publishing Agency, New Delhi, India

14

Analysis of Knowledge Status on Improved Technologies by Groundnut Farmers

Maraddi, G.N., Nagesh, J. Tulasiram, Meti, S.K. and S.N. Bhat

Groundnut is one of the main oilseed crop of India.Unlike many oilseeds, Groundnut can be consumed directly in various farms and hence now it is also being considered a supplementary food crops. In india groundnut is grown in an area of about 6.5 hectares with a production of about 7 million tones of pods per annum (2007). It is one of the few crops that can be cultivated even on marginal lands under low input conditions. Unfortunately average productivity of groundnut in india is quite low (1000kg) compared to (USA-3000 kg/ha, China- 2600 kg/ha, Argentina-2100 kg/ha and Indonesia-1550 kg/ha). A decade ago, India was one of the major exporter of groundnut in the world, now it has lost that position to China. The USA and Argentina are the other competitors.Many new technologies were developed at university and research stations and tested mainly through verification trails and demonstrations in the farmers field before they are released tom the farmers for adoption. But the research results revealed that wide gaps existed between yield on the farmers' field and those recorded at research stations. This indicates the existence of considerable untapped yield potential. With this in Raichur district, Groundnut crop area is being replaced with *Bt*-cotton , Paddy, Chilli, and sunflower due to delayed in Kharif monsoon and non –availability of quality seeds of new varities, lack of knowledge about pest and disease management, non-availability of timely production technologies to combat operational problems. Inspite of availability of drought tolerant, water and input use efficient varieties are available. Hence, there is an urgent need to study the knowledge domain of respondents towards improved technologies in groundnut cultivation and analyze the relationship of independent variables of the respondents with their dependent variable.

Methodology

The Research study was conducted in Raichur district of Karnataka, two taluks like Lingsugur and Devodurga taluks were selected for the study based on maximum area under groundnut cultivation in the district. Three villages from each selected taluks were randomly chosen, in-turn each village 20 groundnut farmers were selected randomly. This constitutes the total sample size of 120 farmers. Jalhalli, B.ganekal, Kerabgera, Deodurga taluka and Neeralkeri, Anahosur and Kasabalingasgur in linasgur taluka were identified for the research

Results and Discussion

With regard to education (Table 1) it was found that, 37.50 percent of the groundnut farmers were educated up to primary school fallowed by illiterate (31.66%), middle school (24.16), secondary school (4.16%) and collegiate (2.50%). Ground nut growers had low literacy rate because majority growers comes in both illiterate and primary school.With respect to farming experience, it was observed that (Table 1) About half of the respondents (49.17%) possessed low experience followed by medium experience (31.67%) and high experience (19.17%) category. Table 1 shows that, About half the (48.33%) ground nut growers belonged to Semi-medium farmers followed by Small farmers (23.33%), Medium farmers (13.33%), Marginal farmers (10.00%) and big farmers (5.00%). This could be attributed to inheritance of land from their ancestors who might have transformed from generation to generation. Regarding risk orientation (Table 1) of the Groundnut farmers low level of risk orientation was noticed in (45.00%), While medium risk orientation was 35.00 percent followed by high risk orientation (20.00%) category. Nearly half of the respondents possessed lower level of scientific orientation (50.83%) followed by medium (36.67%) and High (12.50%) scientific orientation category. Scientific orientation is the orientation of farmer to adopt new technologies in scientific way. The nature of extension contact revealed (Table 2) that, AAO was known to 76.67 per cent of respondents, Around 32 percent of the respondents were found to contact AAO whenever problem arises followed by once in a year (20.00%) and once in a month (9.17%). Amongst 23.33 percent of the respondents who were aware of ADA and very meager percent of respondents were being contacted with ADA. The SMS of the Department of Agriculture was known to 8.33 per cent and respondents was contacted whenever problem arises (5.83%). The scientist of university were known to 28.33 per cent of the respondents, of which 20.00 per cent of respondents contacted scientist whenever problem arises followed by once in year (9.17%). The field officer of the banking sector were known to 65.00 percent of respondents, of which 53.33 per cent of respondents contacted once in a year followed by once in a month (26.67%).

About 74 per cent respondents were known to other extension agencies like Private agriculture input agencies, dealers, Private consultancy, NGOs, etc, of which 31.67 per cent of respondents contacted once in year followed once in month (26.67).

Table 1: Situational attributes of ground nut farmers **(n-120)**

Sl. No.	Situational attributes	Frequency	Percentage
1. Education level			
1.	Illiterate (nil)	38	31.66
2.	Primary (up to 5 years)	45	37.50
3.	Middle (up to 10 years)	29	24.16
4.	Secondary (up to 12 years)	05	4.16
5.	Collegiate (>12 years)	03	2.50
2. Farming experience			
1.	High (> 34.3)	23	19.17
2.	Medium (12.83-34.43)	38	31.67
3.	Low (<12.83)	59	49.17
3. Land holding			
1.	Marginal farmers (Up to 2.5 acers)	12	10.00
2.	Small farmers (2.5 to 5.00 acers)	28	23.33
3.	Semi-medium farmers (5.00 to 10 acers)	58	48.33
4.	Medium farmers (10 to 25 acers)	16	13.33
5.	Big farmers (>25 acers)	06	05.00
4. Risk orientation			
1.	High (> 8.73)	24	20.00
2.	Medium (6.11 -8.73)	42	35.00
3.	Low (<6.11)	54	45.00
5. Scientific orientation			
1.	High (> 10.59)	15	12.50
2.	Medium (8.01-10.59)	44	36.67
3.	Low (<8.01)	61	50.83

Table 2: Distribution of Groundnut growers according to their Extension contact **(n=120)**

S.No	Type of contact	Awareness	Frequency of Contact			
			Once in month	Once in year	Whenever prob. arises	Never
1.	AAO	92 (76.67)	11 (9.17)	24 (20.00)	38 (31.67)	32 (26.67)
2.	ADA	28 (23.33)	00 (00)	04 (3.33)	06 (5.00)	95 (79.17)
3.	Subject Matter Specialist	10 (8.33)	00	00	07 (5.83)	62 (51.67)
4.	Scientist of University	34 (28.33)	04 (3.33)	11 (9.17)	24 (20.00)	89 (74.17)
6.	Bank Officers	78 (65.00)	28 (23.33)	64 (53.3)	10 (00)	28 (23.33)
7.	Others like Private Extn. worker, Input agencies, & Consultancy	88 (73.3)	32 (26.67)	38 (31.6)	12 (10)	10 (8.33)

Knowledge Status of Groundnut Growers about Improved Technologies

More than half (Table 3) of the respondents had (53.33%) partial knowledge about land preparation followed by full knowledge (25.83%) and around 20 percent of ground nut growers had no knowledge at all. It is due to lack of contact with extension officers of department of agriculture and scientists of Agricultural Universities and exposure to TOT system. About 55 percent of groundnut growers had full knowledge about recommended varieties like TMV-2, R-2001-03 followed by partial knowledge category (42.50%). The possible reason why majority of beneficiaries knowledge ranges from partial knowledge to full knowledge. Since seed is basic input, that has being influenced growers to know the knowledge about varieties and source of availability and their rates and available subsidies and its comparison to existing market rates. The results are in line with the findings of Sriram and Palaniswamy (2003) and Maraddi et al., (2006).

Table 3: Knowledge Status of groundnut growers about improved technologies **(n=120)**

SI No	Technologies	Knowledge status					
		FK	%	PK	%	NK	%
1.	Land Preparation	31	25.83	64	53.33	25	20.83
2.	Varieties	65	54.16	51	42.50	04	03.30
3.	Seed rate	34	28.33	65	54.16	21	17.50
4.	Organic manure	06	5.00	72	60.00	42	35.00
5.	Seed treatment (bio-agents)	06	5.00	32	26.00	82	68.33
6.	Fertilizer management	12	10.00	85	70.83	23	19.16
7.	Spacing	31	25.83	78	65.00	11	9.16
8.	Inter cultivation	66	55.00	54	45.00	00	0.00
9.	Pest Management	08	06.67	38	31.67	70	58.33
10.	Disease Management	05	04.16	34	28.33	81	67.50

Note: FK-Full Knowledge, PK-Partial Knowledge, NK-No knowledge

Around half of growers had partial knowledge about spacing (54.16%) followed by full knowledge 28.16% and no knowledge category consists of 17.50 percent growers. Majority of ground nut growers belong to partial to full knowledge category related to seed rate is concerned, this could be due to bulkiness of seed nature, costly seeds and consulting AAO's during purchase of seed material through subsidies. More than half of groundnut growers (60%) had partial knowledge of organic manures followed by no knowledge category (35.00%) and full knowledge category consists of only 5 percentage. Partial knowledge level on land preparation, spacing, seed rate & organic manure, may be due to fact that, these simple practice requires no additional cost. In spite of above reason the results are inconformity with findings of Sriram and Palaniswamy (2000) and Goud (1998). This points out that, a large number of ground nut growers still lack of information about biofertilizers, vermicompost, composting and other organic manures. The possible reason for these situations might be low risk orientation; less contact with extension agencies and poor dissemination system. Majority of ground nut growers (68.33%) belonged to no knowledge category followed by partial knowledge (26.00%) and full knowledge category of 5 percent. This points out that, a large number of ground nut growers still lack of knowledge about seed treatment particularly biofertilizers like Rhizobium, Trichoderma, and PSB etc. Two third portions of groundnut growers had partial knowledge of fertilizer management followed by no knowledge (19.16%) and full knowledge category of (10.00%). Regarding spacing and intercultivation around half of the respondents belonged to partial knowledge category followed by full and no knowledge category. With respect to pest management, Around half of the respondents belonged to no knowledge category followed by partial knowledge category (31.67%) and full knowledge category constitute only (6.67%) percent. Similarly disease management more than half of the respondents belonged to no knowledge category followed by partial knowledge category (28.33%) and full knowledge category (28.33%).

Table 4: Correlation coefficients of independent variables of Groundnut farmers with their knowledge level of Improved Technologies **(n=120)**

Sl. No.	Independent variables	Correlation coeffcients (r) with knowledge level of Improved Technologies
1.	Age	0.068NS
2.	Education	0.477**
3.	Land holding	0.142 NS
4.	Farming experience	0.257*
5.	Risk orientation	0.240*
6.	Scientific orientation	0.134 NS
7.	Management orientation	0.326**
8.	Achievement motivation	0.398**
9.	Innovative proneness	0.256*
10.	Extension contact	0.349**
11.	Organizational participation	0.068 NS
12.	Annual income	0.026 NS

* Significance at 5% level of probability

** Significance at 1% level of probability

The possible reason for the higher use of chemical measures of pests due to the fact that farmers are more relay on pesticides because of its efficiency in controlling pests and its quick in pest control than biological agents, mechanical tool and other ITK's. In spite of conducting training programmes, demonstrations on IPM practice like use of bioagents, collection and destruction of affected plants and other eco-friendly tools were not known to majority of ground nut growers, because the farmers might not have understood the concepts due to low literacy rates, lower extension contact , lower risk orientation, lower scientific orientation and little area coverage by these programmes. The findings of the study are disagreement with the findings of Maraddi et al. and Laxminarayan (1997). Correlation analysis revealed that, education, management orientation, achievement orientation and extension contact of Groundnut farmers showed positive and significant relationship with their knowledge level of improved technologies at 1 percent level of probability. Whereas farming experience, risk orientation, innovative proneness of groundnut farmers exhibited positive and significant relationship with 5 per cent level of probability.

Conclusion

It can be concluded that more than half of the respondents had partial knowledge, which indicates that groundnut growers were in take-off stage regarding knowledge level of improved of improved technologies. It was observed that, majority of respondents found higher knowledge with respect to land preparation, where as lower knowledge was noticed in pest and disease management, Bio-fertilizers & seed treatment with bio-agents. Effective linkage between research, extension and farmers should help in identifying the problems faced by farmers in the adoption of improved technologies, ultimately arriving at the most appropriated solutions through on-farm research and on-farm trials.

References

Sriram, N. and Palaniswamy, A. (2000). Cotton growers extent of awareness about the eco-friendly agriculture practices in cotton. *Journal of Extension Education,* 11(2):27-90.

Lakshminarayan, M. T. (1997). Adoption of sustainable sugarcane farming practices – An analysis. *Ph. D. Thesis,* University of Agricultural Sciences, Bangalore.

Maraddi, G. N. (2006). An Analysis of Sustainable Cultivation Practices followed by Sugarcane Growers in Karnataka. Ph. D. Thesis, University of Agricultural Sciences, Dharwad.

Goud, J. V. (1998). Sugarcane Cultivation Progress and Prospects, Karnataka Institute of Applied Agricultural Research, Sameerwadi, Karnataka, p. 4.

Extension Management Strategies for Sustainable Agriculture, pp 137-147

New India Publishing Agency, New Delhi, India

15

Assessment on Skills of Stakeholders on Precision Farming

Padma S.R. and T. Rathakrishnan

In any work environment the managers, facilitators are equipped with some specific type of skills to perform their activity effectively. In this study, multi stakeholders were involved in the process of Precision Farming from land preparation to marketing. Every stakeholder has a specific task to be performed so as to lead the client in the direction desired. In order to achieve this, the skill components *viz*., Management and Facilitating skills possessed by them and their own perception on those skills were assessed.

Research Methodology

Management skills constitutes strategic skill, influencing skill, leadership skill, implementation skill and personal factors. These skills were assessed on various items in terms of its importance and the competency. The sample stake holders have to place the relative importance of the items under each component and indicate their level of competency. This component was discussed by using Percentage analysis.

As a skilled facilitator, he /she have to demonstrate several practises of Precision Farming to the clients. The extent to which the stake holder performing his activity on the five type of skills viz., Presentation skill, Relationship skill, Learning environment Management skill, Continuous learning attitude skill and Empathic attitude were taken as a measure.The items under each skill were measured based on its usage viz., Rarely, Sometimes, Frequently and Always. Percentage analysis was performed to discuss the skill components.

The items under the said two skills, components were collected from the various management studies and through perusal of literatures. Further discussion was made with the advisory committee members and the faculty of Agricultural

Management and thus the items were finalised and included in the final data collection tool.

Findings and Discussion

Management Skill

In the Precision Farming process multi stake holders viz., Researchers, Extension personnel, Officials of State Department of Agriculture (SDA), State Department of Horticulture (SDH), input dealers, drip marketers, Agri clinic consultants and marketing

personnel have been involved from the planning stage itself. Some of the stakeholders involvement was more in the early stage of the farming practices and as and when required their services have been sought by the sample.

Accordingly, mostly on need basis, their services were made available to the farmers. There are two major skills *viz.*, Management and Facilitation skills are much essential to carry out their tasks, pertaining to Precision Farming. As much as skill they possessed in the said aspect, their competency in tackling the issues related to farming will be higher. Under this section the management skill was assessed in terms of Strategic, Influencing, Leadership, Implementation Skills and Personal factors were assessed.

In this part, the stakeholders themselves have been requested to give their response in terms of the importance on each item, as they felt and also the relative competency he possessed against each. The responses thus collected have been furnished in Table 1.

Strategic Skill

It is quite interesting to observe that all the seven items under this skill were felt as important by the stakeholders. No item was indicated under not important. Among the items, Strives for continuous improvement (48 per cent), Able to manage the Changes effectively, (46 per cent),brings innovative ideas, concepts to the farmers (44 per cent), demonstrate critical thinking (40 per cent) were felt as very important by nearly half of the sample.

Also the items like the stakeholders are oriented towards customers (70 per cent). Keeps abreast of changes in the commercial environment (64 per cent), Innovative and critical thinking (60 per cent) were expressed as important.

In general, it could be concluded that the stakeholders perceived their roles perfectly and hence they felt all the seven items as important in concern with their strategic skill.

With regard to competency level, it has been noticed that more than 60 per cent of them expressed their competency on innovative and tries new approaches, demonstrates critical thinking, strives for continuous improvement, brings innovative ideas, concepts to the farmers and keeps abreast of changes in the commercial environment. However, the competency in the items like, manages changes effectively, keeps abreast of changes and oriented towards customer needs improvement as stated by them.

Influencing Skill

The stakeholders, to what extent they possess the skill in influencing the clients was studied on five items as furnished below.

Here also no one was found under not important category. All the five items were felt as important to very important by the stakeholders in order to build effective relationships, Negotiates win-win outcomes, influencing skills are very important as felt by them.

While observing the competency level, build effective relationships, clear communicator, Negotiates win-win outcomes, they possess competency. Despite exhibits flexibility, influence effectively and directly, needs improvement as stated by more than one-third of the sample.

Leadership Skill

Being a stake holder an individual must have the ability to organise and lead the team towards the direction deemed desirable. For which, he must equipped with the leadership skill. This was assessed on five items.

All the five items assessed under leadership skills were felt as important to very important. However, a few sample had not felt the importance on four items.

Regarding the competency, 94 per cent of the sample had high competency in the item leads a high performing team. And empower the farmers to achieve result (64 per cent). The competency level needs improvement on managing performance (44 per cent) and Creates a sense of shared purpose (26 per cent) as expressed by the sample.

Implementation Skill

Success of any programme depends on the extent to which it is implemented as programmed.In order to bring out the success the stakeholders involved in the process must possess skill on implementation process. Five items were included to assess the implementation skill of stakeholders and the results have been furnished in Table 1.

Table 1: Management Skills of stakeholders **(n=50)**

Sl.No	Skills	Importance %			Competency %		
		Very Important	Not Important	Important Competent	Highly	Competent Development	Needs
I. Strategic Skills							
1.	Brings innovative ideas, concepts to the foarmers	44.00	56.00	—	34.00	62.00	4.00
2.	Demonstrates critical thinking	40.00	60.00	—	6.00	76.00	18.00
3.	Keeps abreast of changes in the commercial environment	36.00	64.00	—	6.00	52.00	42.00
4.	Is oriented towards the customer	30.00	70.00	—	8.00	60.00	32.00
5.	Is innovative and tries new approaches	38.00	62.00	—	8.00	78.00	14.00
6.	Strives for continuous improvement	48.00	52.00	—	34.00	64.00	2.00
7.	Manages change effectively	46.00	54.00	—	8.00	54.00	38.00
II. Influencing Skills							
8.	Is a clear communicator	48.00	52.00	—	32.00	66.00	2.00
9.	Influences effectively, directly, and indirectly	46.00	54.00	—	8.00	54.00	38.00
10.	Exhibits flexibility	44.00	56.00	—	6.00	42.00	52.00
11.	Negotiates win-win outcomes	54.00	46.00	—	24.00	64.00	12.00
12.	Builds effective relationships	70.00	30.00	—	16.00	82.00	2.00
III. Leadership Skills							
13.	Creates a sense of shared purpose	44.00	42.00	14.00	18.00	56.00	26.00
14.	Leads a high performing team	46.00	38.00	16.00	94.00	6.00	—
15.	Empowers the farmers to achieve results	42.00	40.00	18.00	26.00	64.00	10.00
16.	Manages performance	48.00	44.00	8.00	—	56.00	44.00

(Contd.)

Table Contd.

IV. Implementation Skills						
17. Plans ahead to meet the needs of the farmers/clients	36.00	54.00	10.00	24.00	76.00	—
18. Demonstrates effective organising skills	32.00	48.00	20.00	8.00	92.00	—
19. Manages time effectively	42.00	40.00	18.00	34.00	44.00	22.00
20. Demonstrates sound decision-making and problem-solving skills	26.00	48.00	26.00	40.00	48.00	12.00
21. Sees things through despite setbacks	22.00	54.00	24.00	8.00	44.00	48.00
V. Personal Factors						
22. Shows integrity	36.00	56.00	8.00	22.00	74.00	4.00
23. Develops self	40.00	52.00	8.00	18.00	78.00	4.00
24. Shows flexibility	52.00	48.00	—	14.00	50.00	36.00
25. Demonstrates drive and desire to succeed	34.00	52.00	14.00	16.00	72.00	12.00

Among the item studied under implementation skill, all the items were felt as important to very important by 74 per cent to 90 per cent of the sample. Whereas 10 per cent to 26 per cent of the sample in varied degree placed as not important on the five items. While observing the competency level of the sample on implementation skill, it is revealed that 92 per cent of them had the competency in demonstrating effective organisational skills followed by plans ahead to meet the needs of the client.

However, 48 per cent and 22 per cent of the sample respectively needs improvement on the skills *viz*., sees things through, despite setbacks and manages time effectively.

Personal Factors

Among the skills studied under major dimensions of management skill, how a stakeholder assesses his own personal factors in terms of its important and the competency he had on those items were discussed under the personal factors.

As observed in other types of skills, here also the importance of each item *viz*., integrity, flexibility, self development were felt by the sample as important and a negligible portion of the sample found under not important.

Interestingly, except flexibility in all the other 3 items more than 70 per cent of the sample stated their competency level. Highly competent was observed with 14 to 22 per cent of the sample on the items assessed. More than one-third of the sample expressed that the flexibility level is to be improved.

The foregoing discussion on the 25 items studied under five dimensions of management skill revealed that majority of the sample felt the importance of the items studied. Only a meagre portions indicated some of the items as not important. However, no item was felt as not important under strategic and influencing skills. This showed that every stakeholder when performing a function, acting as a manager. He plays the management role in the entire process of a event like Precision Farming.

Under this process the skill, possessed by stakeholders played a major role in influencing the clients and take up the new venture and helping them in implementing the same.

The success totally rest with the skills, the stakeholders possessed. The skill component reflects on their competency level. The study also revealed that majority indicated their competency on various skills and also suggested some of the areas need to be improved. The competency in skill and their technical knowledge have to be complementary. That also served in the technological utilization pattern, perceived effectiveness and market associated activities of the precision farmers stated elsewhere in the report. The result thus indicated that the stakeholders performed their roles as expected in the Precision Farming approach.

Facilitating Skills

In the technology dissemination process, it is unavoidable for the stakeholder to see that the delivered information should reach the clients and adopted in their areas concerned. The continuous personal touch is essential for enabling the clients to take up without any inhibition. Here the facilitating process of stakeholders plays a vital role. In this study, the facilitation skill of the stakeholders were studied under five dimensions viz., Presentation skill, Relationship skill, Learning Environment Management, Continuous learning attitude, Relationship skill Empathic attitude. Totally 20 items were assessed under these major five sub components. The responses were obtained on a four point continuum viz., rarely, sometimes, frequently and always and plotted in Table 2.

Table 2: Facilitating Skills of stakeholders **(n=50)**

Sl.No	Skills	Rarely %	Sometimes %	Frequently %	Always
I. Presentation skill					
1.	Being clear and to the point	8.00	14.00	44.00	34.00
2.	Effectively & accurately capturing participants comments	—	10.00	48.00	42.00
3.	Maintaining eye contact with group	—	—	46.00	54.00
4.	Speaking in easily understandable language	10.00	32.00	36.00	22.00
5.	Ensuring body language and voices are appropriate	38.00	14.00	32.00	16.00
6.	Giving clear instructions	—	4.00	50.00	46.00
II. Relationship skill					
7.	Showing respect for participants diverse experience and perceptions	—	20.00	44.00	36.00
8.	Demonstrating the ability to listen	4.00	16.00	56.00	24.00
9.	Acting in supportive and helpful ways to participants	22.00	24.00	36.00	18.00
10.	Maintaining a high level of positive regard for participants diverse capabilities	—	44.00	38.00	8.00
11.	Showing sensitivity to participants different learning and developmental situations	8.00	24.00	42.00	26.00
12.	Displaying congruence between words, tone & body language	12.00	20.00	46.00	22.00

III. Learning Environment Management				
13. Establishing an environment supportive of individual self-esteem, morale and commitment to the organisation	—	56.00	30.00	14.00
14. Demonstrating the view that mistakes are positive learning opportunities	4.00	30.00	48.00	18.00
15. Encouraging self-directed learning	—	26.00	46.00	28.00
IV. Continuous learning attitude				
16. Demonstrating the capacity to learn from participants	18.00	34.00	32.00	16.00
17. Applying the principle of co-learning.	6.00	18.00	54.00	22.00
18. Demonstrating a non-defensive reaction to feedback	4.00	32.00	38.00	26.00
V. Empathic attitude				
19. Using appropriate language	10.00	22.00	38.00	30.00
20. Maintaining an awareness and respect for another person as he is	8.00	16.00	42.00	34.00

Presentation Skill

Among the presentation skills, Maintaining eye contact with group members was followed by cent per cent of the sample. This was closely followed by effectively and accurately capturing participants comments (90 per cent) and being clear and to the point were also frequently followed. As far as the presentation skill is concerned, speaking in easily understandable language and ensuring body language and speak in appropriate voices were sometimes and rarely followed by more than two-fifths of the sample.

Relationship Skill

The degree of relationship possessed by a stakeholder with the farming community will ensure the guidelines, directions if any passed on by them will easily be accepted by the farmers. This will speed up the Precision Farming activity. Relationship skill of the stakeholders was studied on six items as indicated. It could be observed that Showing respect for participants' diverse experience and perceptions and Demonstrating the ability to listen were always followed by 80 per cent of the sample. Showing sensitivity to participants different learning and developmental situations and displaying congruence between words, toneand body language were also frequently followed by more than two-thirds of the sample. Whereas maintaining a high level of positive regard for participant's diverse capabilities was sometimes followed by 44 per cent of the sample.

Learning Environment Management

The effective learning environment provided by the facilitators will help the participants to learn more and gain much experience on the subject/ skills taught. In order to create an effective environment for learning the stakeholders must possess the facilitating skill to the expected level. The skill was assessed on three items in terms of, self-esteem positive learning and self-directed learning. About 44 per cent to 74 per cent of the sample effectively managed the learning environment. That showed that they might have possessed good facilitation skills. All the three items assessed under this sub component was followed by the sample in varied frequency.

Continuous learning attitude

This part deals with Continuous learning attitude of the stakeholders in terms of learning capacity, co-learning and non-defensive reaction. The result showed that principle of co-learning and non-defensive reaction was frequently followed by the sample whereas half of the sample alone frequently learns from the participants.

Empathic attitude

Establishing and maintaining the relationship with the precision farmers and place empathy on the clients is one of the successful factors required for a facilitator. That will ensure him to play his role according to the level of the clients. Two items studied showed that more than two-thirds of stakeholders frequently followed this items.

It would be inferred among the sub components studied on the facilitation skill of the stakeholders that being clear to the point, maintain eye contact with group members, giving clear instructions, showing respect for participants' diverse experience, ability to listen, sensitivity to participants, self-esteem, self directed learning, Co-learning, non-defensive reaction, respect for another person were frequently followed by the stakeholders.

Being a facilitator in the process of Precision Farming the skills he possessed as stated above certainly helping the precision farmers to learn and gain more from stakeholders. The stakeholders intervention as and when required by the sample was observed to the expected level as stated elsewhere.

The need based and voluntary intervention was also noticed in the study as perceived by the sample. This result could be substantiated for such a type of observation among the sample.

Principle component analysis- skill required by stakeholders

The data on the skill levels of the stakeholders were subjected to Principal Component Analysis (PCA) to identify the underlying structures (components) in the data. For PCA analysis, the data was entered as correlation matrix and the components were extracted. After extraction, the components were rotated using varimax approach.

The PCA estimates Kaiser-Meyer-Olkin (KMO) value of 0.63 was above the minimum level of 0.5, and a significant Bartlett's Chi-square (÷2=217.80, p<0.01) indicated that the sample size chosen for this investigation was adequate (Field, 2000).

The PCA extracted six factors that exceeded the Eigen value of one and explained 71 per cent of the total variance in the data, which is close to the minimum level of 60 per cent variance necessary for social science research (Hair *et al.*, 1998). Since the sample size is 50, a factor loading of 0.72 was used as a lower cut-off value for selection of variables for each factor (Field, 2000). The rotated component matrix showing the principal components along with loaded variables is given in the following table.

Table 3: Rotated component matrix of stakeholder skills

S.No	Skills	Principle Component Loadings*		
		PC1	PC2	PC3
1.	Strategic Skill	**0.855**	-0.052	0.099
2.	Influencing Skill	**0.874**	-0.061	0.107
3.	Leadership Skill	**0.818**	0.210	-0.192
4.	Implementation Skill	**0.751**	0.136	-0.190
5.	Personal Factors	**0.765**	-0.171	0.429
6.	Presentation Skill	-0.105	0.281	**0.819**
7.	Relationship Skill	-0.056	0.398	**0.738**
8.	Learning environment	-0.036	**0.877**	0.180
9.	Continuous Learning Attitude	0.140	**0.771**	0.125
10.	Empathic attitude	0.128	0.293	**0.771**
	Eigen Value	**3.31**	**2.65**	**1.06**
	Variance explained	**32.28 %**	**20.30 %**	**17.50 %**

* Values shown in bold letters indicate significant loading (>0.72)

The data displayed in the above table shows that the Strategic, Influencing, Leadership, Implementation skills and Personal factors loaded significantly on PC1, which explained 32.28 per cent variation in the data. Based on the factors loaded, this component may be named as *"Management skills"*. Learning environment and continuous learning attitude have loaded in PC2 (20.30per cent) which may be termed as "*Learning skills*". The presentation, relationship and emphatic skills loaded in PC3 (17.50 per cent) which may be named as "*Human relation skills*".

Conclusion

A stakeholder is any entity with a declared or conceivable interest or stake in a policy concern. The range of stakeholders relevant to consider for analysis varies according to the complexity of the reform area targeted and the type of reform proposed and, where the stakeholders are not organized, the incentive to include them. By analysing Management and Facilitating skills possessed by them and their own perception on those skills, could identify the areas need to be strengthened for effective implementation of precision farming.

Extension Management Strategies for Sustainable Agriculture, pp 149-159

New India Publishing Agency, New Delhi, India

16

Impact of Polambadi (Farmer Field School) on Knowledge Level of Groundnut Farmers Regarding Improved Cultivation Practices

S. Sreenivasulu, P.K. Jain and T.P. Sastry

Farmer Field School (FFS) is a capacity building method based on adult education principles for group of farmers. FFS is best described as a *'school without walls'*, where farmers learn through observation and experimentation in their own fields. FFS methodology is potential extension approach to enhance farmer's knowledge which is a basis for adoption of improved practices (Manoj and Vijayaragavan, 2014). In Andhra Pradesh Department of Agriculture initiated FFS programme in the name of Polambadi during 2006. Groundnut was major crop in Chittoor district of Andhra Pradesh grown in area of 1.62 lakh ha with a production of 1.91 lakh tonnes and with an average yield of 864 kg per ha. If the farmers have proper knowledge about the improved production technologies of groundnut, it would be possible to achieve their targeted yields. About 805 polambadis (FFSs)were organized from 2009 to 2012 in groundnut in Chittoor district of Andhra Pradesh to improve the knowledge of improved cultivation practices (*Source*: Department of Agriculture, Chittoor). It is essential to assess the impact of polambadi (FFS) on knowledge level of groundnut farmers with regard to improved cultivation practices. Hence, the present study was conducted with the following objectives:

1. To assess the knowledge level of participant and non participantfarmers of groundnut polambadi (FFS) regarding improved cultivation practices.
2. To know the detailed and comprehensive analysis of knowledge items of participant and non participant farmers of groundnut polambadi (FFS) about improved cultivation practices.

Material and Methods

The study was conducted in the year 2012-13 in Chittoor district of Andhra Pradesh. *Expost-facto* research design was followed in the present study. Twelve mandals, four each from three revenue divisions of the district were selected randomly for the study. Twelve villages from twelve mandals were selected randomly. Ten participant and ten non participantfarmers were selected from each selected village by random sampling method.Thus, making a sample size of 240 farmers i.e. 120 participant and 120 non participant farmers. The relevant information pertaining to knowledge on improved groundnut cultivation practices was collected, scored, quantified, categorized, tabulated and interpreted using statistical methods like mean, standard deviation, and 't' test. To measure the knowledge of participant and non participantfarmers about improved groundnut cultivation practices a knowledge test was developed based on the following steps:

Item collection: Forty nine knowledge test items of groundnut were arranged in different types as Yes/No, True/False, Multiple Choice and Fill in the blanks with the help of relevant experts.

Selection of items: Selection was carried out by applying standard criteria such as item should promote thinking rather than memorization, it should differentiate well informed respondent from the less informed and items included should cover all areas of knowledge about improved groundnut cultivation practices.

Item analysis: Forty nine items were administered to thirty randomly selected non sample respondents for their response. Item analysis yielded item difficulty index, item discrimination index and point biserial correlation.

Final selection of the items for the test: Items with difficulty indices ranging from 25 to 80, discrimination indices ranging from 0.20 to 0.80 and items having point biserial correlation values > 0.25 were considered for inclusion in the final knowledge test. Thus, final knowledge test consisted of 40 items related to improved groundnut cultivation practices.

Results and Discussion

The overall knowledge level of participant and non participant farmers regarding improved groundnut cultivation practices presented in the Table 1. The results indicated that more than half (59.17%) of participant farmers of groundnut Polambadi (FFS)had medium level of knowledge followed by high (25.83%) and low (15.00%) level of knowledge about improved groundnut cultivation practices. Whereas, less than half (44.17%) of non participant farmers had medium knowledge level followed by low (34.17%) and high (21.67%) knowledge levels.

Table 1: Distribution of participant and non participantfarmers of groundnut polambadi (FFS) based on their knowledge level **(n=120)**

S.No.	Category	Participant farmers		Non Participant farmers		t-value	p-value
		F	%	F	%		
1.	Low knowledge	18	15.00	41	34.17		
2.	Medium knowledge	71	59.17	53	44.17	8.483**	0.000
3.	High knowledge	31	25.83	26	21.67		
	Total	120	100	120	100		
	Mean	27.48		13.68			
	S.D.	11.81		13.34			

** Significant at 0.01 probability level

The study revealed that the participation of farmers in polambadi (FFS) helped them to gain significant level of knowledge with regard to improved groundnut cultivation practices. The polambadi (FFS) methodology is based on adult education principles and a learner centered education process has empowered the groundnut farmers in increasing their knowledge with regard to improved cultivation practices. The low knowledge level of particpantfarmers may be due to lack of participation in polambadi (FFS), low extension participation and low mass media exposure when compared to non participant farmers of groundnut polambadi (FFS). The findings are in confirmation with those of Manoj and Vijayaragavan (2014), Sunitha *et al.* (2013), Sreenivasulu (2011) and Yeshwant Kumar Naik (2008).

Independent 't' test was carried out to assess the significant difference in mean knowledge between participant and non participant farmers of groundnut polambadi (FFS). It is observed from the Table 1 that calculated 't' value is greater than that of 't' table value at 0.01 level of probability. Thus it is clear that there was a significant mean difference between knowledge of participant and non participant farmers about improved cultivation practices of groundnut. When the respective means were compared it was observed that mean knowledge score of participant farmers (27.48) was higher than that of non participant farmers (13.68). This indicates that participant farmers had higher level of knowledge than the non participant farmers about improved groundnut cultivation practices.

To know the detailed and comprehensive idea on the knowledge level of participant and non participant farmers about improved groundnut cultivation practices, an item wise response analysis was carried out and the results were presented in the Table 2. The analysis of 120 participant farmers compared with 120 non participant farmers on 40 specific improved cultivation practices of groundnut.

Majority (94.17%) of participant farmers had correct knowledge on summer ploughing practice whereas, 56.67 per cent of non participant farmers had correct knowledge. With regard to seed treatment with Mancozeb reduces seed borne diseases, 97.50 per cent of participant farmersand 61.67 per cent of non participant farmers had correct knowledge. Regarding seed rate required for kharif groundnut majority (81.67%) of participant and fifty per cent of non participant farmers had correct knowledge. More than half (54.17%) of the participant farmers had correct knowledge about pre emergence herbicide used in groundnut crop, but only 21.67 per cent of non participant farmers had correct knowledge. Regarding application of phosphorous fertilizers 60.00 per cent of participant farmers had correct knowledge, whereas only 35.00 per cent of non participant farmers had correct knowledge. Majority (90.83%) of participant farmers and 60.83 per cent non participant farmers had correct knowledge about sprinkler irrigation, respectively. Eight five per cent of participant farmers and more than half (54.17%) of non participant farmers had correct knowledge on red gram as an intercrop in groundnut gives additional returns. With respect to red hairy caterpillar and its egg laying nature, 57.50 per cent of participant farmers and 31.67 per cent of non participant farmers had correct knowledge. Regarding Lady Bird beetle, majority (87.50%) of participant farmers had correct knowledge and less than half (48.33%) of the non participant farmers had correct knowledge. Regarding transmission of PSND by thrips, majority (92.50%) of participant farmers had correct knowledge but only 45.83 per cent of non participant farmers had correct knowledge.

With respect to Imidachloprid used as seed treatment chemical to control sucking pests in groundnut, majority (89.17%) of participant farmers had correct knowledge but only 40.00 per cent of non participant farmers had correct knowledge. Majority (81.67%) of participant farmers had correct knowledge on treat the seed first with bio agent followed by insecticide and fungicide whereas, only 29.17 per cent of non participant farmers had correct knowledge. More than half (54.17%) of the participant farmers had correct knowledge on Kalahasthi variety is resistant to nematodes whereas, in case of non participant farmers only 30.00 per cent had correct knowledge. About 89.17 per cent of participant farmers had correct knowledge on usage of complex fertilizers increases cost of cultivation whereas, only 48.33 per cent had correct knowledge in case of non participant farmers. Eighty five per cent and fifty per cent of participant and non participant farmers had correct knowledge on drought tolerant varieties in groundnut, respectively. Regarding trap crop for Spodoptera, majority

(82.50%) of participant farmers had correct knowledge and only 45.83 per cent of non participant farmers had correct knowledge. Ninety per cent and 59.17 per cent of participant and nonparticipant farmers had correct knowledge on major stored grain pest in groundnut, respectively.

With respect to polambadi (FFS) and its aims, majority (92.50%) of participant farmers had correct knowledge but only 21.67 per cent of nonparticipant farmers had correct knowledge. Regarding symptoms of iron deficiency in groundnut, 45.83 per cent and 14.17 per cent of participant and nonparticipant farmers had correct knowledge, respectively. With regard to optimum time of sowing for *rabi* groundnut, majority of participant (95.83%) and nonparticipant (55.83%) farmers had correct knowledge. Majority (72.50%) of participant farmers had correct knowledge on optimum plant population per sq m in *kharif* groundnut whereas, only 20.00 per cent of nonparticipant farmers had correct knowledge. With respect to quantity of zinc sulphate used per acre to overcome zinc deficiency, majority (85.83%) of participant farmers had correct knowledge but only 45.83 per cent of nonparticipant farmers had correct knowledge. With respect to quantity of nitrogen, phosphorus and potassium used per acre in groundnut crop, 40.00 per cent of participant farmers had correct knowledge. In case of nonparticipant farmers, only 11.67 per cent had correct knowledge on quantity of nitrogen, phosphorus and potassium used per acre in groundnut crop. Majority (90.00%) of participant farmers had correct knowledge on quantity of gypsum used per acre whereas, in case of nonparticipant farmers it is only forty per cent. Majority (84.17%) of participant farmers had correct knowledge on the role of azadirachtin in pest management whereas, only 48.33 per cent of nonparticipant farmers had correct knowledge. With regard to pest defender ratio followed in IPM, one fourth of participant (25.83%) and less number of nonparticipant (4.17%) farmers had correct knowledge. With respect to Economic Threshold Level (ETL) of Red Hairy Caterpillar, leaf miner and tikka leaf spot only 28.33, 31.67 and 22.50 per cent of participant farmers had correct knowledge, respectively. In case of nonparticipant farmers none of them had correct knowledge on Economic Threshold Level of Red Hairy Caterpillar, leaf miner and tikka leaf spot.In case of egg parasitoid used for control of Spodoptera, only 20.83 per cent of participant farmers had correct knowledge and none of the nonparticipant farmers had correct knowledge.

More than half (55.00%) of the participant farmers had correct knowledge on post emergence herbicide used for control of grasses in groundnut but, only 21.67 per cent of nonparticipant farmers had correct knowledge. Forty five per cent of participant farmers had correct knowledge on quantity of borax used to control boron deficiency and only five per cent of nonparticipant farmers had correct knowledge. Sixty per cent of participant and forty per cent of nonparticipant farmers had correct knowledge on bacteria present in root nodules helps in fixation of nitrogen. Majority (80.83%) of participant and less than half (46.67%) of the nonparticipant farmers had correct knowledge on sowing of 4-8 rows of bajra or jowar or maize as guard crop can reduce the Peanut Stem Necrosis Disease (PSND) in groundnut. More than half (54.17%) of participant farmers had correct knowledge on basal application of 8 kg phorate granules checks root grub whereas, only 28.33 per cent of nonparticipant farmers had correct knowledge. Eighty five per cent of participant and 61.67 per cent nonparticipant farmers had correct knowledge on spraying of urea overcomes drought situation. Majority of participant (81.67%) and nonparticipant (65.83%) farmers had correct knowledge on critical stages for irrigation in groundnut. With respect to monitoring of Spodoptera by using pheromone traps, majority (86.67%) of participant farmers had correct knowledge and only 30.83 per cent of nonparticipant farmers had correct knowledge. Regarding dosage of NPV used for control of Spodoptera, 31.67 per cent of participant farmers had correct knowledge and none of the nonparticipant farmers had correct knowledge. Half (50.83%) of the participant farmers and 37.50 per cent of nonparticipant farmers had correct knowledge on drying of pods.

Item wise response analysis indicated that majority of the participant farmers had more knowledge on improved cultivation practices of groundnut than non participant farmers and it is due to the participation of season long polambadi (FFS) programme.

Table 2: Knowledge level of participant and non participant farmers regarding specific improved cultivation practices of groundnut

S. No.	Knowledge item	FFS farmers **(n=120)**				non-FFS farmers **(n=120)**			
		Incorrect		Correct		Incorrect		Correct	
		F	%	F	%	F	%	F	%
1.	Deep summer ploughing helps to reduce pest and disease incidencein groundnut crop YES/NO	113	94.17	7	5.83	68	56.67	52	43.33
2.	Seed treatment with Mancozeb reduces seed borne diseases YES/NO	117	97.50	3	2.50	74	61.67	46	38.33
3.	Seed rate required for *kharif* groundnut is 50-60 kg per acre YES/NO	98	81.67	22	18.33	60	50.00	60	50.00
4.	*Imazethapyr* is used as pre emergence herbicide in groundnut YES/NO	65	54.17	55	45.83	26	21.67	94	78.33
5.	Phosphorus can be applied in split doses (basal and top dressing) YES/NO	72	60.00	48	40.00	42	35.00	78	65.00
6.	Sprinkler irrigation saves 20-30% of irrigation water YES/NO	109	90.83	11	9.17	73	60.83	47	39.17
7.	Red gram sown as intercrop in groundnut (1:7 ratio) to get additional returns YES/NO	102	85.00	18	15.00	65	54.17	55	45.83
8.	Red Hairy Caterpillar lays eggs in single YES/NO	69	57.50	51	42.50	38	31.67	82	68.33
9.	Lady bird beetle is harmful insect in groundnut YES/NO	105	87.50	15	12.50	58	48.33	62	51.67
10.	Thrips transmit Peanut Stem Necrosis virus Disease (PSND) in groundnut YES/NO	111	92.50	9	7.50	55	45.83	65	54.17
11.	Imidachloprid is used as seed treatment chemical to control sucking pests in groundnut T/F	107	89.17	13	10.83	48	40.00	72	60.00

S. No.	Knowledge item	FFS farmers **(n=120)**				non-FFS farmers **(n=120)**			
		Incorrect		Correct		Incorrect		Correct	
		F	%	F	%	F	%	F	%
12.	Treat the seed first with bio agent followed by insecticide and fungicide T/F	98	81.67	22	18.33	35	29.17	85	70.83
13.	Kalahasthi (TCGS-320) variety is susceptible for nematodes or Kalahasthi malady T/F	65	54.17	55	45.83	36	30.00	84	70.00
14.	Usage of complex fertilizers reduces cost of cultivation T/F	107	89.17	13	10.83	58	48.33	62	51.67
15.	Kadiri-9, Abhaya, ICGV-91114, Vemana are tolerant varieties to drought T/F	102	85.00	18	15.00	60	50.00	60	50.00
16.	Bajra grown as trap crop for *Spodoptera* T/F	99	82.50	21	17.50	55	45.83	65	54.17
17.	Pod borer is major pest in groundnut during storage T/F	108	90.00	12	10.00	71	59.17	49	40.83
18.	FFS aims at a) To increase decision making capacity of farmers b) To grow healthy crop c) More yield with low cost of Cultivation d) All	111	92.50	9	7.50	26	21.67	94	78.33
19.	Leaves turn pale and white due to the deficiency of which micro nutrient a) Iron b) Zinc c) Manganese d) Boron	55	45.83	65	54.17	17	14.17	103	85.83
20.	Optimum time of sowing for groundnut crop in *rabi*a) September 15th to October 15th b) October 15th to November 30th c) November to December 15th d) December to January	115	95.83	5	4.17	67	55.83	53	44.17
21.	Optimum plant population per sq m for realizing higher yields in *kharif* groundnut a) 22 b) 33 c) 44 d) 55	87	72.50	33	27.50	24	20.00	96	80.00

S. No.	Knowledge item	FFS farmers **(n=120)**				non-FFS farmers **(n=120)**			
		Incorrect		Correct		Incorrect		Correct	
		F	%	F	%	F	%	F	%
22.	Quantity of Zinc sulphate used per acre to avoid Zinc deficiency in Groundnut a) 20Kg b) 30Kg c) 40Kg d) 50 Kg	103	85.83	17	14.17	55	45.83	65	54.17
23.	The recommended dose of nitrogen, phosphorus and potassium per acre for *kharif* groundnut a) 8:16:16 kg b) 8:16:20 kg c) 12:20:16 kg d) 16:16:8 kg	48	40.00	80	66.67	14	11.67	106	88.33
24.	Quantity of gypsum used per acre for groundnut a) 100 Kg b) 200 Kg c) 300 Kg d) 500 Kg	108	90.00	12	10.00	48	40.00	72	60.00
25.	Azadirachtin acts as a) Repellent b) Anti-feedant c) Growth regulator d) All	101	84.17	19	15.83	58	48.33	62	51.67
26.	Pest defender ratio followed in IPM a) 1:1 b) 2:1 c) 3:1 d) 4:1	31	25.83	89	74.17	5	4.17	115	95.83
27.	The Economic Threshold Level (ETL) for Red Hairy Caterpillar is a) One caterpillar per plant b) 20 – 25 % leaf damage c) Both d) None	34	28.33	86	71.67	0	0.00	120	100.00
28.	Egg parasitoid used for control of Spodoptera a) *Trichogramma* b) *Trichoderm* c) *Telenomusremus* d) *Geocorid*	25	20.83	95	79.17	0	0.00	120	100.00
29.	The ETL for leaf miner is a) 5 leaf mines or webs per plant b) 10% leaves damage c) Either a or b d) None	38	31.67	82	68.33	0	0.00	120	100.00
30.	The ETL for Tikka leaf spot is a) 8% incidence per leaf b) 4% incidence per leaf c) 2% incidence per leaf d) 1% incidence per leaf	27	22.50	93	77.50	0	0.00	120	100.00

S. No.	Knowledge item	FFS farmers (n=120)				non-FFS farmers (n=120)			
		Incorrect		Correct		Incorrect		Correct	
		F	%	F	%	F	%	F	%
31.	Post emergence herbicide used to control monocot weeds or grasses in groundnut ———.	66	55.00	54	45.00	26	21.67	94	78.33
32.	To control boron deficiency, quantity of Borax used for one acre of groundnut crop ———.	54	45.00	66	55.00	6	5.00	114	95.00
33.	Bacteria present in groundnut root nodules helps in the fixation of ——— nutrient.	72	60.00	48	40.00	48	40.00	72	60.00
34.	Sowing of 4-8 rows of bajra or jowar or maize as guard crop can reduce incidence of --disease in groundnut.	97	80.83	23	19.17	56	46.67	64	53.33
35.	Basal application of 8 kg phorate granules kept check in ——— insect in groundnut	65	54.17	55	45.83	34	28.33	86	71.67
36.	——— is sprayed to overcome drought situation in groundnut.	102	85.00	18	15.00	74	61.67	46	38.33
37.	Critical stages for irrigation in groundnut crop are ——— and ———.	98	81.67	22	18.33	79	65.83	41	34.17
38.	For monitoring of *Spodoptera* in groundnut ——— traps are used.	104	86.67	16	13.33	37	30.83	83	69.17
39.	Dosage of NPV for control of *Spodoptera* in groundnut is ——— L.E. per acre.	38	31.67	82	68.33	0	0.00	120	100.00
40.	Groundnut pods should be dried up to ——— % moisture content.	61	50.83	59	49.17	45	37.50	75	62.50

Conclusion

From the study it is concluded that participant farmers had more knowledge on improved groundnut cultivation practices when compared to nonparticipant farmers. Significant level of knowledge was observed in participant farmers with respect to improved cultivation practices of groundnut. It was due to participation of farmers in weekly training classes, regular observations, field demonstrations, regular touch with extension personnel and subject matter specialists and more exposure to mass media sources during season long programme of polambadi (FFS). Therefore it is concluded that polambadi (FFS) is an effective and innovative extension approach to improve the knowledge level of groundnut farmers and it can be used as a potential tool for transfer technology.

References

Manoj, A. and Vijayaragavan K. (2014). Impact of Farmers' Field School on Farmer's Knowledge of Integrated Crop Management Practices in Paddy. *Indian Res. J. Ext. Edu.* 14 (1): 5-10.

Sreenivasulu, M. (2011). Empowerment of cotton farmers through farmer field school in Andhra Pradesh. Ph.D. Thesis submitted to Acharya N.G. Ranga Agricultural University, Hyderabad.

Sunitha, A. B., Lalitha, K. C., Preethi and Surendra, H. S. (2013). Comparative study on knowledge level of participant and non participant farmers of farmers field schools. *International J. of Engineering and Management Sciences*. 4 (3): 352-356.

Yeshwanth Kumar Naik, L.G. (2008). A study on knowledge and adoption of integrated crop management practices by the participants of farmers field school in Bellary district. *M. Sc. (Agri.) Thesis*, Univ. Agric. Sci., Dharwad, Karnataka, India.

Extension Management Strategies for Sustainable Agriculture, pp 161-168

New India Publishing Agency, New Delhi, India

17

Food Security Army – Extension Management Strategy for Quick Deliverance of Technology Packages

U. Jaikumaran, A. Latha, Shyla Joseph and P.S. Preman

Food security is an essential feature of a country's independence and sustenance. Rapid growing population and increasing purchase capacity, both play an influential role on food security. The cultivation or agriculture has declined the world over, with India and Kerala being no exceptions mainly due to the lack of human resources to work with the soil. Uneconomic crop pricing, lack of social accreditation, poor and unstable income, insecurity to life and health etc. are the main factors which discourage the youngsters in our country from becoming 'farmers'.

If food security is to be attained, we need a core of people to work in the soil and to transfer the advanced technologies to farmers and people through service providers. Here comes the need of a 'Food Security Army' (FSA) to protect us from this food insecurity. Society must recognize them as the 'Service Personnel' who work in the agricultural fields for sustaining humanity by producing enough food for all. Agricultural Research Station, Kerala Agricultural University, Mannuthy has conceptualized and is building up a 'Food Security Army' for the entire country for attaining Food Security

Design of 'Food Security Army' (FSA)

Agricultural Research Station, Mannuthy under Kerala Agricultural University acquired a Chinese made YANJI SAKTHI 8 row mechanical transplanter. On acquiring the machine, it was operated at ARS farm fields and all the operational details were acquainted with and experienced. The machine was then taken to different farmers fields in Thrissur district (Kerala) for interacting with farmers and to collect their opinion and apprehensions. The demonstration of the machine in farmers field led to several questions. These queries were taken the machine back to the farm and following research projects were formulated and conducted to answer the questions scientifically.

1. Production technology for rice under mechanized transplanting
2. Standardization of mat nursery for rice
3. Identification and development of rice genotypes suitable for mechanized farming.

The projects were taken up at the paddy fields of ARS, Mannuthy. Projects were conducted for three years and the resultant data were pooled and analyzed.

The data on number of missing hills and number of seedlings per hill with Transplanter revealed that mechanized transplanting using mat nursery through 8 row mechanical transplanter was an economic alternative to manual transplanting using conventional nursery (Table 1). The mechanized transplanting needed raising of mat nursery. However only 20-25 kg seeds were required for preparing this mat nursery for transplanting in an acre of main field. Nursery area required for this is 40 sq.m. i.e., 1/10th of conventional nursery. Seedlings can be cut as mats and fed to tray directly. Eight rows are planted simultaneously and depth of planting is 2-5 cm. (Jaikumaran and Beena, 1997).

Table 1: Mean no. of seedlings per hill and percentage of missing hills in mechanically transplanted crop as influenced by density of sowing in mat nursery

Density of sowing in the mat nursery(Kg / m^2)	Mean number of seedlings per hill	Percent of missing hills
0.4	3	13.5
0.6	3	15.6
0.8	4	9.4
1.0	3	13.6

The data also showed that mechanized transplanting has the definite monetary advantage over manual transplanting (Table 2). Saving in total cost by mechanically transplanted crop over manual transplanting varied from Rs.1950 to Rs. 3500 per hectare depending upon the mat nursery used. Considering the overall merit of mechanized transplanting seed density between 0.4 to 0.8 kg m^{-2} in the mat nursery is appropriate.

Table 2: Yield and yield attributes of mechanically transplanted rice at the farmers' field in Kole land in Thrissur

Variety	No. of panicles/m^2		No. of filled grains/panicle		1000 seed weight (g)		Yield of grain (kg/ha)		Yield of straw (kg/ha)	
	MT	LT	MT	LT	MT	LT	MT	LT	MT	LT
Annapurna	748	651	60	54	225	217	6797	6753	6750	6300
Jyothi	792	555	65	47	277	243	7678	5913	7900	5625

MT = Mechanized transplanted; LT = Labour transplanted

Conceptualisation of Technical Service core - The "Food Security Army"

When the research findings of mechanical paddy transplanting were empowered to the trainees through work experiential training, the "soft ware" of technology was loaded to them for application. The hardware "the machine" to be handled by them is also exposed to them by imparting training on repair, service and maintenance apart from the training on operation of the machine. Then their confidence and esteem were boosted. They were empowered to operate both hardware and software part of the technology.

The groups trained had to work for larger area in short span of time as demanded by the farmer to the sustenance of the trained persons. So it was decided to form Societies under the Indian Charitable societies Act. Draft Bye Law was formulated under 1955, 12th Act Travancore Cochin Literary Science and Charitable Trust Registration Act and issued to trained groups based on Panchayath and persuaded them to form society. Twelve such societies have already been formed in Thrissur District. They started contract work under the name of these societies. As per the Bye-law, they are named as AMOSC (Agro Machinery Operation Service Centre). As the Indian Army is for Territorial Security so is the "Food Security Army" (FSA) for the 'Food Security' activities of the Nation. A logo was also created to FSA.

Agricultural Research Station received a lot of demand for agro machinery service, operators, operation, machinery and technical guidance. Considering the demand, the Agro Machinery Operation Service Centre was started at the station. Youths were selected and trained and developed as 'Agro Machinery Operations Service Executives' ie. (FSA Officers). Their service was utilized as Master Trainer cum Service Providers. Machinery were procured and rented out. Machinery repairs were also attended at farmer's field. Farmers' enquiry regarding machinery was addressed. All got a relief; there is a machinery solution centre at Agricultural Research Station, Mannuthy.

Activities of the AMOSC include training and building up a corpus of AMOSE (Agro Machinery Operations Service Exicutives), offer efficient agro machinery operations service for crops,offer efficient repair and servicing for all agro machinery,offer franchise training for all agro machinery, take up all agro technology specialised works,undertake KAU Agro products distribution at homesteads,undertake agro micro enterprises establishment works, take up agro trading and marketing

Services of AMOSE

The Food Security Army training is rendered by skillfully trained, highly professsional and motivated group of manpower known as Agro Machinery Operations Service Executives (AMOSE). In the FSA training programmes, for every five trainees, there is a master trainer to train them. The duties of the AMOSE are not confined in the FSA training, but they form the mainstay of mechanized agriculture in the state by providing technically perfect agricultural mechanization services. They provide service for 365 X 24 to meet any "Agricultural Emergency". Other services rendered are Agro machinery operations service for all crops, repair and maintenance services for all machinery, franchise training for all agro machinery, formation of self help groups (AMOSE) to take up all agricultural mechanization works at farmers' fields, establishment of agro micro enterprises, specialised works and consulting on agro technology. When he/she successfully completed the training programme and is awarded a certificate.

Farm Machinery Facilitation Centre (FMFC)

A lot of farmers visit the centre every day to understand the operation of different farm machineries, its rental services, repair services and procurement details. They also need contractual details of service providers (AMOSCs), details of agricultural machinery and its manufacturers. Considering the overwhelming demand, the station started 'Farm Machinery Facilitation Centre' (FMFC) to provide service to the farmers.

Mobile Agro Machinery Mobile Training Unit (MAMTU)

Frequent contact with farmers and workers led to the understanding that apart from in house trainings at Mannuthy, training was needed at their own places to avoid the difficulty in traveling. Hence this Station launched Mobile Agro Machinery Mobile Training Unit (MAMTU) Each unit is commanded by one

Commandant (Chief Master Trainer) and 4 Master Trainers (Technical and service core). One unit trains 20 trainees, i.e. for every five trainee, service of one master trainee is available. Each unit is provided with vehicle and one piece each of all farm machinery On demand, the training will be conducted in the farmers field and Master trainers stay there along with trainees, for 20 days and impart work experimental training programme. Training schedule depends upon the subject of training, viz."Mechanised Paddy Transplanting", "Total Paddy Mechanisation", "Total Farm Mechanisation", "Small Farm Mechanisation", "Garden land mechanisation" etc. The day to day schedule is well planned and documented. The training includes morning drill, parade, physical training, briefing on a subject, theory and practical of the machine. Working lunch is provided at the spot. FSA uniform *viz*., Army Green Pants, 3/4th, Shirt, Cap and Shoe and Tool bags are provided. A pre-evaluation before training and post-evaluation after training are conducted. After training, a passing out parade is arranged and certificate is distributed to those who have successfully completed the training. The trainees are given draft bye laws and directed to form 'AMOSC' at Panchayath level.

Mobile Agro Machinery Repair & Service Unit (MAMRSU)

The basic problem with machines was lack of adequate timely service. Hence when FSA training was conducted, a Moble Agro machinery Repair and Service Unit (MAMRSU) was established at ARS, Mannuthy. Spare parts were stored and AMOSEs were sent to field on call to repair the machines. This served as greatest relief to AMOSCs since the break down problems were timely addressed. This added strength to AMOSC and confidence to undertake the work in larger area and to go for joint operation.

Reach and social acceptability of FSA

With the tremendous strength achieved through research findings and vast stretch of experience acquired in operation, aftercare, repair and maintenance of the machine, the Station started 22 days vocational training programme. The schedule of training programme consisted of 20 hours of class room lecture and 155 hours of machine operation. The trainees receive a certificate from KAU for participating in the work experiencing vocational training programme. After the training, the trainees become the referral persons of the Station for machine transplanting and they themselves contract for mechanical transplanting. Once their contract work is going on, the Station interacted with them to remove their problems if any and to add confidence and strength.The first batch of training was given to two groups of Kudumbasree women, 10 members of Kodakara Panchayath and 9 members of Tholur Panchayath.

On seeing the success rolling, the Thrissur District Panchayath has decided to sponsor the training programme and to supply machines to Grama Panchayaths free of cost. Wadakkanchery Block Panchayath got trained their 18 farmers/ farm workers from 9 Panchayaths at this Station Thrissur District Panchayath, further sponsored 60 trainees from twelve Panchayaths They started engaging in the work of mechanical transplanting with the technical assistance and guidance of ARS, Mannuthy. Now the station completed training for 176 sessions including trainees from Andaman and Nicobar Islands and officials from Andhra Pradeshincluding .

The AMOSE is working under three models

1. Co-operative / group AMOSC Model

Among this most successful example is the "Wadakkanchery Green Army". Nearly 800 persons have been trained and deployed as the members of Green Army. They are organized as Block Panchayath level Green Army. This AMOSC is registered under the Charitable Society Act and its functioning is overseen by the Peringandoor Co-operative Service Bank. All the members are covered under insurance, pension fund and provident fund scheme.

2. Individual AMOSC Model

AMOSC has fixed service charge for every farm machinery operation as decided by the leader and approved by the members of AMOSC. "Sivasakthi" is an example of an individual-based AMOSC. It has 43 members working in it. This AMOSC is currently able to contract agro machinery operations worth more than Rs. 50 lakhs per annum. The profit of operations comes to the leader.

3. SHG Model AMOSC

This model of FSA Service Centres, based at Grama Panchayath level, operate as Self Help Groups. Such AMOSCs are also registered under the Charitable Society Act and managed by the elected officials of the AMOSC. "Janasakthi" is an AMOSC functioning in this mode. At the end of each financial year, the AMOSc office bearers prepare the balance sheet and distribute profit equally among all members.

Success Stories

"Operation Ponnamutha 300/5"

The mission, "Operation Ponnamutha 300/5" conceived by Agricultural Research Station, Mannuthy, was executed jointly by Kerala Agricultural University,

Department of Agriculture and Ponnamutha Kole Padavu Committee. The 'mission' transplanting of paddy in 300 acres in 5 days using 8 row mechanized transplanter was conducted with the aim of 'double cropping' of Ponnamutha Kole Padavu. 'Food Security Army' personnel trained by the Station at different Panchayaths as well as those formed Agro Machinery Operation Service Centers after training constituted the core team to execute the work and they are grouped under nine regiments comprising of 24 battalions and covered 2.5 acres of planting per day. The mission was successful.

"Mission Elamutha 1000"

Mission Elamutha 1000 was launched to tap the yield level of 8 tones of paddy per hectare from 304 acres of Elamutha Kole Padavu. The whole area of 304 acres was transplanted mechanically utilizing service of 6 groups of Food Security Army. The transplanting started on 12th December 2009 and completed on 16th.

Round Paddy Straw Baler – A Boon to paddy farmers

Agricultural Research Station, Mannuthy introduced a "Ronud Paddy Straw Baler" which led to straw revolution in rice cultivation of Kerala especially in Kole lands.

Green Cadet Corps

Agricultural Research Station is introduced a new wing of Food Security Army named "Green Cadet Corps" among the school students. The purpose of GCC is to present the basic theory and practices of agriculture and cultivation practices of different crops, thereby creating interest among school and college students in agriculture and motivating them to adopt farming as a career. They will be made to understand the need, scope and importance of mechanized agriculture. This programme focuses on familiarizing and experiencing various crops, its cultivation and management, starting from land preparation up to harvesting. So they will be physically and mentally prepared for agricultural works with a positive attitude towards it and will encourage the youth to take up agriculture.

Upscaling of the technology delivery

Govt. of Kerala, appreciating this unique FSA programme, launched a massive project of establishing 35 Agro Service Centres throughout the state. Among this, 14 Model Agro Service Centres, one each in every districts of Kerala is under the control of Agricultural Research Station. The members of these Model Agro Service Centres were given first and second level FSA training and now handholding is done by the station with funding by Government of Kerala. Department of Co-operation, Government of Kerala is also establishing 60 Farmer Service Centres throughout the State through Service Co-operative Banks. Among these FSA trainings were given to 20 Farmer Service Centres.

Conclusion

An efficient food security system should withstand challenges posed by calamitous weather conditions, unfavorable for agricultural production over successive seasons. Food Security Army – Green Cadet Corps implies the younger generation with complete knowledge about agriculture who will never stretch their hands for borrowing food. The Nation's food resources will be secure in their hand. The growing young generation will understand their social responsibility to keep the Nation 'food secure'. They will commit themselves to the nation's growing pride of food independence. Finally, from each and every home, one will join the core of Food Security Army to build an independent and sovereign Nation with plentiful of food and even offer food to the starving millions of other Nations. The concept and activities are socially accepted and developed as a successful programme in the 'ToT' in agricultural mechanization with prime aim of immortalization of slogan 'Jai Jawan Jai Kisan'.

Extension Management Strategies for Sustainable Agriculture, pp 169-181

New India Publishing Agency, New Delhi, India

18

Farmers' Perception on The Usage of Chemicals in Agriculture: A Study Conducted in Nadia District of West Bengal

J.K. Das, Jayasree Datta and Tapas Kumar Mondal

Agricultural practices have been undergoing changes over a period of time throughout the world. The intensive agricultural practices followed during the past four decades in our country resulted in food security and self-sufficiency. This was achieved through development of input responsive varieties coupled with use of chemical fertilizers and plant protection chemicals. Agriculture is an important sector for economic development in India. Crop production is an important income source for millions of farmers in some cases with few cropping alternatives due to ecological constraints. All over the world majority offarmers rely on pesticides and fertilizers to increase yields. Pesticide use in most Indian agriculture is an essential part of production technology. Therefore farmers in developing countries are perceived as overusing pesticides, both in quantity and quality, with mixtures of chemicals, known locally as cocktails, being the favoured form of application (Crissman et al., 1994).

In India pesticides were introduced in 1948 but production was started in the year 1952 mainly benzene hexa chloride (BHC) to use against locust incidence. Farmers were then becoming dependent to pesticides and to some extent they used excessively and inappropriately until today. India is the second largest user of pesticides after China in Asia. India, 51% of foods commodities are contaminated with pesticide residues and out of these, 20% have pesticides residues above the maximum residue level values (Gupta, 2004). In another survey based on market basket samples in Kolkata was found out of 62 samples; 19 samples were contaminated with organ chlorine pesticides. Residues of endosulfan were reported in 14 samples, residues of BHC in 6 samples and residues of heptachlor in two samples. Interestingly in all the samples the levels of pesticide residues were below permissible level. It has been observed that their long-term, low dose exposure are increasingly linked to human health effects such as immune suppression, hormone disruption, diminished intelligence, reproductive abnormalities, and cancer. In this light, problems of pesticide safety, regulation of pesticide use, use of biotechnology, and bio pesticides, and use of pesticides obtained from natural plant sources such as neem extracts are some of the future strategies for minimizing human exposure to pesticides.

Food and water samples collected across many countries were found to be contaminated with various organochlorine pesticide residues of DDT and its metabolites, HCH and its isomers, heptachlor and its expoxide and aldrin. The amount of pepsticide detected was higher than the permissible limits prescribed by FAO/WHO (2005). In drinking water only a few pesticides exceeded the permissible limits. Seasonal variations of pesticides residues were also observed during the study period. The tested samples showed 100% contamination with low but measurable amounts of residues. Among the four chemical groups, the organophosphates were dominant followed by organochlorines, synthetic pyrethroids and carbamates. About 23% of the samples showed contamination with organophosphorous compounds above the irrespective MRL values. More extensive studies covering different regions of Haryana state are suggested to get a clear idea of the magnitude of vegetable contamination with pesticide residue (Kumari *et al.*, 2002; John and Bhatnagar 2004).

Farmer's survey is an important data-gathering process for assessing the needs of intended beneficiaries to determine their knowledge and perceptions of pest problem, their constraints in dealing with the problem and their attitudes in pest management. Such works related to rice farmers of Asia have been documented by Heong and Escalada 1997; Katti *et al.*, 1999). On an average, 33% of crop loss in India occurs exclusively due to crop pests (Puri *et al.*, 1999).

In a recent study on pre and post harvest loss by pests has been estimated to be around 4,000 million dollars. The loss due to only one insect species, Helicoverpa armigera, which attacks many crops such as cotton, tomato, brinjal, cabbage, cauliflower, bhendi, strawberry and apple, is estimated to exceed 500 million dollars in a year. In order to meet the demand of a growing population we need to protect our crops from pests as India needs to increase the annual food production by 5 million tons per year as against the current rate of 3.1 million tones. Pesticides have become a weapon in the farmers struggle to protect crops. The consumption of pesticide in India has been increased recently to 0.5 kg ha^{-1} from as low as 0.28 kg ha^{-1} a pentad ago. This quantity is very low when compared to 7.0 kg ha^{-1} by the US, 2.5 kg ha^{-1} in Europe, 12 kg ha^{-1} in Japan and 6.6 kg ha^{-1} in Korea (JPCR, 2004). The unscientific and indiscriminate use of agrochemicals has bad affect on the soil health and brought down the fertility of soil and productivity of crops. The study has been designed to know the farmers' perception on bad effects of the chemical usage in agriculture on the environment.

Methodology

Nadia district is a district of the state of West Bengal, in the north east of India. Nadia is situated between 22°53" and 24°11" North latitude and 88°09" and 88°48"East longitude and about 3927 Sq. Km. in Area, this District is linear in shape with orientation of North-South. The District is Approximately 46 ft. above the mean sea level. The Tropic of cancer divides the district in two parts. The soil of the district is composed of recent alluvium and the surface consists of sandy clay and sand along the course of the rivers and fine silt consolidating into clay in the flatter portions of the plains, such as in Kalantar tract between the Bhagirathi and the Jalangi. The soil varies but little over the district except in this tract and a portion of Ranaghat sub-division. It is almost universally a light sandy loam with low fertility status and incapable of retaining adequate moisture.

Locale of Research

The village namely Fatepur of the Haringhata block of Nadia district in West Bengal was purposively selected for the study. The area had been selected for the study because of (a) The availability of appropriate respondents who are using chemicals in their crop field (b) Acquaintance with the local people as well as the local language (c) The concern area was easily accessible to the researcher in terms of place of residence and (d) The closure familiarities of the student researcher with the area, people, officials and local dialects.

Sampling Techniques: Purposive as well as simple random sampling techniques were adopted for the study. For selection of state and district purposive sampling techniques was adopted because the area was ideal with respect to the problem, convenient for researcher and having the infrastructural facilities and in case of selection of block, villages and farmers or respondents simple random sampling technique was taken up.

Diagrammatic Representation of Sampling Technique and Sampling Design

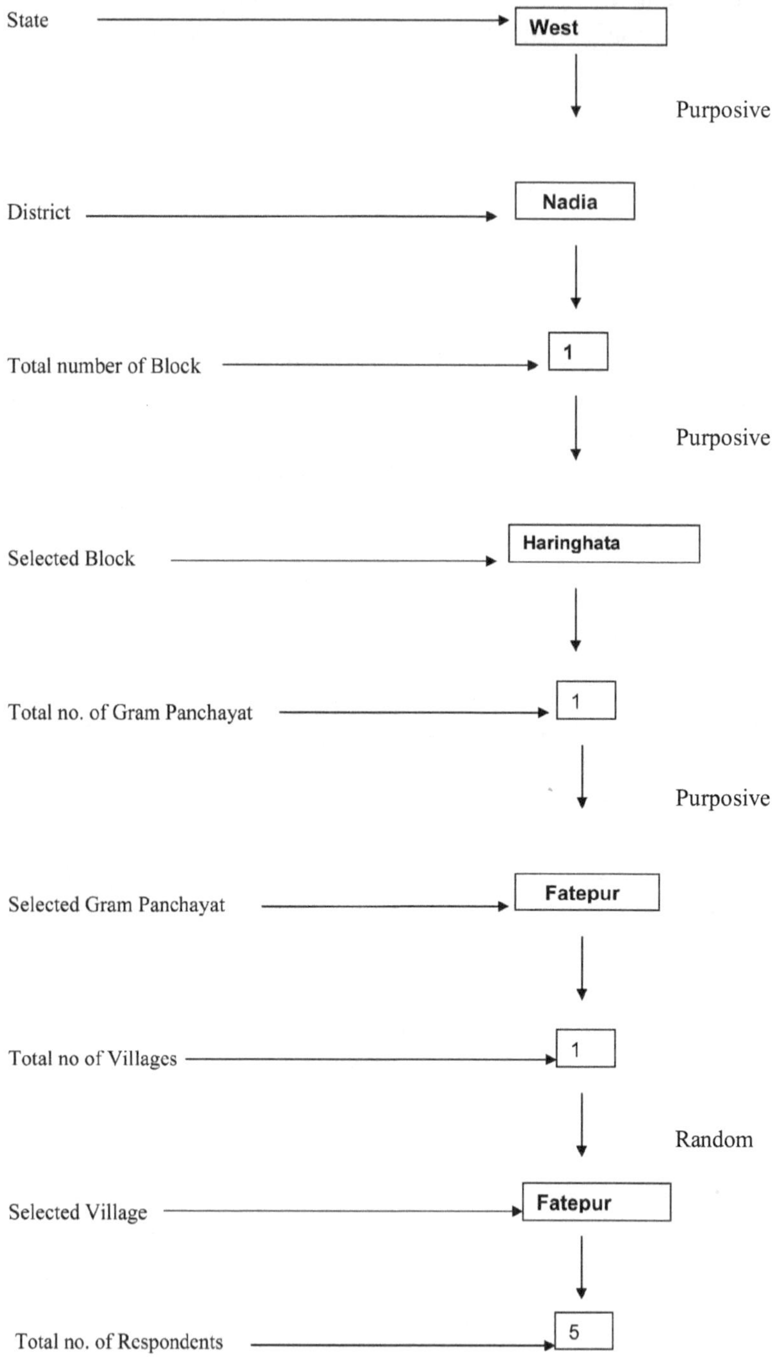

Result and discussion

Table 1: The socio-economic profile of selected respondents

Items	Category	Frequency	Percentage
Age (X_1)	18-24 Years	8	16
	25-30 Years	9	18
	31-36 Years	3	6
	37-42 Years	5	10
	43-48 Years	11	22
	49-54 Years	9	18
	55-60 Years	3	6
	Above 60 Years	2	4
Community Affiliation (X_2)	Hindu	28	56
	Muslim	22	44
	Christian	0	0
	Budhist	0	0
	Others	0	0
Family Type (X_3)	Single	29	58
	Joint	21	42
Family Size (X_4)	Upto 5 members	30	60
	Above 5 members	20	40
Education (X_5)	Illiterate	2	4
	Primary	15	30
	Middel School	16	32
	Matriculation	9	18
	Higher Secondary	5	10
	College	3	6
House Type (X_6)	No house	0	0
	Hut	10	20
	Kutcha House	7	14
	Mixed House	24	48
	Pucca House	7	14
	Mansion	2	4
Farm Power (X_7)	No drought animal	9	18
	1-2 drought animal	27	54
	3-4 drought animals/one or more prestige animal	12	24
	5-6 drought animals/tractor/power tiller	2	4
Material Possession (X_8)	Furniture	40	80
	TV	25	50
	Improved Agril, implements	10	20
	Vehicle	20	40
	None of these	5	10
Primary Occupation (X_9)	Farmer	19	38
	Day Labour	12	24
	Business	6	12
	Mason	3	6

Items	Category	Frequency	Percentage
	MGNREGA	6	12
	Others	4	8
Secondary Occupation (X_{10})	None	0	0
	Farmer	18	36
	Day Labour	3	6
	Business	29	58
	Mason	0	0
	MGNREGA	0	0
	Others	0	0
Annual Income (X_{11})	Below 24,000	18	36
	24,000-60,000	8	16
	60,000-1,20,000	20	40
	1,20,000-2,40,000	4	8
	2,40,000 and above	0	0
Social Participation (X_{12})	No any Participation	35	70
	Member of an organization	15	30
	Office bearer in 1 organization	0	0
	Office bearer in more than 1 organization	0	0

Table 1 presents the frequency distribution of the respondents according to their Age, Community affiliation, Family type, Family size, Education, House type, Farm power, Material possession, Primary occupation, Secondary occupation, Annual income, Social participation.

Age

The attribute Age has categorized into eight distinct categories viz. 18 to 24 years, 25 to 30 years, 31 to 36 years, 37 to 42 years, 43 to 48 years, 49 to 54 years, 55 to 60 years and above 60 years. In case of the selected respondents majority percentage are in 43 to 48 years (22%), then 49 to 54 years and 25 to 30 years (18% each); then 18 to 24 years (16%); then 37 to 42 years (10%); then 31 to 36 years and 55 to 60 years (6% each) and very less in above 60 years (4%).

Community Affiliation

Community affiliation is basically a process of cognitive personality wherein community has become a social reference. It includes both belongingness and affinity towards community. The attribute Community affiliation has categorized into five distinct categories viz. Hindu, Muslim, Christian, Buddhist and others. In case of the selected respondents majority percentage is in Hindu (56%) and less percentage in Muslim (44%).

Family type

The attribute Family type has categorized into two distinct categories viz. Single and Joint family. In the selected area the majority percentage of the respondent is Nuclear family (58%) and Joint family (42%).

Family size

The attribute Family size has two distinct categories *viz*. Up to 5 members, above 5 members. In the selected area the majority percentage of the respondent is up to 5 members (60%) and above 5 members (40%).

Education

The attribute Education has categorized into seven distinct categories *viz*. Illiterate, Primary, Middle School, Matriculation, Higher secondary and College. The majorpercentage of the respondent is Middle School (32%) the category followed byPrimary (30%), Matriculation (18%), Higher Secondary (10%), College (6%) and Illiterate (4%).

House type

The attribute House type has categorized into five distinct categories *viz*. No house, Kutcha house, Mixed house, Pucca house, Mansion. In the selected area the majority percentage of the respondent is Mixed house (48%), followed by Hut (20%), Kutcha house and Pucca house (14%) and Mansion (4%). No one is in No house category.

Farm power

The attribute Farm power has categorized into four distinct categories *viz*. No drought animal, 1-2 drought animals, 3-4 drought animal/one or more prestige animal, 5-6 drought animal or tractor or power tiller. In the selected area the majority percentage of the respondent is 1-2 draught animals (54%), followed by 3-4 drought animals (24%), no drought (18%) and 5-6 drought animals or tractor or power tiller (4%).

Material possession

The attribute Material possession has categorized into 5 distinct categories *viz*. Furniture, TV, Improved agricultural implements, vehicle and None of these. The majority percentage of the respondent is Furniture (80%), followed by TV (50%), Vehicle (40%), Improved agricultural implements (20%) and None of these (10%) respectively.

Primary occupation

The attribute Primary occupation has categorized into six distinct categories viz. Farmer, Day labour, Business, Mason, MGNREGA and others. The major percentage of the respondent is Farmer (38%), followed by Day labour (24%), MGNREGA (12%), Business (12%), others (8%) and Mason (6%) respectively.

Secondary occupation

The attribute Secondary occupation has categorized into six distinct categories *viz*. Farmer, Day labour, Business, Mason, MGNREGA and others. The major percentageof the respondents is in business (58%), followed by farmer (36%) and day labour (6%).

Annual income

The attribute Annual income has categorized into three distinct categories *viz*. Below Rs. 24000, Rs. 24,000 to 60,000, Rs. 60,000 to 1,20,000, Rs.1,20,000 to 2,40,000, Rs. 2,40,000 and above. The major percentage of the respondent is Rs. 60,000 to 1,20,000(40%), followed by Below Rs. 24,000 (36%), Rs. 24,000 to 60,000 (16%), Rs. 1,20,000 to 2,40,000 (8%) respectively. No respondent in the category of annual income of Rs. 2, 40,000 and above.

Social participation

The attribute Social participation has been three distinct categories *viz*. None, Member of an organization, Office bearer in one Organisation and Office bearer in more than 1 Organisation. In case of the selected respondents majority percentage is in none category (70%) and less in the category of Member of 1 organization (30%). No respondent in the category of Office bearer in 1 organization and Office bearer in more than 1 organization.

Table2 : Correlation coefficient between the Independent variables and 'Awareness about bad effects on environment' (Y^1)

Variables	Pearson	Spearman
Age (X_1)	0.455**	0.480**
Community Affiliation (X_2)	0.319*	0.321*
Family type (X_3)	0.061NS	0.06NS
Family size (X_4)	0.092NS	0.09NS
Education (X_5)	0.883**	0.919**
House Type (X_6)	0.209NS	0.206NS
Farm Power (X_7)	0.192NS	0.217NS
Material Possession (X_8)	0.339*	0.344*
Primary Occupation (X_9)	-0.078NS	-0.114NS
Secondary Occupation (X_{10})	0.371**	0.367**
Annual Income (X_{11})	0.319*	0.321*
Social Participation (X_{12})	0.643**	0.647**

**Significant at 1% level of significance; *Significant at 5% level of significance; NS-Non significance

Table 2 revealed that the 'Awareness of bad effects on environment' (Y_1) of the selected respondents is positively and significantly correlated with their Age (X_1), Education (X_5), Secondary Occupation (X_{10}) and Social Participation (X_{12}) at 1% level of significance in both the Pearson Correlation and Spearman Correlation. It has revealed that 'Awareness about bad effects on environment' (Y_1) of the selected respondents is positively and significantly correlated with their Community Affiliation (X_6), Material Possession (X_8) and Annual Income (X_{11}) at 5% level of significance in both the Pearson Correlation and Spearman Correlation. It means that there exists high community affiliation of the selected respondents irrespective of their community wise classification with the awareness.

Table 3: Correlation coefficient between the Independent variables and 'Awareness about cost increase of crop production practices (Y^2)

Variables	Pearson	Spearman
Age (X_1)	0.905**	0.911**
Community Affiliation (X_2)	-0.19NS	-0.185NS
Family type (X_3)	0.047NS	0.02NS
Family size (X_4)	0.101NS	0.084NS
Education (X_5)	0.216NS	0.252NS
House Type (X_6)	0.494**	0.441**
Farm Power (X_7)	0.310*	0.337*
Material Possession (X_8)	0.344*	0.366**
Primary Occupation (X_9)	0.094NS	0.047NS
Secondary Occupation (X_{10})	0.811**	0.819**
Annual Income (X_{11})	0.867**	0.905**
Social Participation (X_{12})	0.467**	0.482**

**Significant at 1% level of significance; *Significant at 5% level of significance; NS-Non significance

It has revealed that 'Awareness about cost increase of crop production practices'(Y_2) of the selected respondents is positively and highly correlated with their Age (X_1), House Type (X_6), Secondary Occupation (X_{10}), Annual Income (X_{11}), and Social Participation (X_{12}) at 1% level of significance in Pearson Correlation and withthese Material Possession (X_8) is also significant in Spearman Correlation. It has also revealed that 'Awareness about cost increase of crop production practices' (Y_2) of the selected respondents is positively and significantly correlatedwith their Farm Power (X_7) at 5% level of significance in Spearman Correlation and with this Material Possession (X_8) is also significant in Pearson Correlation.

It means that therc cxists high Community Affiliation of the selected respondents irrespective of their community wise classification with the 'awareness on cost increase'.

Table 4: Correlation coefficient between the Independent variables and 'Awareness about crop yield increase' (Y_3)

Variables	Pearson	Spearman
Age (X_1)	0.816 **	0.826**
Community Affiliation (X_2)	-0.197NS	-0.195NS
Family Type (X_3)	0.041NS	0.003NS
Family Size (X_4)	0.097NS	0.094NS
Education (S_5)	0.238NS	0.293*
House Type (X_6)	0.474**	0.417**
Farm Power (X_7)	0.313*	0.346*
Material Possession (X_8)	0.359*	0.367**
Primary Occupation (X_9)	0.097NS	0.041NS
Secondary Occupation (X_{10})	0.736**	0.742**
Annual Income (X_{11})	0.745**	0.774**
Social Participation (X_{12})	0.429**	0.446**

**Significant at 1% level of significance; *significant at 5% level of significance; NS. Nons significance

Table 5: Correlation coefficient between the Independent variablesand 'Awareness about profit increase' (Y4)

Variables	Pearson	Spearman
Age (X_1)	0.940 **	0.917**
Community Affiliation (X_2)	-0.161NS	-0.159NS
Family Type (X_3)	0.066NS	0.058NS
Family Size (X_4)	0.099NS	0.099NS
Education (S_5)	0.323*	0.320*
House Type (X_6)	0.576**	0.552**
Farm Power (X_7)	0.394**	0.444*
Material Possession (X_8)	0.485**	0.494**
Primary Occupation (X_9)	0.036NS	0.011NS
Secondary Occupation (X_{10})	0.948**	0.936**
Annual Income (X_{11})	0.967**	0.974**
Social Participation (X_{12})	0.490**	0.495**

**Significant at 1% level of significance; *significant at 5% level of significance; NS. Non significance

It has revealed 'Awareness about profit increase' (Y_4) that of the selected respondents of the programme is positively and highly correlated with their Age (X_1), House Type (X_6), Farm Power (X_7), Material Possession (X_8), Secondary Occupation (X_{10}), Annual Income (X_{11}) and Social Participation (X_{12}) at 1% level of significance in both the Pearson Correlation and Spearman Correlation. It has also revealed that 'Awareness about profit increase' (Y_4) of the selected respondents of the programme is positively and significantly correlated with their Education (X_5) at 5% level of significance in both the Pearson Correlation and Spearman Correlation. Others 4 variables are found to be non significant.

Table 6 : Factor analysis for clubbing of variables into factor based on factor loading

Factors	Variables Accounted	Factor Loading	% of Variance	Comulative %	Factor Rename
Factor 1	Secondary Occupation ($X_{10)}$)	0.933			
	Annual Income (11)	0.919			
	Age (X_1)	0.895			
	House Type (X_6)	0.758	35.843	35.843	Family Resource
	Material Procession (X_8)	0.680			
	Farm Power (X_7)	0.606			
Factor 2	Family Type (X_3)	0.956	16.295	52.139	Family Background
	Family Size (X_4)	0.934			
Factor 3	Education (X_5)	0.773			
	Community	0.707	15.37	67.510	Social
	Affiliation (X_2)	0.707	15.37	67.510	Status
	Social Participation (X_{12})	0.720			
Factor 4	Primary Occupation (X_9)	-0.765	11.232	78.743	Occupation

Factor analysis has been carried out for the static conglomeration of variables based on Eigen roots that is derived from coefficient of correlation. So, a recombination types of agglomeration results which can be trenched as factor. In Table 3 Factor 1 has accounted for 35.84% of total data variance, six variables with high factor loading i.e. Secondary Occupation (X_{10}), Annual Income (X_{11}), Age (X_1), House Type (X_6), Material Possession (X_8) and Farm Power (X_7) chosen to interpret this factor. The factor loading of all the variables is positive.

Factor 1 has been renamed as Family Resource. The Factor 2 has accounted for 16.29% of total data variance, two variables with high factor loading i.e. Family Type (X_3) and Family Size (X_4) chosen to interpret this factor. The factor loading of both the variables is positive. Factor 2 has been renamedas Family Background. The Factor 3 has accounted for 15.37% of total data variance, three variables with high factor loading i.e. Education (X_5), Community Affiliation (X_2) and Social Participation (X_{12}) chosen to interpret this factor. The factor loading of all the variables is positive. Factor 3 has been renamed as Social Status. The Factor 4 has accounted for 11.23% of total data variance; one variable with high factor loading i.e. Primary occupation (X_9) chosen to interpret this factor. The factor loading of the variable was positive. Factor 4 has been renamed as Occupation.

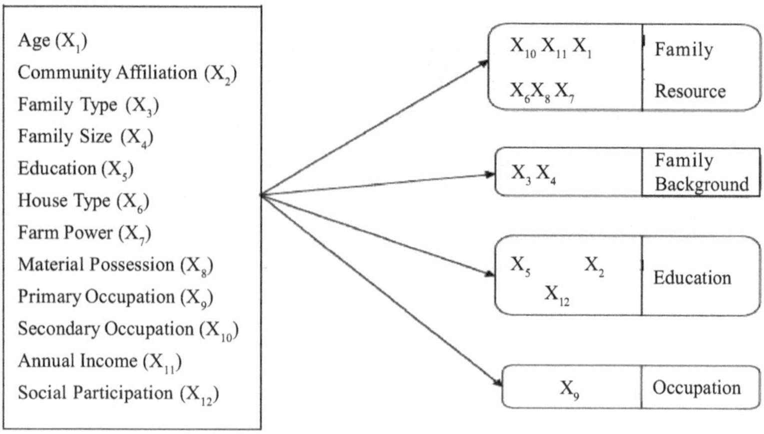

Fig 1: Factor analysis paradigm of Independent variables

Conclusion

The study was conducted to know the farmers' perception on the usage of chemicals in the form of fertilizers, herbicides; plant protection chemicals and organic matter have also been considered. Environmental effect of chemicals has been noticed by the farmers; 70% farmers in the study area have expressed their views of degradation on soil, air, underground water, surface water, livestock and fisheries etc. Though they recognized the environmental effects of chemicals and cost of production has also increased, still they are using it for the enhancement of productivity and good net return of profit. Bad effects of chemicals and the degradation of natural resources have been strongly coupled with age, education, secondary occupation, social participation and somewhat a positive relationship with community affiliation, material possession and annual income of the respondents in the study area.The study have also identified and explained the contribution of different variables on the overall perception of the farmers on the usage of chemicals. On the basis of the contribution of the variables out of 12 independent variables, only 4 factors have been identified clubbing all the 12 variables. These 4 factors have been renamed as Family Resource, Family Background, Social Status and occupation. Among these 4 factors the most important factor is 'Family Resource'.

References

Kumari, B., Madan, V.K., Kumar, R. and Kathpal, T.S. (2002). Monitoring of seasonal vegetables forpesticide residues. *Environmental Monitoring and Assessment*, 74: 263-270.

Katti, I.C. Pasalu; N.R.G. Varma and N. Dhandapani (2001). Quantification of natural biological control in rice ecosystem for possible exploitation in rice IPM. *Indian Journal of Entomology* 63: 439-448.

Crissman CC, Cole DC, Carpio, F. (1994). Pesticide Use and FarmWorker Health in Ecuadorian Potato Production, *Am. J. Agric. Econ.*76: 593-597.

Heong, K.L. and Escalada, M.M. (Eds.) (1997b). Pest management of rice farmers in asia. Los Baños, Philippines: International Rice Research Institute.

Puri, S.N., Murthy, K.S. and Sharma, O.P. (1999). 'IPM for sustainable crop production,' Management in Sundaram V. *et al.*, Handbook of cotton in India, ISCI, Bombay.

John, B.N. and Bhatnagar, P.J. (2004). Organochlorine pesticide residues in wheat and drinking water samples from Jaipur, Rajasthan, India. *Earth and Environ. Sci.*, 98 (1-3):381-389.

JPCR (2004). Report. Pesticides and good agriculture practises *In: Joint Parliamentary Committee on Pesticide Residues in and Safety Standards for Soft Drinks*, Fruit Juice and Other Beverages, New Delhi, pp 98-123.

Gupta, P. K. (2004). Pesticide exposure-Indian Scene, *Toxicology*. 198(1-3):83-90.

Extension Management Strategies for Sustainable Agriculture, pp 183-187

New India Publishing Agency, New Delhi, India

19

Extension Strategies to Combat the Environmental Issues in Grapes Cultivation

D. Jebapreetha and Rexlin Selvin

The cropping pattern in India is changing fast, particularly in favour of export-oriented commercial crops. These crops pose threat to the environment. Grapes is one such crop which accelerates environmental degradation. Borkar (2007) reported that with the advancement of technologies, the cultivation practices of grapes had improved increasing the yield and quality of grapes. However, to control the tremendous losses occurred due to pests and diseases in grapes, thousands of hundreds of kilograms of pesticides were dumped into the environment causing pollution.

The area and production of grapes in Tamil Nadu is 27,000 ha and 53,000 MT in 2010-11 (*Source*: National Horticulture Board, Government of India). The production of grapes in Cumbum valley of Theni district was 90,000 tonnes of Paneer grapes and 10,000 tonnes of Thomson seedless grapes. The unique feature in this region is that the fruits are harvested throughout the year and grapes is cultivated in the same land for many decades.

The farmers were using pesticides and fertilizers unscrupulously to get maximum yield. The indiscriminate use of pesticides resulted in water pollution and residues in berries/ fruits. Moreover it endangers human health and kills non- targeted species. The continuous cultivation and increases use of tractors and other machineries would lead to destruction of organic matters, depletion of soil fertility besides soil compaction. Hence, the grapes growers were contacted to document the environmental issues related to soil, water and produce residues.

Methodology

The study was conducted in Theni and Dindigul district of Tamil Nadu. The study area was selected purposively since intensive cropping is practiced vigorously and cash crops like grapes were cultivated intensively in a larger area and for longer years (25-30 yrs). Proportionate allocation was employed to draw the 150 respondents from the three blocks of the two districts.

Findings

The following issues were documented on enquiry with the grapes growers, crop specialists and Extension personnel of the State Department of Horticulture and Agricultures, input dealers, marketing agents and other progressive farmers of the study area.

A. Soil related issues

Worn out and weak soil: Perennial nature of grapes result in continuous extraction of nutrients without enrichment weakened the soil.

Loss of inherent fertility of the soil: Unscrupulous, indiscriminate and excessive use of chemicals had degraded the soil.

Compaction of the soil : The soil remained undisturbed for many years as the crop was perennial in nature and mechanical weeding, hand weeding which loosened the soil was neglected.

Human activities and mismanagement: Intensive cultivation with excessive use of chemicals over the years had turned the cultivable land into barren patches of land not suitable for cultivation.

B. Water related issues

Depleting the aquifers and lowering of the water table: Continuous extraction of groundwater depleted the aquifer and lowered the water table.

Delayed monsoon and unpredicted weather condition: Prolonged dry weather, less number of rainy days, erratic rainfall, unpredicted and frequent heavy down pour, affected the grape cultivation.

C. Produce residues related issues

Pesticide residues in crop produce: Grapes growers in the study area avoided eating grapes harvested from their own field in fear of chemical residues in berries which showed the presence of residues in the produce.

Reduction in keeping quality and taste: More chemicals were used without maintaining post harvest interval. Berries harvested decayed immediately as the physical maturity was not attained naturally.

D. Other issues

Loss of biodiversity: Insecticides used to target pests also kill the natural predators and pollinators available in the eco-system.

Health hazards: Farmers handled chemicals without necessary precautions, suffered from skin allergies, nausea, irritation of eyes, giddiness, vomiting and even fainting while spraying.

Mismanagement of agricultural practices are the primary reasons for the above documented environmental issues in the study area. They are presented below. In general the grapes growers had deviated from the recommended package of practices.

Table 1. Distribution of respondents according to their practice-wise adoption of recommended grapes growing technologies **(n=150)**

Sl. No	Items	Adopted	%	Non -adopted	%
1.	Spacing	68	45.33	82	54.67
2.	Manures	3	2.00	147	98.00
3.	Fertilizers	5	3.30	145	96.70
4.	Disease management	2	1.30	148	98.70
5.	Pest management	2	1.30	148	98.70
6.	Weed management	3	2.00	148	98.00
7.	Time of harvesting	2	1.30	148	98.70

The recommended spacing was not followed by the grapes growers. The application of FYM and green manures was much less but the application of chemical fertilizers were manifold than the scientific practice. In order to ward off the pests and diseases a large quantity of pesticides was used. The safety measures were also totally ignored. Harvesting was done earlier because of the unhealthy competition among the grapes growers to get more yield and high remuneration. Though the grapes growers expressed concern about the environment, they pleaded the inability to follow the recommended practices as they were more concerned about getting immediate profit.

Factors influencing the negligence of environment and health are

- Lack of technical knowledge.
- Lack of education on keeping the environment safe.

- Lack of awareness on various forms of toxicity of pesticides and chemicals and health hazards.
- Lack of awareness on importance of keeping the flora and fauna of the soil and beneficial insects and
- Poor natural resource management.

Hence, to protect the natural resource base i.e., soil, water and environment for the survival of future generation, it is the need of the hour to educate the farmers about the environment.

Strategies to combat environmental issues and to preserve the natural resources

1. Research should be carried out to increase the observability of botanicals and other organic manures.
2. Effective and feasible eco-friendly technologies should be identified with yields comparable to the conventional agricultural practices.
3. Campaigns should be organized to demonstrate the health, economic and environmental consequences associated with the use of highly toxic chemicals to farmers.
4. Awareness workshops could be organized for creating awareness in safe use and handling of pesticides and protective gadgets.
5. Farming community should be educated about the negative impact of pesticides on the environment and health of human beings.
6. Continuous learning should be promoted on organically based alternatives to pesticides by popularizing eco-friendly farming alternatives.
7. By conducting result demonstrations farmers should be made aware of the comparative results of pure organic farming, integrated organic farming and conventional inorganic farming.
8. Farmers might be encouraged to maintain written record of pesticides applied in a particular season, including chemical names, dose applied and total number of applications which will be helpful for the future planning of pesticides.
9. Trainings could be given to pesticide dealers about the impact of chemical pesticides on health and environment.
10. Legislations should be passed which will make the pesticide producers to print the labels in local languages with pictograms which will helps in

clearly identifying the product hazards, appropriate mode of storage and handling, giving warning indications on repeated and cocktail use of pesticides.

11. While protecting the environment is our concern, the plea of the farmers to make profit cannot be totally ignored. They should be guaranteed with good price for their produce. Hence, certification for grapes grown following the recommended package of practices could be thought of which might bring good prices/ remuneration from consumers who are concerned with environment and health.

References

Borkar, S.G (2007). Diseases of Grapes their Forecasting and Control. Pointer Publishers, Agrawal Printing Press, Jaipur.

http://agriexchange.apeda.gov.in/Market%20Profile/one/GRAPES.aspx

Extension Management Strategies for Sustainable Agriculture, pp 189-203
© 2016, Editors, T. Rathakrishnan, H. Philip, P.P. Murgan and Jagadeesan
New India Publishing Agency, New Delhi, India

20

A Study on Fish Consumption Behaviour of Consumers

K. Veerabhadran, G. Arul Oli and D. Mahalakshmi

Fish has a vital role to play in mitigating the protein deficiency in developing countries. Fish has been described as the meat of the third world, and fish contributes significantly in developing and developed countries as a source of employment, income generation and foreign exchange earnings. The minimum per capita consumption of fish of an average Indian should be 11 kg per annum, as recommended by the ICMR. However, the per-capita consumption of fish in India is only 5 kg per annum, whereas, per capita consumption of fish in Japan was 62.6 kg, China - 30 kg and Bangladesh - 35 kg (Goswami *et al.,* 2012). Fish consumption pattern was dependent on the economic conditions of individual households and per capita fish consumption increased with increase in income. The percentage share of expenditure on fish over animal protein expenditure is higher for lower income groups, which clearly indicates that the poor people (lower income group) depend more on fish as a source of animal protein than the rich (high income group). The low-priced fishes were consumed more by the low income groups than by the high income groups and the high income groups spent a significant portion of budget on expensive fishes. The per capita consumption of fish was substantially higher in rural areas than in urban areas (Dey *et al.,* 2005).

According to FAO, the estimation about 84 per cent of total fish production (121.8 million tonnes in 2009) was used for direct human consumption. The remaining 16 per cent or 22.8 million tonnes were destined for non-food products, mainly for the manufacture of fish meal and fish oil. In India, about 28 per cent households consume fish, about 600 lakh families consume fish on a regular basis, and the estimated monthly per capita consumption of fish is 0.204 kg. The sociological factors such as the influence of group/family behaviour upon the individual behaviour, customs, traditions and culture of the family have their effect on the buying behaviour of the consumers. Then there are economical factors such as the income of the buyers, its distribution, savings, etc. which also contribute towards the consumer behaviour to certain extent. In order to specify the fish consumption level, this study conducted on the following aims: (i) the profile of the fish consumers with different social status, (ii) awareness and attitude of fish consumers, (iii) specific fish varieties as preferred by fish consumers with reasons, (iv) S ources of fish supply and mode of purchase, (v) characteristics of fish consumers and their consumption behaviour.

Approach Methodology

The Virudhunagar taluk of Tamilnadu selected for the study which has 36 wards and the sample unit was the non-vegetarian. The study population covered all the non-vegetarian families. Fifteen non-vegetarian heads of families consisting of five from each occupational category were randomly selected from each ward. Thus, the total sample numbered 150 for 10 wards together equally distributed among the three occupational categories. A total of 12 independent variables were selected for the study with single dependant variable fish consumption behavior. The interview schedule was constructed including all the selected dependent and independent variables. The structured interview schedule was presented in non-sampling area. Based on the results of the pre-testing, modifications were made in the schedule, the final interview schedule was prepared for collecting information from the selected respondents.The data were collected through personal interview. The following statistical tools were used for analysis. Percentage analysis was done to make simple comparisons, Chi-square test was applied to find out the significant difference between three categories namely labourers, businessmen, professionals and to find out the correlation between any two sets of variables viz., independent and dependent variables.

Results and Discussion

It may be concluded that the age factor had not differentiated the respondents belonging to different occupational groups. The educational status of fish consumers is a critical factor. The consumer categories of different occupational levels differed in their literacy level. The analysis was also found to be significant. Similarly the highest percentages in each occupational group of the respondents were Hindus followed by Christians and Muslims respectively. Thus there was no difference among different groups of consumers with references to religion. There was no other community in the labourers category. Similarly among the businessmen and the professionals the highest percentage of the consumers belonged to backward caste followed by most backward caste, schedule caste/

schedule tribe and then by other caste respectively. Thus there was no difference among different groups of consumers with reference to caste. There was a predominance of married respondents in the total sample and also in each occupational groups. There was no difference among the different group of consumers with reference to family structure.

It is found that there was difference between different occupational levels of consumers with reference to their annual income. The difference is obvious from the fact that labourers get low annual income compared to business people and professionals. Total annual income also differed in each occupational level. There was difference existed among the different occupational levels of consumers with reference to their annual expenditure. This difference is obvious from the fact that labourers spend low annual expenditure compared to business people and professionals.

Awareness and Attitude of Fish Consumers

Awareness of fish consumers

The respondents (91 per cent) were aware of the fact that fish contain more vitamins and minerals. About 84 per cent of the respondents were aware that eating fish is easy to digest followed by 81 per cent of them felt that fish is good for pregnant women, 74 per cent of them fish is good for heart disease, 67 per cent of them aware that fish is good for development of child brain and contain low cholesterol. About 59 per cent of the respondents known that fish strengthen the immune system of the body and 2 to 4 per cent of them felt that eating fish

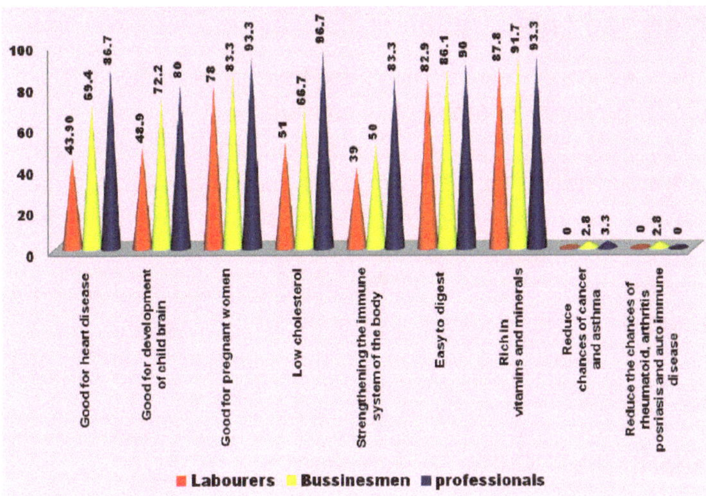

Fig. 1: Awareness on health benefit on fish consumption

reduce the chances of cancer, asthma, rheumatoid and auto immune diseases. Among the labourers and businessmen about 87 to 92 per cent of the respondents aware fish contains rich vitamins and minerals 97 per cent of the professional respondents aware fish contain low cholesterol. It was found that the majority of the professional respondents know more about the benefits of fish consumption to other two occupational groups.

Awareness of the quality of fish

About 86 per cent of the respondent among the total sample identified the quality of fish through appearance of eyes followed by 66 per of them identify the quality with gills appearance. About 56 per cent of the respondents identified the quality of fish with the appearance of body shape followed by 52 per cent of them identified freshness by seaweed odour smell and 48 per cent of them identify the quality of fish by softening nature of fish flesh at the time of fish purchase. Among the three occupational groups, professional respondents gave utmost importance to the fish quality at the time of fish purchase.

Sources for awareness of fish price in the market

The respondents get fish price information from the street vendors. About 30 per cent of the respondents got fish price information from their friends followed by 9 per cent of them read newspaper and get fish price information. It may be concluded that majority of the respondents in three occupational groups got fish price information first from street vendors followed by their friends and News Papers.

Awareness and preparation of fish dishes

A total fourteen value added fishery products as explained to the respondents to find out the awareness among fish consumers. About 21 per cent of the total sample respondents were aware of the masala dry fish followed by 10 per cent of them were aware of the fish pajji and Masmin and 9 per cent of them were aware of the fish pickles. The other dishes like fish cutlet, fish vadai were known to few of them. Fish consumers were not aware dishes like fish pakkoda, fish pups, fish rolls, fish noodles, fish macroni, fish balls, fish burger and fish fingers. Majority of the respondents among the three occupational groups aware about masala dry fish and majority of the respondents' aware preparation of fish pajji and fish vadai.

Training on preparation of fish dishes

The knowledge and skill of the fish consumers in the preparation of various fish dishes was found to be poor in general. They were asked whether they needed any training in the preparation of fish dishes. Among the 107 respondents only about 19 per cent wanted training and others did not want training on the fish/ dishes preparation. The respondents were mainly interested in the preparation of masmin followed by other products.

Attitude of Fish Consumers

Perception related to fish consumption

The defined statements relating to nutritive value, tasty, easy preparation, availability and expensive price for the study were read out and the respondents asked to indicate fish which statement was more applicable. Around 87.80 per cent of the consumers perceived fish is tasty followed by 59.80 per cent of them perceived fish is more nutritive value, 47.60 per cent of the consumers perceived fish price is expensive, 40.20 per cent perceived fish preparation is easy and 30.80 per cent of them perceived fish is easily available. Among the three occupational groups as well as in total sample most of the fish consumers perceived fish is tasty and high nutritive value.

People willing to consume fish dishes

About 91.60 per cent of the fish consumers in total sample willing to consume fresh fish dishes followed by 73 per cent of them consume boneless fish dishes. Parents are mostly influenced in all the occupational groups of the respondents for fish consumption.

Thrust about health benefits of fish consumption

About 92 per cent of the consumers thrust family as a source for health benefits followed by 77.60 per cent as friends, 52.30 per cent as doctors, 34.60 per cent as Newspaper/magazines, 30.80 per cent as Nutrition/ Dietician and Radio respectively. Among the three occupational groups as well as in total sample family is a source for most of the respondents thrust about health benefits of fish consumption.

Specific Fish Varieties as Preferred by Fish Consumers with Reasons

Consumers preference for fish compared to other non-vegetarian food items

The findings of the study on consumers preference for fish compared to other non-vegetarian food items data were collected and analysed.

Distribution of respondents by different non-vegetarian items

S.No.	Non-vegetarian items	Labourers (n=50)		Businessmen (n =50)		Professionals (n =50)		Total (n =150)	
		No.	%	No.	%	No.	%	No.	%
1.	Fish	41	82.00	36	72.00	30	60.00	107	71.30
2.	Mutton	2	4.00	4	8.00	5	10.00	11	7.30
3.	Chicken	4	8.00	5	10.00	10	20.00	19	12.70
4.	Beef	-	-	-	-	1	2.00	1	0.70
5.	Pork	1	2.00	2	4.00	1	2.00	4	2.70
6.	Egg	2	4.00	4	8.00	5	10.00	11	7.30
	Total	**50**	**100.00**	**50**	**100.00**	**50**	**100.00**	**150**	**100.00**

X^2 = 9.386 Non significant at 5% level

It is evident from the above table that the majority (71.30 per cent) of the respondents had chosen fish as their first choice. About 13 per cent of respondents had ranked chicken as their first choice followed by 7 per cent of the respondents ranked egg as their first choice, 5 per cent of them preferred mutton as their first choice. Only few respondents in the total sample had ranked either pork (or) beef as their first choice. Among the labourers, a vast majority of the respondents (82 per cent) had ranked fish as their first choice. Among the businessmen, 72 per cent of the respondents had ranked fish as their first choice and among the professionals, 60 per cent of the respondents preferred fish as their first choice. Thus it may be concluded that preference of fish among the labourers found to be higher compared to other two occupational groups. Evidently the chi-square value was also not significant at 5 per cent level.

Reasons for consumption of more fish

The less cost factor was the primary reason for the highest proportion of the labourers (92.70 per cent) preference of fish as their first choice. Businessmen and professionals (92 per cent to 87 per cent) choose fish as first choice for its easy digestibility and high nutritional values.

Reasons for not consuming fish

The bad smell was invariably quoted by the highest percentage in each occupation groups for non-consuming of fish. The fish price is expensive was the least of all the reasons among labourers, the reasons of difficult work involved in cleaning among the businessmen and the reason of non-availability of particular fish variety among the professionals.

Fish variety preference

There was no differential preference among the three occupational groups for selecting the fish variety either inland or marine fishes. About 51% of the respondents preferred inland fish varieties and 34.80% of them preferred marine fish and 11.20% of them liked to have both the fish varieties inland and marine.

Reasons for preference of inland fish varieties

Among 73.80 per cent of the consumers told inland fish variety available at all seasons followed by 65.60 per cent consumers told that inland fish was cheaper and 50.80 per cent of them told that inland fish was non allergic to skin. The total five reasons, particularly two reasons namely freshness and taste were found to be very strong points for preferring the inland fish compared to marine fish by a higher proportion of the respondents in each occupational group as well as in the total sample.

Reasons for preference of marine fish varieties

About 94.10 per cent of the consumers said marine fish varieties are tasty in comparison with inland fish varieties followed by 82.40 per cent of consumers viewed that there was no muddy smell in marine fish varieties, 70.60 per cent of consumers mentioned of less bony nature, 52.90 per cent of the respondents prefer as they are non allergic to skin, 47.10 per cent mention that marine fish available at all seasons and only 2.90 per cent of the consumers said reason for marine fish is low cost. Particularly two reasons namely absence of muddy smell and taste were found to be the very strong points for preferring marine fish compared to inland fish by a higher proportion of respondents in each occupational group as well as in the total sample.

Preference of specific fish varieties

Fish consumers of 35.50 per cent preferred murrels, about 17 per cent of the consumers liked Seer fish, 13 per cent of the consumers preferred Carps, 10 per cent of them preferred Sardines and Mackerel, 8 per cent of the consumers preferred Barracuda. Those who expressed their desire for other varieties were only a few not exceeding 3 per cent in any case. The same trend was also observed among occupational groups. It can be concluded that Murrels, Seer fish and Carps were the most preferred varieties. There is a habit to eat almost all fish variety except Tuna, Prawns and Shrimp among the labourers. These facts further strengthen the earlier conclusions that the fish is very popular among the labourers.

Table 1: Preference of specific fish varieties by the consumer respondents

S.No.	Specific fish varieties preference	Labourers (n = 41)		Businessmen (n = 36)		Professionals (n = 30)		Total (n =107)	
		No.	%	No.	%	No.	%	No.	%
1.	Murrel	15	36.60	14	38.90	9	30.00	38	35.50
2.	Carps	6	14.60	5	13.80	3	10.00	14	13.10
3.	Tilapia	1	2.40	1	2.80	-	-	2	1.80
4.	Cat fish	2	4.90	1	2.80	-	-	3	2.80
5.	Mackeral	2	4.90	1	2.80	2	6.70	5	4.60
6.	Eel	1	2.40	-	-	-	-	1	1.00
7.	Seer fish	6	14.60	7	19.40	5	16.70	18	16.80
8.	Pomfret	1	2.40	1	2.80	-	-	2	1.88
9.	Sardine	3	7.30	1	2.80	1	3.30	5	4.60
10.	Anchovie	1	2.40	2	5.50	-	-	3	2.80
11.	Barracuda	2	4.90	1	2.80	6	2.00	9	8.40
12.	Tuna	-	-	1	-	1	3.30	1	1.00
13.	Lethirines	1	2.40	-	-	2	6.70	3	2.80
14.	Crab	-	-	-	2.80	-	-	1	1.00
15.	Prawn	-	-	1	-	1	3.30	1	1.00
16.	Shrimp	-	-	-	2.80	-	-	1	1.00
	Total	**41**	**100.00**	**36**	**100.00**	**30**	**100.00**	**107**	**100.00**

Reasons for the preference of specific fish varieties

Taste is a first reason for 87.85 per cent of consumers preferring specific fish variety. Thus it can be concluded that Murrels, Seer fish and Carps were the three most preferred varieties and main reasons were tasty and less bony for consuption.

Common type of fish preparation at home

About 56.10 per cent of the consumers in total sample fish curry were the mostly prepared at home. About 30 per cent of them mostly prepare fish fry at home and 14 per cent prepared both (fish curry and fish fry). The sequential order of fish dishes like fish curry, fish fry and the both are prepared at each occupational group home. This means that there was no differential preparation among the three occupational groups.

Time for fish consumers prefer to eat fish

About 30 per cent of the consumers prefer to eat fish at dinner and 14 per cent of consumers eat both in lunch and dinner time. Only one per cent of the consumers eat fish at breakfast. The sequential order of fish prefer to eat consumers like lunch, dinner and both time was same for each occupational groups except in labourers only one consumers prefer to eat fish in breakfast.

Form of fish variety preferred by consumers

About 72 per cent of the consumers in total sample prefer to eat fresh fish followed by 21 per cent of consumers eat dry fish and 7 per cent of them eat iced fish. Only one per cent of the consumers prefer to eat canned fish. Sequential order of fish prefer to eat consumers like fresh fish, iced fish, dry fish and canned fish was same for each occupational groups except in professional only one consumer prefer to eat canned fish.

Reasons for preference of fresh fish

About 94.80 per cent of the consumers in total sample like fresh fish because of tastiness'. More or less of the same percentage of consumers mentioned about its hardness while touching and about 71 per cent of the consumers said that they could not get any bad smell and 32 per cent of them preferred fresh fish because it was not damaged and spoiled. There were very least numbers of the respondents liked to have iced fish.

Reasons for preference of Dry fish

About 86.40 per cent of the consumers in total sample like dry fish because of its taste. About 72.70 per cent of them preferred dry fish because it was stored for long days and 59.10 per cent of the consumers told that dry fish is easily available in all places with low cost.

Sources of fish supply and mode of purchase

Purchasing point

It may be concluded that the cycle vendor was the main source from purchase of fish in each occupational group followed by the fish rearing place, market and landing centres.

Frequency of fish purchasing behaviour

About 44 per cent of the consumers in total sample for purchase of fish once in fortnight. About 32 per cent of the consumers purchase fish once a week followed by 21 per cent of them purchase monthly once and 3 of them purchase fish twice a week. Among three occupational groups labourers were found to purchase fish more frequently compared to businessmen and professionals.

Fish consumption level

About 29 per cent of the respondents in total sample purchased fish ranging from 2 to 4 kg per month. About 23 per cent of them purchased less than 2 kg followed by 22 per cent of them purchased 4 to 6 kg, 21 per cent of them purchased 6 to 8 kg and 6 per cent of the respondents purchased fish on highest quantity like above 8 kg. This consumption trend also existed among the individual occupational groups. Thus the different occupational groups did not differ with reference to monthly intake of fish. This was supported by chi-square value also not significant.

Decision making

About 43 percent of the families wife was the decision makers in purchasing fish followed by Husband and wife (26 per cent) and husband alone (20 per cent) respectively. Otherwise, in the true sense they could not be the decision makers taking the cost and other factors into consideration as it would be. Among the three occupational groups most of the families, wife was the decision maker in purchasing of fish, because they were looking after most of the home activities

Purchaser

Decision makers may not be always the actual purchasers. At time there may be the different family members playing the role of decision making and the purchasing separately. About 53% of Husbands and 20% of wives were the purchaser of fish in the family. The three occupational groups did not different among themselves with reference to the purchaser.

Distance of fish purchasing point

The distance to which the highest percentage of each occupational group had to walk, was within one km and such there were no differences with reference to distance among the occupational groups.

Non – consumption of Fish during certain periods

About 80 per cent to 92 per cent of the respondents in total sample did not consume the fish either on "Kiruthikai" or in "New moon days". About 69 per cent of the respondents were the strict vegetarians on the day of "Adi Friday". About 50 to 56 per cent of the respondents avoided fish eating in the month of "Pankuni" or "Sastiviradam" and "Navarathri" days. Also 43 per cent of the respondents avoided fish eating in the month of "Puratasi" and 20 per cent to 25 per cent of respondent avoided fish consumption in "Thai pusam" or "Pongal" and "Maha sivarathri" days. Among Christians of fish consumers to avoid fish on certain religious days. About 81 per cent of the respondents do not eat fish from Good friday and 62 per cent of them avoid fish from "Vibuthi thirunal". About 38 per cent of then do not eat fish during "New year" or "Christ birth day. Among Muslims of fish consumers also were not the exception to this custom. But of 9 Muslims, 7 did not consume fish on "Ramzan fasting days" and "Miladinabi", 3 on "Muharam".

More fish consumption on certain periods

About 94 per cent of the respondents in total sample used to purchase more fish during holidays and during the visit of the guests to their home. Also 33 per cent consumed more fish during social festivals. As met with any market commodity fluctuation in fish price is not uncommon. Whenever market price had gone down about 27 per cent of them motivated to buy more fish and 23 per cent had bought more fish on the days when they had higher income. About 5 to 9 per cent of them purchase more fish in Deepavali, Ramzan and Christmas days.

Regular fish consumption days

Among Labourers higher 71 per cent of them consume fish on regular week days compared to other two occupational groups. Among the three occupational groups, majority of the respondents consume fish regularly on Sundays and Saturdays.

Place to consume fish

Majority (56 per cent) of the consumers in total sample wish to consume fish at home followed by 16 of them consume fish at restaurants and 12 per cent of the respondents consume fish at home and restaurants. About 8 per cent of the respondents chosen to consume fish at mobile stalls. Only few of them consume fish in home & restaurants, restaurants & mobile stall and home, restaurants & mobile stall. Thus it may be concluded that majority of the respondents in three occupational group wish to consume fish at home.

Fish consumption practices

About 14 per cent of them consumed next day of without keeping it in fridge followed by 4 per cent of the respondents consumed fish in next day preparation after keeping it in fridge. Among the three occupational groups of the respondents prefered to consume first day of fish preparation.

Non-availability of fish varieties

About 80 per cent reported that varieties preferred by them were available in market and only 20 per cent on the non-availability of chosen fish variety.

Totally eight varieties of fish were not available to them. About 29 per cent of the respondents told that Goat fish were the non-available variety and 14 per cent told that Eel and Prawn were non-available variety of them. About 10 per cent of them told that Tuna, Valai, Uluvai and Crab varieties were the non-available variety.

Suggestion to improve fish consumption

The ultimate purpose of research would be to suggest the ways and needs to promote fish consumption among the non-vegetarian families. The respondents themselves may have some suggestions on this issue.

Distribution of fish consumers by their suggestion to improve fish consumption level

No.	Suggestions	Labourers (n = 41)		Businessmen (n = 36)		Professionals (n = 30)		Total (n =107)	
		No.	%	No.	%	No.	%	No.	%
1.	Makes reservoir fishes like catla, rohu, mrigal, silvercarp easily available	14	34.10	23	63.90	9	30.00	46	43.00
2.	Making other inland fishes available throughout the year	18	44.00	20	55.60	7	23.30	45	42.00
3.	Making available fresh marine fish in the market	15	36.60	31	86.10	11	36.70	57	53.30
4.	Avoiding fluctuation in the fish price by retailers convenience	10	24.40	29	80.60	15	50.00	54	50.50
5.	Avoidance of adulteration like mixing with spoiled fish	26	63.40	22	61.10	20	66.70	68	63.60
6.	Opening more number of fresh fish stall by Tamilnadu Fisheries Development Corporation in all places	28	68.30	30	83.30	27	90.00	85	79.40
7.	Fish market centers in nearby places to easy the purchase	38	92.70	34	94.40	25	83.30	97	90.70
8.	Improving the sanitation of the fish market	7	17.10	16	44.40	21	70.00	44	41.10
9.	Fish consumption awareness and benefits to be created	18	43.90	14	38.90	10	33.30	42	39.30

Characteristics of fish consumers and their consumption behavior

The socio-personal and socio psychological characteristics and their association and contribution towards the fish consumption behaviour of consumers have been studied the result are presented in the following table.

Characteristics of fish consumers and their consumption behaviour

Variable No.	Constraints	Correlation co-efficient	
		Fish consumers	Non – fish consumers
X_1	Gender	0.0892 NS	- 0.4094 **
X_2	Age	- 0.2921 **	0.2096 *
X_3	Educational status	0.08570 NS	0.4200 **
X_4	Religion	- 0.0901 NS	- 0.0554 NS
X_5	Caste	0.3015 **	0.5034 **
X_6	Marital status	- 0.1305 NS	- 0.1938 NS
X_7	Family structure	0.0849 NS	0.0102 NS
X_9	Annual income	0.2282 *	0.7652 **
X_{10}	Annual expenditure	0.2671 **	0.7160 **
X_{11}	Awareness of fish consumer	0.2282 *	0.4570 **
X_{12}	Attitude of fish consumer	0.2088 *	0.7188 **

** Significant at 1 percent level * Significant at 5 per cent level NS – Non-significant

It could be seen from the above Table that Eight of the independent variables had positive relationship with the dependent variable and the remaining three variables exhibited negative relationship in fish consumers level. Independent variable viz., Gender (X_1), Educational status (X_3), caste (X_5), Family structure (X_7), Annual income (X_9), Annual expenditure (X_{10}), Awareness of fish consumers (X_{11}), Attitude of fish consumers (X_{12}) were to have positive relationship with dependent variable, extent of fish consumption behaviour. Among eight variables that exhibited positive relationship to the fish consumption behaviour two independent variables viz., Caste (X_5) and Annual expenditure (X_{10}) had significant relationship at one per cent level. While another independent variable namely Annual income (X_9) had significant relationship at 5 per cent level. Remaining three independent variables namely Gender (X_1), Educational status (X_3) and Family structure (X_7) had non-significant relationship with the dependent variable.

Among the three variables that exhibited negative relationship to the fish consumption behaviour two independent variable viz., Religion (X_4) and Marital status (X_6) had non-significant relationship with the dependent variable. Remaining one variable namely Gender (X_1) had significant at 1 per cent level respectively.

Summary and Conclusion

The Problem of 'protein gap' is more acute in India and therefore there is an urgent need for massive production of animal protein for which 'fish culture' could contribute considerably. Fish culture, a relatively new and developing science, has been receiving great deal of attention in recent years because of its potential role in integrated rural development programmes, gainful employment and sustained income. Fish Farmers Development Agencies have been set up in Tamil Nadu with the general objective of popularizing the improved techniques of fish culture so as to step up inland fish production.

1. Since lack of fish market in nearby places was a constraint, it is essential to open more fish markets in different places of the Taluk to promote fish consumption.
2. The government should open a cold storage in Virudhunagar taluk to store the fresh fish and the rent should also be at reasonable price for the private fish traders so that they can avoid the spoilage of fishes which would promote fish consumption.
3. Quality control of fish should be done every day by municipal authority as in the case fish in order to restrict the selling of spoiled fish.
4. The metric system of weighment in selling fish should be imposed so that we can rationalise fish price.
5. Fish is one of the most delicious and perishable of all foods. There is a diversity of dishes that can be prepared with fish, an assortment which takes in sandwiches, soups, and salads as well as appetizers and entre features. Certain amount of knowledge on the above and related aspects should be imparted to fish consumers through mass media.
6. Once purchased, the fish must be kept refrigerated until preparation time. Such expensive facilities are rarely available in all households. There is a need to suggest cheaper methods of preserving the purchased fish for sometime till it is cooked.

References

Ali Hassan (2006). Identification of training needs of women shelf help groups of Tamilnadu. Unpub. M.F.Sc.Thesis. FC&RI. TANUVAS, Thoothukudi.

Arpita Sharma. (2003). Valuing women's work in fisheries: *Fishing Chimes*, 23 (7): 46-47.

Brunso, K. (2009). Trend in consumer attitude and selection. *FAO Fisheries and Aquaculture Proceedings*, 13: 115-121.

*Dey, M., Rab, M.A., Paraguas, F.J., Piumsombum, S., Bhatta, R., Alam M.F., and Ahmed, M. (2005). Fish consumption and food security: A disaggregated analysis by types of fish and classes of consumers in selected Asian countries. *International Journal of Aquaculture Economics and Management,* 9(1): 89-111.

Edun, O.M., O.A. Akinrotimi, A. Uka and K.N. Owhonda, (2010). Patterns of mudskipper consumption in selected fishing communities of rivers state. *Journal of Agriculture and Social Research,* 10 (2): 100-108.

Ergonul, B. (2011). Meat consumption and buying behaviors of consumers living in Manisa city center, Turkey. *Journal of Animal and Veterinary Advances,* 10 (3): 286-290.

Esakkias, Y. (2007). Impact of fisheries training programme on the socioeconomic status of women self help group in Thoothukudi District. Unpub. M.F.Sc.Thesis. FC&RI. TANUVAS, Thoothukudi.

FAO (2009). Fishery and Aquaculture Statistics. Food and Agriculture Organization of the United National, Rome: pp18.

Ganesh Kumar., Engle, C., Quagrainie, K. (2008). Household preferences and consumption patterns for Farm-Raised catfish in the U.S. Department of Agriculture and Country government cooperating. Co-operative extension program, University of Arkansas at Pine Bluff, U.S. PP 15.

Goswami, C., and Bhuyan, P.C. (2012). Segmental variations in fish consumption behaviour: A case study of Sonitpur district of Assam. *Indian Journal of Marketing*, 42(4): 50-57.

http://dahd.nic.in/dahd/division/fisheries.aspx

http://nfdb.ap.nic.in/html/aboutus.htm

Iyangbe, C.O. and Orewa, S.I. (2009). Assessment of the calorie-protein consumption pattern among Rural and low-income urban households in Nigeria. *Middle-East Journal of Scientific Research*, 4(4): 288-296.

Jackie Singh, Y. (2009). Adoption of hygienic fish handling practices by fishermen. Unpub. M.F.Sc.Thesis. FC& RI. TANUVAS, Thoothukudi.

Kavitha, S. (2005). Consumption pattern of fish and fishery products in the selected towns of Thoothukudi district, Tamilnadu. Unpub. M.F. Sc. Thesis. FC&RI. TANUVAS, Thoothukudi.

Mahesh Babu, Y. (2011). Impact of Training Programmes on Ornamental Fish Culture on the Empowerment Status of the Trainees. Unpub. M.F.Sc.Thesis. FC&RI. TANUVAS, Thoothukudi.

Marine Products Export Development Agency (MPEA) (2012). Ministry of Commerce & Industry, Govt. of India. XXI (2): 20.

Santhakumar, R. (1988). Study on the characteristics of fish consumers and their consumption behaviour. Unpub. M.Sc.(Ag.) Thesis. Dept.of.Agrl. Exten., TNAU, Coimbatore.

Senthil Kumar, K. (2008). Adoption of composite fish culture technology among the fish farmers of Thanjavur District. Unpub. M.F.Sc.Thesis. FC&RI. TANUVAS, Thoothukudi.

Sundararajan, A. and Renuka, T. (1998). Palm leaf artisans in Manapad, Tamilnadu. *Social Welfare*, 45(1): 35-37.

Thriveni Kasukurthi (2011). Retail Fish marketing of commercially important marine fishes in Thoothukudi city. Unpub. M.F.Sc.Thesis. FC&RI. TANUVAS, Thoothukudi.

Verbeke, W., Sioen, I., Pieniak, Z., *et al.* (2005). Consumer perception versus scientific evidence about health benefits and safety risks from fish consumption. *Public Health Nutritn*, 8(4): 422-429.

Vinod, K.and Sharma O.P. (2011). Socio-economic status and consumption pattern of fish consumers in Udaipur (Rajasthan). *Progressive Agriculture*, 11(1): 89-94.

Extension Management Strategies for Sustainable Agriculture, pp 205-213

New India Publishing Agency, New Delhi, India

21

New Dimension of Sugarcane Technology Delivery Through Open and Distance Learning Mode and Its Usefulness Among the Learners

R. Sureshverma, R. Rajasekaran, Venkatapirabu and Guru

Agriculture is the key force in India and nearly two-third of the work force is directly or indirectly depends on agriculture to support their livelihood. In every day situation, farmers are facing complex problems viz., pest hazards, weed control, moisture insufficiency, soil fertility, farm credit, labor shortage, soil erosion, crop production technologies and so forth to improve the production and productivity. It is not sufficient to give agricultural related information to farmers alone and it is very essential to give information on newly released technologies to the people involving in the extension activities or extension personnel's. The best method of educating working group of population is distance education. Distance education or distance learning is a mode of delivering education and instruction, often on an individual basis, to students or employees of various fields who are not physically present in a traditional setting such as a classroom. Since 2005, TNAU is offering distance learning programme on various courses. The major objective of this study is to identify the utilization behaviour of the ODL course on Sugarcane Technology. The study conducted among the learners of PG course on Sugarcane technology offered through ODL mode by the Directorate of ODL, Coimbatore. The major findings of this study as perceived by the learners are job satisfaction, adequacy of the course content, usefulness of the course for skill development and usefulness of the course for attitude change were found to have a positive and significant relationship with the 'r value' of 0.364, 0.266, 0.296 and 0.393 respectively with the learners utilization behaviour.

Introduction

The projected growth of the worldwide population of people qualified to proceed from high school to higher education will yield a significant increase in demand that cannot be met by existing capacity or infrastructure. It is generally acknowledged that open and distance education is a good way of reaching out to large numbers of students. India accounts for a quarter of the developing world's population and has the third largest higher education system in the world as reported by ICDE 2009. Approximately 24 per cent of all higher education students in India are enrolled in distance education institutions, specifically in the 13 national and state open universities and the 106 institutions, mostly public, which offer both on-campus and correspondence programs (Daniel *et al.*, 2007).Even it is a herculean task, it is the right time to identify the utilization behaviour of the Open and Distance Learning course on Sugarcane Technology among the learners.

Research Methodology

Tamil Nadu Agricultural University offers different Post Graduate courses through Open and Distance Learning mode to the needy people. Post graduate course on Sugarcane Technology is one among them. For the present study, ex-post facto research design was employed. The entire population who registered and completed the Post Graduate course on Sugarcane Technology through Open and Distance mode of learning was selected for this study. Totally, 89 members were registered for this course.

The entire population has been taken for the investigation. A well structured questionnaire was prepared and the informations were collected by sending questionnaire through registered post and in-person to the registered members. After giving considerable period of time for the learners to send back their responses, 60 registered learners were responded. These 60 responses were taken as the sample for the study.

The utilization behaviour was studied with reference to different components as mentioned in the Table 5. The utilization behaviour was analysed and interpreted using the procedure adopted by Chandrasekar (2013) had been used for this study.

Findings and Discussion

The results of the study have been presented in the following tables along with their interpretations. The discussion is taken up simultaneously along with the results.

Job Satisfaction

The respondents were satisfied or dissatisfied with different aspects of their job after completion of the post graduate course on sugarcane technology. Therefore, the job satisfaction of the respondents was studied and the results are depicted in Table 1 and Figure 1.

Table 1: Distribution of the learners according to their job satisfaction **(n=60)**

Sl. No.	Category	Number	Per cent
1.	Low	12	20.00
2.	Medium	35	58.33
3.	High	13	21.67
Total		**60**	**100**

From Table 1, it could be interpreted that majority of the respondents (58.33 per cent) had medium level of job satisfaction, followed by high (21.67 per cent) and low (20.00 per cent) levels respectively.

Among the respondents very few were doing agriculture and unemployed. A meagre section felt dissatisfaction in their job, due to the health reasons, nature of the job, intensive travelling and other administration tangles, might be the probable reasons for the medium level of satisfaction, even though respondents were satisfied to the extent of getting promotion and financial increment, immediately after completion of the post graduate course through distance learning mode. The findings were in accordance with Mishra *et al.* (2007).

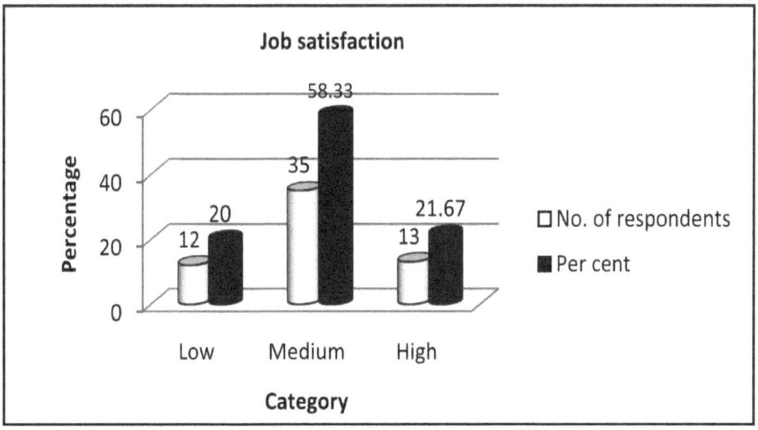

Fig. 1: Distribution of the learners according to their job satisfaction

Perception about the adequacy of course content

The perusal of the Table 2 and Figure 2 explains the adequacy of the course contents as perceived by the learners.

Table 2: Distribution of the learners according to the adequacy of the course content **(n=60)**

Sl. No.	Category	Number	Percentage
1.	More adequate	13	21.67
2.	Adequate	45	75.00
3.	Inadequate	2	3.33
Total		**60**	**100**

It could be very well diagnosed from the table 2 that exactly 75.00 per cent of the learners have responded that the course content was adequate, whereas, 21.67 per cent have stated that the contents were more adequate and only 3.33 per cent expressed as inadequate.

Most of the respondents expressed that the course content was adequate. This might be due to the reason that post graduate programme covered most of the recent sugarcane production technologies which may satisfy the learners' expectation. Hence, more than two-third of the learners were of the opinion that the course content was adequate. The result derives support from the result of Naika (1999) and Palaniswamy (1994).

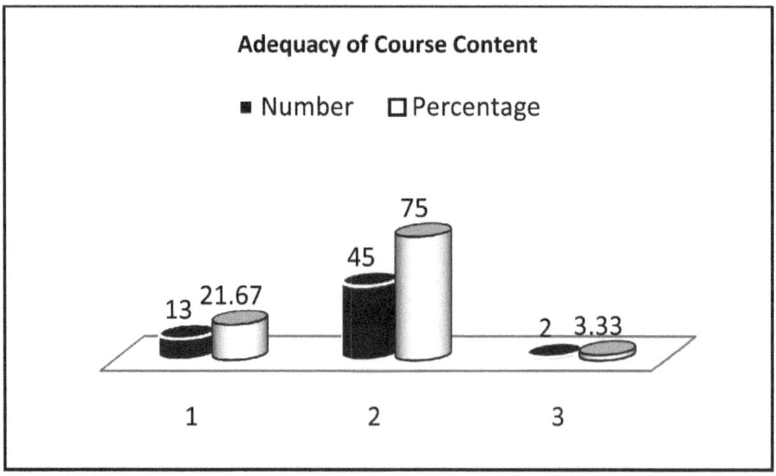

Fig. 2: Distribution of the learners according to the adequacy of the course content

Perception on usefulness of the course for skill development

It is vivid from Table 3 and Figure 3 that nearly 81.67 per cent responded that the courses were useful in skill development and 13.33 per cent reported as most useful, followed by negligible opined that it was not at all useful.

Table 3: Distribution of the learners according to their perception on usefulness of the course for skill development **(n=60)**

Sl. No.	Category	Number	Percentage
1.	More useful	8	13.33
2.	Useful	49	81.67
3.	Not useful	3	5.00
Total		**60**	**100**

It is very interesting to observe that the courses were useful (81.67 per cent) in skill development on various sugarcane production technologies, which were mostly expected by the learners while making enrolment in the Post Graduate Course on sugarcane technology. The results are in agreement with the findings of Naika (1999).

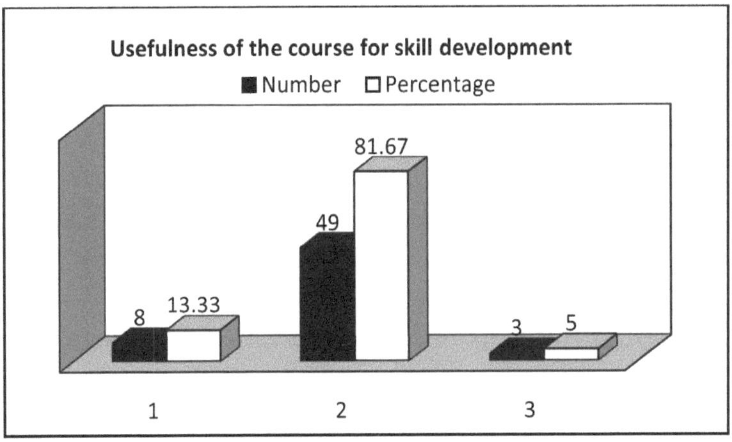

Fig. 3: Distribution of the learners according to their perception on usefulness of the course for skill development

Perception about the Change in Attitude

From Table 4 and Figure 4, it could be exonerated that majority (55.00 per cent) of the respondents stated that the courses were useful for change in the attitude of the learners, whereas 40.00 per cent were not useful in changing learners attitude and very meager (5.00 per cent) were more useful in changing the attitude of the learners.

Table 4: Distribution of the learners according to their perception on usefulness of the course for attitude change **(n=60)**

Sl. No.	Category	Number	Percentage
1.	More useful	3	5.00
2.	Useful	33	55.00
3.	Not useful	24	40.00
Total		**60**	**100**

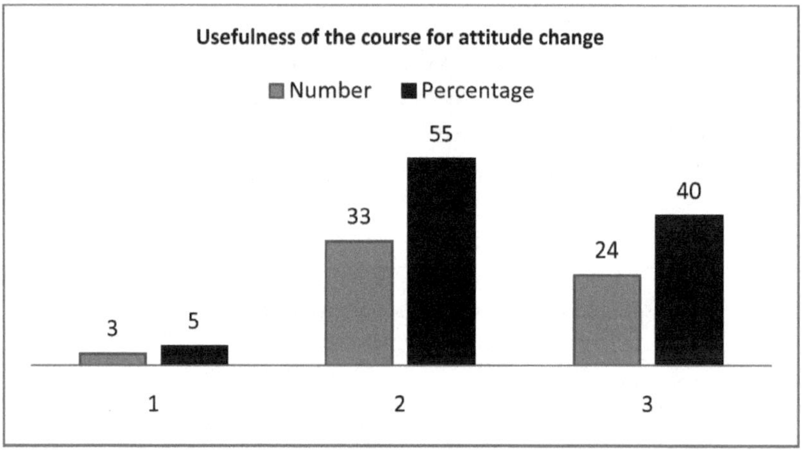

Fig. 4: Distribution of the learners according to their perception on usefulness of the course for attitude change

Perception about the Utilization behaviour of the Course among the Learners

It is observed from the Table 5 and Figure 5, that a majority of the beneficiaries have been impacted much on the utilization behaviour, since they were impacted skill and knowledge through sugarcane technology course, which in turn had lead them to improve their writing skills, recognition and approval from the peer groups and super ordinates in their work accomplishments.

Table 5: Distribution of learners based on the perceived impact on utilization behaviour of the ODL course on Sugarcane Technology **(n=60)**

Sl. No.	Particulars	Decreased	No	Increased change	Increased impact percentage	Overall percentage of increased impact
	Utilization Behaviour					
1.	Use of technical knowledge in sugarcane production technology	0	4	56	93.33	71.04
2.	Team work	0	19	41	68.33	
3.	Extension activities	0	19	41	68.33	
4.	Writing research articles related to sugarcane technology	0	39	20	33.33	
5.	Educating colleagues/ co-workers	0	6	54	90.00	
6.	Research activities undertaken	0	38	22	36.67	
7.	Sharing of knowledge with others	0	2	58	96.67	
8.	Status attained	0	11	49	81.67	
9.	Leadership quality	0	21	39	65.00	
10.	Technical Communication with superiors	0	2	58	96.67	
11.	Technical Communication with coworkers	0	1	59	98.33	
12.	Motivation towards your fixed goal	0	18	42	70.00	
13.	Job responsibility	0	20	40	66.67	
14.	Job satisfaction	0	7	53	88.33	
15.	Positive employees relation	0	18	42	70.00	
16.	Negative employees relation	8	44	8	13.33	

* (Multiple responses obtained)

In addition they were able to build team, strengthen the extension activities and became prolific writers, through their penmanship. Apart, the learners were engaged in educating colleagues (90.00 per cent) through the inputs they gained through the course under ODL.

Beside that, they were able to shape their research ability (33.33 per cent), shared the knowledge on demand (96.67 per cent), and gained special status from others (81.67 per cent). More so, their leadership quality (65.00 per cent), technical communication with the superiors (96.67 per cent) and coworkers (98.33 per cent), high amount of motivation shouldering job responsibility (66.67 per cent) and job satisfaction (88.33 per cent) and maintaining positive employee relationship were found to be high.

Less amount of impact has seen in the utilization behaviour especially in research activity. It is very much encouraging to note that the technical communication with the coworker on sugarcane technology is highly increased among all the learners.It could be understood that Post Graduate course on Sugarcane Technology had increased impact on utilization behaviour among the learners.

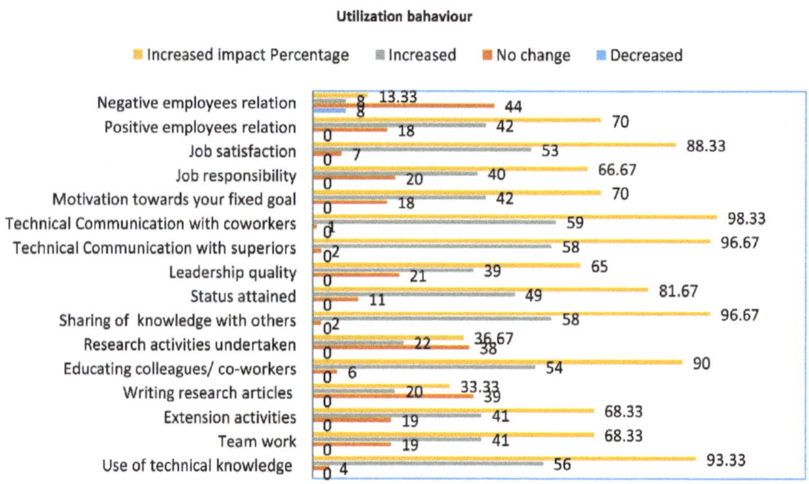

Fig. 5: Distribution of learners based on the perceived impact on utilization behaviour of the ODL course on Sugarcane Technology

Conclusion

The present study revealed that,majority of the respondents (58.33 per cent) had medium level of job satisfaction, followed by high and low levels respectively with their value of 0.364 and significant relationship at one per cent level. With regard to the perception about the adequacy of course content, more than 2/3rd (75.00 per cent) of the learners have responded the course content was adequate, whereas 21.67 per cent have stated that the contents were more adequate and very meager per cent expressed as inadequate. Perception about the adequacy of course content was found to have a positive and significant relationship at five per cent level with the value of 0.266.

It is very interesting to observe majority of the respondents (81.67%) replied that the courses were useful in skill development. Perception on usefulness of the course for skill development was found to have a positive and significant relationship with the value of 0.296 at five per cent level. Perception about the change in attitude, majority (55.00%) of the respondents stated that the courses were useful for change in the attitude of the learners, whereas 40.00% responded that they were not useful in changing learners' attitude and very meager (5.00%) said that it was more useful in changing the attitude of the learners. Perception on usefulness of the course for attitude change was found to have a positive and significant relationship at five per cent level with the value of 0.393 respectively with the learners utilization behaviour of Post Graduate Course on Sugarcane Technology.

It could be conclude from the findings that, adequacy of the course content was very much useful for skill development and attitude change towards new sugarcane production technologies which leads to increased utilization behaviour and job satisfaction of the learners finally.

References

Chandrasekar, K. (2013). Impact analysis of Agricultural Management Agency (ATMA) programme amongst end-users in the process of Agriculture and Rural Development. M.Sc. (Ag.) Thesis, AC & RI, TNAU, Coimbatore.

Daniel, John, Asha Kanwar, and Stamenka Uvalic-Trumbic (2007). Mass Tertiary Education in the Developing World: Distant Prospect or Distinct Possibility? Commonwealth of Learning. http://www.col.org/colweb/site/pid/4605.

ICDE (International Council for Open and Distance Education). (2009). Global Trends in Higher Education, Adult and Distance Learning.

Mishra, D.M. Chandargi and I.V. Hirevenkanagoudar (2007). A comparative study on job performance and job satisfaction of men and women extension Officers. *Karnataka. J. Agric.Sci.,* 20 (2): (432-434).

Naika (1999). A study on the impact of Educational Technology on Students' Development in a State Agricultural University of Karnataka, Ph.D Thesis, TNAU, Coimbatore.

Palaniswamy (1994). Human Resource Development in Agricultural Sector in Tamil Nadu. M.Sc. (Ag.) Thesis, TNAU, Coimbatore.

Extension Management Strategies for Sustainable Agriculture, pp 215-222

New India Publishing Agency, New Delhi, India

22

A Study on Assessing the Effectiveness of the Distribution Channel of Parle Agro Pvt. Ltd. in Palakkad District

M. Israel Thomas, Shipin, P. and R. Sendilkumar

Distribution channels are behind every product and service that consumers and business buyers purchase everywhere. Product distribution is one of the four elements of the marketing mix. Usually, combination on institutions specializing in manufacturing, wholesaling, retailing and many other areas join force in Distribution channels. A Distribution channels are a set of inter-dependent organization involved in the process of making a product or service available for use or consumption. Distribution channels decisions play a role of Strategic importance in the overall presence and success a company enjoys in the market palace.

India is the second-largest producer of food, and has the potential to be the largest on a global food and agriculture canvas. The food industry in India comprising of food-producing and processing industries. The food processing industry is one of the largest in India – it is ranked fifth in terms of production, consumption, export and expected growth. Maharashtra is the leading state in India as far as food processing is concerned. Annually, India produces 205 million tonne fruit and vegetables, and is the second-largest country in the world as far as farm production is concerned. However, it processes just 4.6 per cent of its output. In contrast, countries like the United States (65 per cent), China (23 per cent) and Philippines (78 per cent) are far ahead of India in reducing wastage and enhancing the value addition and shelf-life of the farm products. In India, the processed food industry contributing to 6.3 per cent of GDP. It accounts for 13 per cent of the country's exports and 6 per cent of total industrial investment.

Parle Agro Pvt Ltd one of the largest Indian food and Beverage Company. Their brands that have won the hearts of consumers everywhere, they are in almost every home across the length and breadth of India. Parle Products was founded in 1929 in British India. It was owned by the Chauhan family of Vile Parle, Mumbai. The Parle brand became well known in India due to the success of products such as the Parle-G biscuits and Thums Up soft drink. The original Parle company was split into three separate companies owned by the different factions of the original Chauhan family: Parle Products, Parle Agro and Parle Bisleri, All three companies continue to use the family trademark name “Parle”. Parle products have wide acceptance, like Frooti India’s first mango drink in a tetra pack, Appy Fizz, the country’s first sparkling apple drink, their beverages have over the decades acquired iconic stature in one of the most competitive markets in the world.

Statement of the Problem

Frooti which was launched way back in 1985 and still upfront in the fruit beverage market. But in the present scenario of cut-throat competition from domestic and multinational giants, it’s very difficult to remain at the top. To be at the front, it’s very necessary that products are readily available to the consumers as and when they demand for it. There should not be any delay in that, further this should be one of the most important objectives of the firm. This objective can be achieved only through effective distribution channel.

This study examined the effectiveness of the distribution channel of Parle Agro Pvt. Ltd. With reference to fruit drink called Frooti in Palakkad district. It studied whether the product is being distributed efficiently and effectively or not in the market, to make sure that the customer is satisfied. It also studied whether there is any service gap in relation with the distribution channel of Frooti and if it’s there how it can be solved. It examined three levels of distribution, through which the product is actually reached in the hands of consumers. The study took into account of all the elements included in the distribution channel like frequency of order, period, route, cost etc. to check whether the distribution system is effective enough to meet the changing needs of customers.

Objective

To study the effectiveness of the distribution channel of Parle Agro Pvt. Ltd. in Palakkad with reference to Frooti.

Methology

The primary data for the study were collected from distributors, retailers and from consumers. A sample of 10 distributors, 20 retailers and 50 customers of Frooti were selected from Thrissur district.The study used both primary and secondary data. Primary data were collected by administering structured interview schedules for distributors, retailers and consumers. Secondary data were collected through books, journals and websites related to the subject domain area. Statistical tools such as Percentage analysis and Satisfaction.

Results and Discussions

Reason for adopting distribution of Parle Agro

The different factors which made the distributors to adopt the distribution of Parle Agro were depicted in the Table 1. It covered the area like margin, promotional schemes etc. The index scores have been obtained for each factor.

Table 1: Reason for adopting distribution of Parle Agro

Sl No.	Particulars	Strongly Agree	Agree	No opinion	Disagree	Strongly Disagree	Index
a.	Margin	2	7	0	1	0	80
b.	Promotional Schemes	1	5	2	2	0	70
c.	Availability	1	6	0	3	0	70
d.	Dealer relationship	3	6	0	1	0	82
e.	Credit Policy	1	2	1	5	1	54
f.	Quality	2	7	1	0	0	82

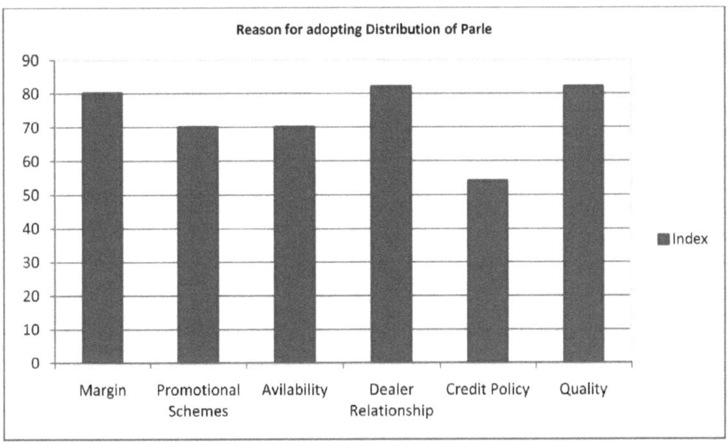

It is very clear from the Table 1 that the reasons for the dealers to adopted the distribution of Parle Agro. In factors studied, dealer relationship and quality scored with the index of 82. Distributors have good opinion in margin (80) provided by the Parle Agro. Credit policy provided by the Parle Agro does not satisfy the distributor which scored the least (54).

Timely delivery

It is very essential that customers should get the products when they actually want it. Otherwise they will shift to other product. So there should be timely delivery of the product. Table 2 showed the timely delivery of Parle.

Table 2: Timely delivery of Parle Agro

Sl No	Particulars	Opinion	Respondents	Percentages
1	Timely delivery of Parle	Excellent	1	10
		Very Good	4	40
		Good	3	30
		Average	2	20
		Poor	0	0

Regarding the timely delivery of Parle Agro, 40 per cent of the distributor says that they are very good at delivering the product timely. Whereas 30 per cent felt it's good only. 10 per cent says they are excellent in the timely delivery of products. But even though there is good impression about the delivery, there should be continuous improvement so as to remain leader in the market.

Time Required for Placing the Delivery to Retailer

Table 3 showed that how much time it requires to deliver the order placed by the retailer. Because if the delivery has not taken place in time, it will be problem for both distributor and also the retailer. Both will be losing their business

Table 3: Time required for placing the delivery to Retailer

Sl No	Particulars	Opinion	Respondents	Percentages
1	Time it requires to deliver the order placed by the retailer	Within 1 day	0	0
		1 to 2 day	7	70
		3 to 5 day	3	30
		More than 5 day	0	0

In case of delivering the order to the retailer, most of the order will be placed in a period of 1 to 2 day (70 per cent). And maximum time period is 3 to 5 days. Distributors try to deliver the order on time.

Reason for Association with Parle Agro by the Retailer

The Table 4 showed the reason for association with Parle. It's been studied with different parameters and the index was calculated.

Table 4: Reason for association with Parle Agro

SI No.	Particulars	Strongly Agree	Agree	No opinion	Disagree	Strongly Disagree	Index
a.	Margin	2	14	1	3	0	75
b.	Promotional Schemes	4	11	1	4	0	71
c.	Availability	5	12	1	2	0	80
d.	Dealer relationship	4	11	2	3	0	76
e.	Credit Policy	0	6	7	5	2	64
f.	Quality	7	9	2	2	0	81

Reason for association with Parle Agro

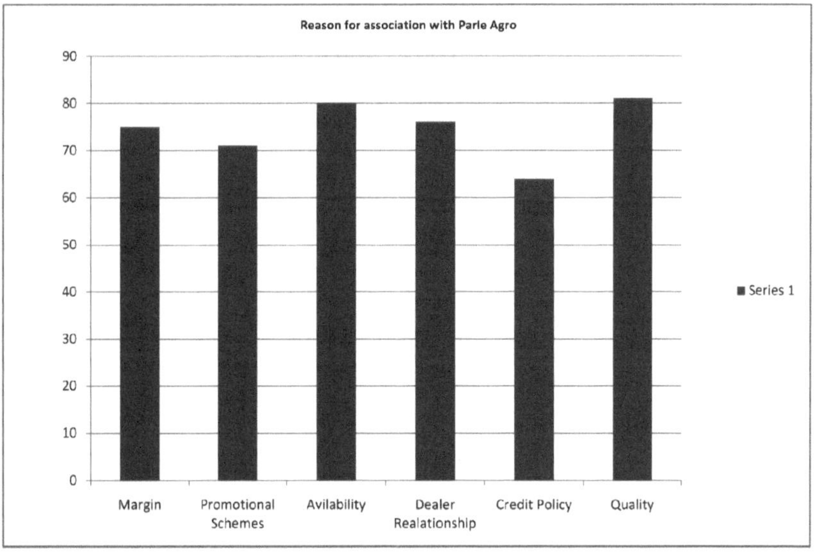

The Table 4 showed the reasons why retailors are doing business with Parle Agro and important factors which have greater influence is been discussed with table. All the parameters scored above 70 index except credit policy. This gives clear idea that retailors are not satisfied with credit policy followed by the company and also the distributor. The main reason with the collaboration of Parle Agro is its quality, which scored the maximum index of 81, followed by availability with 80. Margin, promotional schemes and dealer relationship have scored 75, 71 and 76 respectively.

Timely delivery

It is very essential that customers should get the products when they actually want it. Otherwise they will shift to other product. So there should be timely delivery of the product. Table 5 showed the timely delivery of Parle.

Table 5: Timely delivery of Parle Agro

Sl No	Particulars	Opinion	Respondents	Percentages
1	Time delivery of Parle	Excellent	4	20
		Very Good	9	45
		Good	5	25
		Average	2	10
		Poor	0	0

Regarding the timely delivery of Parle Agro, 45 per cent of the distributor says that they are very good at delivering the product timely. Whereas 25 per cent feels it's good. 20 per cent says they are excellent in the timely delivery of products. But even though there is good impression about the delivery, there should be continuous improvement so as to remain leader in the market.

Time Required for Receiving the Delivery

The Table 6 showed that time it required to receive the order that been placed with the distributor. Because if the delivery has not taken place in time, it will be problem for both distributor and also the retailer, both will be losing their business.

Table 6: Time required for receiving the delivery

Sl No	Particulars	Opinion	Respondents	Percentages
1	Time it required to deliver the order placed by the retailer	Within 1 day	0	0
		1 to 2 day	16	80
		3 to 5 day	4	20
		More than 5 day	0	0

In case of receiving the order, most of the order will be placed in a period of 1 to 2 day (80 per cent). And maximum time taken is 3 to 5 days. Distributors try to deliver the order in time.

Provision for Cold Storage Facility

The Table 7 shows whether Parle Agro provide any provision for cold storage facility. As it is a fruit beverage and most of consumers wish to have it as chilled, so the company should provide cold storage facility.

Table 7: Provision for cold storage facility

Sl. No.	Particular	Opinion	Respondents	Percentages
1.	Provision for cold storage facility	Yes	0	0
		No	20	100

The response of the retailers clearly shows that Parle Agro does not provide any provision for cold storage facility. Something has to be done in this case; otherwise it will affect the sale. The main reason is that the competitors are providing it almost freely.

Uniqueness in the Distribution of Parle Agro

The Table 8 showed the uniqueness found in the distribution of Parle Agro Frooti.

Table 8: Uniqueness in the distribution of Parle Agro

Sl No.	Particulars	Strongly Agree	Agree	No opinion	Disagree	Strongly Disagree	Index
a.	Order taking system	3	14	2	1	0	79
b.	Timely delivery	4	15	0	1	0	82
c.	Frequent visit by the staff	5	11	2	2	0	79
d.	Payment basis	4	13	0	3	0	78
e.	Behaviour of staff	2	15	1	2	0	77

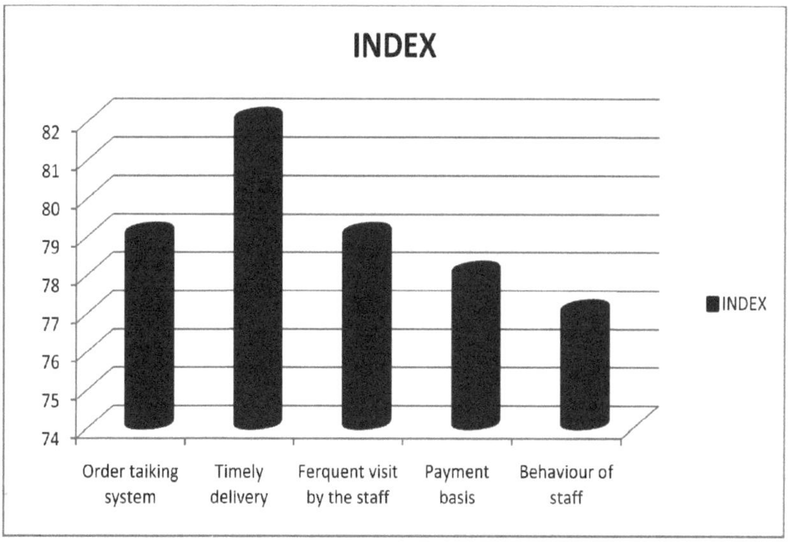

Timely delivery on Customer point of view

It is very essential that customers should get the products when they actually want it. Otherwise they will shift to other product. So there should be timely delivery of the product. Table 9 showed the timely delivery of Parle.

Table 9: Timely delivery of Parle Agro

Sl No	Particulars	Respondents	Percentages
1.	Excellent	9	18
2.	Very Good	14	28
3.	Good	18	36
4.	Average	7	14
5.	Poor	2	4

Among the total respondents 18 per cent feels that delivery of Parle is excellent. But the majority feels its good (36 per cent) only. 28 per cent feels it's very good.

Conclusion

The fruit beverage market is in phase of expansion. Being much conscious about the health and having a clear picture about what ultimately carbonated soft drinks will give, consumers are shifting towards fruit beverages. Even with competing with international giants, Parle Agro has always had their market share in the fruit beverage with their most successful product of all time- Frooti.

The study points the need of a developing meticulous marketing strategy to reach out to more channel participants and to maintain consistent relationship with them. The prior motive must be to strengthen the relationship with both distributors and retailers and develop schemes which will be beneficial to the entire distribution channel. Since the product and price of Parle Agro is comparatively good, improvement in promotion and physical distribution would extend the reach of its products substantively. From this study on effectiveness of distribution channel of Parle Agro reveals the fact that company have a well-structured distribution system. But as mentioned in the beginning, there is an immense potential for the expansion.

References

Philip Kotler (1996). "*Channels of Distribution*", *Principles of Marketing*, seventh edition, p.336.

Neves, Zuurbier, Campomar (2000). "*Chain-Management in Agribusiness and the Food Industry Proceedings*", A model for the Distribution Channels Planning Process, pp. 245-255.

Baijal and Mardsen (2005). "FMCG Distribution in India- Challenges and Opportunities for Manufactures and Retailers", *Journal of Business & Industrial Marketing*, 20(10): 200-213.

Extension Management Strategies for Sustainable Agriculture, pp 223-227

New India Publishing Agency, New Delhi, India

23

Importance of Transfer of Animal Nutrition Knowledge in The Upliftment of End Users

Chellapandian, M., Suresh, C., Arulnathan, N. and Thirumeignanam, D.

Animal husbandry contributes 30 percent of agricultural gross domestic products in India and hence remains an important livelihood activity to the growing population of India. Mixed and diversified farming provides significant profit to the farmers than either only with agricultural or livestock farming. One of the biggest challenges in animal production is to produce animal products for human consumption through sustainable technologies. In the southern districts of Tamilnadu, dairy farming, sheep and goat rearing and back yard poultry form the major livestock livelihood activities. Commercial poultry production, mainly broiler farms is also undertaken under integration with poultry companies, however on a small scale in certain rural and semi urban areas. Animal nutrition is an integral part of livestock and poultry production. Feeding costs of livestock and poultry are about 60-70 percent of the total cost. Nutritional improvements are continuously used for further improvements in production efficiency. Solutions for problems in animal production can be partly achieved by the developments in the discipline of animal nutrition (Reddy, 2011). Hence, animal nutrition is an important thrust area to be concentrated upon and latest technologies should to be educated to farmers by subject matter specialists to enable themselves to obtain maximum profit.

Transfer of Animal Nutrition Knowledge

i) Assessment of needs of the farmers

Tamilnadu Veterinary and Animal Sciences University continuously endeavors to improve the production potential of livestock sector by identifying, refining and developing new or modified technologies, methods and practices useful for livestock and poultry farmers through its research institutions and various departments of constituent colleges. Whatever the technologies developed, these are to be communicated and transferred to the farmers for ultimate adoption at field level. This can be achieved by imparting scientific knowledge to the farmers through extension activity like conducting training programmes on specific topics. But the actual needs or gaps in knowledge have to be identified at the beginning stage itself before venturing into extension activities. Hence the training needs were assessed by conducting surveys in the villages of southern districts on livestock and poultry farming activities by following well designed and structured interview schedules. In one such survey conducted by Senthilkumar and Thanaseelaan (2013), it was revealed that "nutritional aspect" was the most needed area for training and the respondents mainly small ruminant farmers perceived training needs in nutritional aspects like feeding mineral mixture, feeding colostrum to lambs and kids, cultivation and feeding of azolla etc.

ii) Institutional extension of animal nutrition techniques

Department of Animal Nutrition, established since the inception of the new Veterinary College and Research Institute in Tirunelveli in 2012, with the main objective of providing quality education to the undergraduate veterinary students, also has the mandate on transfer of technology through organizing extension programmes on livestock and poultry nutrition for the benefit of farming community. Many programmes for the transfer of knowledge in animal nutrition have been successfully organized/conducted and participated as resource person by the staff of this department in association with extension workers to the farmers of Southern districts of Tamilnadu (Table 1). The extension programmes thus conducted can be grouped as follows.

i. Training programmes - Feeding management of various livestock and poultry, fodder cultivation techniques, fodder seed production, least cost ration formulations etc.

ii. Demonstration camps - Silage making, Urea enrichment of paddy straw, Feed mixing.

iii. Farm Consultancy services – technical advice to the farmers through telephone, email, in person and also on-farm nutrient based farming solutions.

iv. Popularization of proven technology - feeding mineral mixture, feed supplements etc.

v. Technical lectures on animal nutrition in farmers' forums.

vi. Exhibitions - Display of publications, fodder production models, fodder seeds, fodder varieties, feed ingredients etc.

vii. Mass media publications – Popular articles in newspaper/magazines, Radio/ TV talks.

viii. Feed analysis, quality control of livestock/poultry feeds and ration formulation.

ix. Supply of fodder seeds – Fodder Sorghum, Fodder Maize, Hedge lucerne

Table 1: Extension activities of Department of Animal Nutrition, VC&RI, Tirunelveli

S.No. Category	No. of Programmes organized /conducted/ participated as resource person	No. of Beneficiaries
1. Training Programmes		
i) Feeding management of Dairy, Sheep and Goat, Piggery, Poultry – Broilers, Country chicken	23	1166
ii) Fodder Cultivation techniques	15	227
iii) Fodder seed production	03	86
2. Method Demonstrations –Azolla cultivation Silage making, Urea enrichment of paddy straw Ration formulation and feed mixing	25	482
3. Popularisation of Proven Technology camps – TANUVAS Mineral Mixture	05	250
4. Consultancy Services		
i) Technical advice in person, through telephone, email	—	867
ii) On farm nutrition solutions	77	129
5. Guest lectures	32	1280
6. Mass media		
i) Radio/TV talks	12	—
ii) Popular articles in newspaper/ magazines	45	—
iii) Pamphlets	10	5460
7. Exhibitions	22	6221
8. Feed analysis and Ration formulation	53	44
9. Supply of fodder seeds	—	265

iii) Impact analysis

The impact of the training programmes, demonstrations, popularization of technology was assessed by studying the rate of adoption of practices by the beneficiary farmers. The feedback of the trainees obtained by well structured questionnaire also assisted in assessing the impact of the transferred animal nutrition technologies on the production parameters under field conditions. Majority of the farmers have a tendency to adopt a practice if it is found beneficial with a relative advantage over the existing practices (Narmatha *et al.,* 2010). Continuous rise in feed costs leading to increased cost of milk, meat and egg production was perceived as the major constraint for adoption of better feeding practices learnt through training and other extension programmes. Lack of availability of human labour for urea treatment just after the harvest of paddy crop, fungal growth on the top layer and sticky dung in animals fed urea treated paddy straw were cited as the reasons for poor adoption of "urea treatment of paddy straw" technique among farmers (Jaiswal *et al.,* 2011). However, a high rate of adoption was observed in techniques that were demonstrated to farmers viz. azolla cultivation, silage making, feeding mineral mixture, feeding concentrates to lambs, kids. Similar high level of adoption (56.81%) of cattle feed formulation and feed mixing technique was observed among dairy farmers by Rajesh kumar *et al.* (2013) and the cross bred cows also showed a marked increase in milk yield due to the adoption of the new technology. In southern districts, a positive impact on milk yield and milk quality was reported by many farmers on adoption of mineral mixture feeding technique to dairy cows. Similarly more farmers were inspired on fodder cultivation and more land was brought under fodder crops in Tirunelveli and Thoothukudi districts than ever before. Ensuring the availability of a variety of fodder seeds at affordable costs in nearby locality further added a new dimension to the improved fodder cultivation scenario.

Conclusion

The sustained efforts of transfer of technology and knowledge in animal nutrition to the rural farmers of southern districts resulted in better adoption thereby showed a perceptible improvement in fodder production and milk yield. Hence, it could be concluded that extension education of farmers on animal nutrition techniques is highly essential to fulfill the dream of "Lab to Land" in order to achieve maximum production. As demonstration of animal nutrition techniques achieved comparatively better results, method demonstration, frontline demonstration and on farm trials should always be considered as suitable means of transfer of animal nutrition knowledge and technology to the rural masses.

References

Jaiswal, R.S., Chetna Bhatt and Richa Pathak (2011). Techniques for better utilization of drop residues for ruminant diet. In. Animal feeding and production in India. Daya Publishing House, Delhi, 111-116.

Narmatha,N., A.Manivannan, V.Uma and C.Pandiyan (2010). Socio economic and psychological problems associated with poor adoption of livestock and poultry enterprise. *Tamilnadu J. Veterinary & Animal Sciences* 6 (5) : 210 – 214.

Rajesh Kumar,B., D.Baskaran, S. Saraswathi and C.Theophilus Anand Kumar (2013). Impact of training program in adoption of cattle feed computation by farmer interest groups (FIGs) of Tamilnadu. *Tamilnadu J. Veterinary & Animal Sciences,* 9 (4):267- 271.

Reddy, D.V. (2011). Discovery of nutrients briefly. In. Advanced Animal Nutrition. Oxford & IBH Publishing Co. Pvt. Ltd., New Delhi, 3-5.

Senthilkumar, S. and V. Thanaseelaan (2013). Training needs assessment of small ruminant farmers in southern Tamilnadu. *The Indian J. Small Ruminants,* 19(2):232–234.

Extension Management Strategies for Sustainable Agriculture, pp 229-235

New India Publishing Agency, New Delhi, India

24

Perception of Farmers and Extension Personnel Regarding Various Aspects of Participatory Extension Approach in Uttarakhand State

Anbarasan, P.

Agricultural sector worldwide is faced with several serious challenges as follows the spiralling demand of food, declining agricultural productivity due to natural resources degradation and increasing competition in international markets. The new millennium throws up new challenges in the Indian agriculture sector that are different from those in the past. Continuous pressure of producing more food from the same land with shrinking natural resources and marketing the product is a tough task for farmers. Public agricultural extension system with a mandate and mission, beyond transfer of technology, is needed to develop the capacities of the farming community. It is expected to change itself to respond to global changes and act as an integral component of extension-farmer-research-education-market linkage chain. The emphasis on farmer-centered and participatory approach, usage of information technology must overtake the top-down centralized approach. The present extension scenario calls for private-public partnership, diversification, intensification, natural resource management, research on consumer preferences, and continuous capacity building of all the stakeholders in the agricultural development. Despite the acknowledged need for participation of the world's poor in exerting greater influence over decisions that affect their lives, there is still a significant need for mechanisms that ensure genuine participation of the citizenry (Dill, 2009). Supporting empowerment and enhancing the voice of rural population in the development process has increasingly become a central element in poverty reduction strategies (GCARD, 2010). However, our understanding of the impact of participatory extension programs on poor farmer's ability to break with conservative traditions and norms and to engage with many constraints facing them in an innovative manner is limited. Extension will need to involve farmers themselves in the process of research and development in such a way that their participation is highly interactive and empowering. This implies changes in values, attitudes, and behaviour in order to ensure that significant learning takes place among all actors: researchers, extensionists and farmers (Roling & Pretty, 1997).

In Uttarakhand the extension agent to farmers' ratio is very low. Mostly interpersonal communication means are used for information dissemination. This warrants an immediate attention for implementing PEA (Participatory Extension Approach) where individuals will be active in development programmes and processes. Thus, feasibility of using this approach for strengthening agriculture needs to be explored in Uttarakhand.

Research Methodology

Selection of farmers

All the farmers who were availing the extension services rendered by the KVK situated in the *Bhimtal* block of Nainital district constituted as population of the present investigation. Probability Sampling Method had been applied for the selection of sample. A list of 600 farmers from the adopted village of KVK Jeolikote was taken and a sample of ten per cent farmers was selected based on simple random technique. Thus, 60 farmers were selected from the adopted village of KVK Jeolikote for the present study.

Selection of extension personnel

The KVK is a grass root level institution designed and devoted to impart need-based and skill-oriented vocational training to the practicing farmers, in-service extension personnel and to those who wish to go in for self-employment through "learning by doing". These KVKs are established in mind to accelerate the agricultural production and also to improve the socio-economic conditions of the farming community of various districts of India. There are total thirteen KVKs in Uttarakhand state. These KVKs are situated in two different regions of Uttarakhand, i.e. Kumaon region (seven KVK) and Garhwal region (six KVK). The primary responsibility of extension personnel in KVK is Transfer of Technology through both formal and informal means. The opinion of extension personnel regarding the present study had been taken. As each district has KVK, therefore it was difficult for investigator to approach them personally visiting each district. Therefore, investigator selected twenty extension personnel through incidental sampling technique during the *kisanmela* organized by directorate of extension, Gobind Ballabh Pant University of Agriculture and Technology, Pantnagar, which was held at university campus on 9th to 12th March, 2011. In addition to that twenty mailed questionnaires had been sent to extension personnel working in KVKs located in various region of Uttarakhand. Only 15 questionnaires were received by investigator. Five extension personnel from KVK Jeolikote were also selected because it was located in close vicinity of Gobind Ballabh Pant University of Agriculture and Technology, Pantnagar. Thus, 40 extension personnel were selected from KVK located in Kumaon and Garhwal region respectively for the present study.

Results

Opinion of farmers and extension personnel about various characteristics of PEA

Table 1: Distribution of respondents on the basis of their opinion on various characteristics of Participatory extension approach

S. No.	Statement Category	Opinion of respondents			
		Farmers **(N=60)**		Extension personnel **(N=40)**	
		Mean score value	Rank	Mean score value	Rank
1.	Ensuring people participation is an efficient method for participants to reach an higher income.	03.18	III	2.66	VI
2.	Participation of people contributes to reaching the targets set in the programme.	02.82	V	3.25	IV
3.	Assigning responsibility to people make programme more effective.	02.58	VII	2.46	VIII
4.	Most of the time people give irrational ideas which cannot be applied.	01.42	X	2.46	VII
5.	Involving people, sharing responsibilities and people's control over extension make programme effective.	03.32	I	01.42	X
6.	Most of the times people do not take initiative themselves.	02.20	VIII	3.05	V
7.	Only few people in any extension programme/activity take initiative.	02.10	IX	4.20	II
8.	Participation of people has proven to be an efficient tool to increase the problem solving skills of the participants.	03.05	IV	4.32	I
9.	Participation among different stakeholders. helps to reach the target groups.	03.20	II	2.15	IX
10.	Local people's traditional knowledge is recognized or valued.	02.77	VI	3.88	III

The data in Table 1 reveals that Farmers had agreed that participation was a way of involving people, sharing responsibilities and people's control over extension activities made programme effective and subsequently perceived rank I (3.32). Farmers supported that involvement, sharing and control in any activities ensured active participation of the members which would be pre-requisite condition for Participatory Extension Approach. They also perceived that participation among different stakeholders helped to reach the target groups and thus assigned it rank II (3.20). Farmers supported that ensuring people participation was an efficient method for participants to reach a higher income with a rank III (3.18). Participation of people had proven to be an efficient tool to increase the problem solving skills of the participants had fourth ranking with a rank IV (3.05). Reaching the target sets, traditional knowledge and assigning responsibility were also perceived as important aspects of PEA with rank V (2.82), VI (2.77) and VII (2.58) respectively. This gives the sense of equal partnership among all stakeholders. Farmers denied that they did not take initiative in extension activities as told by extension personnel, so they perceived the aspects like not taking initiative, only few people took initiative and irrational ideas given by them with rank VIII (2.20), IX (2.10) and X (01.42) respectively.

Extension personnel admitted that they were unaware about this new approach. So, they could not give appropriate responses for new approach. But with great effort their opinion were recorded to identify aspects of PEA. When extension personnel came to know about the major elements of PEA, they supported its implementation in present situation. But at the same time they did not agree that only people initiative could make PEA feasible, they identified important role of organisation in this regard. Subsequently they assigned rank I (4.32) that problem solving skill of people would become major element in PEA. They perceived rank II (4.20) that only few people took initiative and perceived rank III (3.88) for recognizing their valuable knowledge as they explained that many times farmers innovation and experimentation was fruitful for the society. They perceived reaching target groups as rank IV (3.25) while lack of people participation would be barrier in execution of PEA and perceived it as rank V (3.05). Lowest three perceived ranks were higher income of participant, irrational ideas of people and assigning them responsibility and allotted rank VI (2.66), VII (2.46) and VIII (2.46) respectively. Contradictory opinions were noted down in case of extension personnel against farmers' opinion. They perceived that merely participation of different stakeholders and involving people with responsibility and their control over extension programmes were not important aspect for PEA. They perceived rank IX (2.15) and rank X (1.42) to stakeholders' involvement and involving people, sharing ideas and control respectively. They thought that if involving people, sharing ideas and control became the key characteristics of PEA then we followed such approach since independence of India with a nationwide Community Development project.

Majority of respondents were unaware about Participatory Extension Approach and could not provide concise and particular views regarding Participatory Extension Approach. However The statement "Involving people, sharing responsibilities and people's control over extension make programme effective" was perceived rank I by farmers and the statement that participation could enhance problem solving skill was perceived rank I by extension personnel. It is recommended to create awareness before initiation of this holistic approach so that every stakeholder clearly identifies key characteristics of PEA and prepare community to initiate PEA accordingly.

Opinion of Farmers and Extension Personnel for Initiating Participatory Extension Approach

Influential people are those people who give direction to community towards sustainable development in the village were categorized in the below Table 2.

Table 2: Distribution of responses of respondents towards distinguish qualities of influential people

Quality Category	Responses of respondents			
	Farmers **(N=60)**		Extension personnel **(N=40)**	
	Frequency	Percentage	Frequency	Percentage
Motivating others	26	43.33	18	45.00
Coordination	42	70.00	22	55.00
Innovativeness	20	33.33	32	80.00
Decision making	28	46.67	31	77.50
Leadership	46	76.67	29	72.50

Note: Multiple responses were allowed

The perusal of data in the above table 2reveals the responses of both farmers and extension personnel. It is important to note that when respondents were asked to enlist qualities of influential people necessary to initiate Participatory Extension Approach, majority of farmers (76.67%) affirmed that leadership quality was prerequisite condition while majority of extension personnel (80.00%) supported innovativeness quality would be necessary to initiate PEA. According to farmers, a leader can bring revolution. Thus, a good leader is essential to initiate PEA. But extension personnel said that innovativeness would be essential before start of any holistic approach. Seventy per cent farmers perceived influential person should have co-ordinating abilities where as more than 3/4

(77.50%) of the extension personnel perceived decision making as important quality followed by leadership (72.50%). Farmers also presumed decision making (46.67%), motivating others (43.33%) and innovativeness (33.33%) with substantial percentage. When it came for extension personnel they supported motivation and coordination with 55.00 per cent and 45.00 per cent respectively.

Thus, findings reveals that leadership as well as innovativeness were most preferred distinguished qualities of influential people that could prepare community to initiate PEA in the prevailing social situation.

Methods of Selection of Influential People

Category for methods of selection had been prepared and consequently responses of respondents are presented in the below Table 3.

Table 3: Distribution of responses of respondents towards methods of selecting influential people

Category for methods of selection	Response of respondents			
	Farmers **(N=60)**		Extension personnel **(N=40)**	
	Frequency	Percentage	Frequency	Percentage
Participatory Rural Appraisal (PRA)	36	60.00	36	90.00
Personal contact	14	23.33	28	70.00
Observation	18	30.00	30	75.00
Sharing views of local people	52	86.67	11	27.50

Note: Multiple responses were allowed

It is clear from the table that majority of farmers and extension personnel supported sharing views of local people (86.67%) and PRA (90.00%) would be most appropriate method for selecting influential people. It can be supported with the fact that both these methods involve high level of participation. Thus, they thought that participation of local people must be incorporated during selection of influential people. Extension personnel suggested observation (75.00%) and personnel contact (70.00%) as other preferred methods for selecting influential people. When they were asked to give reason for such option, they replied that KVK works on pre-determined mandate formulated by Indian Council of Agricultural Research (ICAR) and they could not go beyond prescribed mandate. Farmers said that PRA (60.00%) became also a better method applied for selecting influential people. Some of the farmers preferred least the observation (36%) and personal contact (23.33%) as methods for selecting influential people. The reason for such opinions perceived by them was prevalence of biasness in adopting such methods for selecting local leaders.

Contradictory to the response of farmers, extension personnel supported sharing of views with least response (27.50%). It was due to prevalence of social stratification which could create biasness in selection of influential people as perceived by extension personnel.Thus, it can be concluded that preferences in selecting influential people are given to those methods which incorporates elements of participation.

Conclusion

It is clear from the study that different types of organisation have different roles. Perceptions about institutional roles do not differ considerably between extension personnel and farmers. They identified village community as major element that can mobilize community for adopting such approach (PEA) which can lead community towards sustainable development and consequently provide moral and social support in sustaining such activities. Both respondents supported KVKs for providing technical knowledge and identified government agency as key institution which support financial assistance. Use of these complementarities would not only make farmer participatory approach more effective, it would also help in spreading the approach to different areas.

References

Dill, B. (2009). The Paradoxes of Community-based Participation in Dar es Salaam. *Development and Change*.40(4):717-743.

Global Conferences on Agricultural Research for Development (2010). Transforming Agricultural Research for Development : Report for GCARD 2010 from the Global Author Team. FAO : Rome. pp. 156-196.

Roling, N. and Pretty, J.N. (1997). Extension's role in sustainable agriculture development. In: B. E. Swanson, R. P. Bentz, and A. J. Sofranko (eds.), Improving Agricultural Extenison: A Reference Manual. FAO. 18: 181-191.

Extension Management Strategies for Sustainable Agriculture, pp 237-243

New India Publishing Agency, New Delhi, India

25

Impact of Climate Change and Adaptation Strategies to Sustain Livestock Production in Telangana

D.Sreenivas

Climate change is a complex problem involving varied interactions between the environment, natural resources (land, crops, animals and water) and people. These interactions are likely to change the ecological and agricultural landscape, and therefore influence agricultural and animal production. The Intergovernmental Panel on Climate Change (IPCC) has dictated that global warming/climate change could lead to many environmental problems i.e., Drought, floods, crop and animal farming, food production, and health hazards etc. Agriculture is closely dependent on the endowment of natural resources and environmental conditions of soil and climate. India is divided into many agro—climatic zones depending upon rainfall, agriculture and soil fertility. The agro climatic conditions of India and the various other factors that largely determine the cropping patterns vary throughout the country. Warangal district falls into two agro climatic regions namely Southern Telangana Zone and North Telangana Zone.

India is very much vulnerable to climatic change due to its large crops sector, very warm spring (Mendelsohn, 2014) and huge dependency ratio on agriculture it is ranked 18 in Global Climate Rick Index 2014 (Kreft and Eckstein, 2014). Grazing and mixed rain-fed system of livestock production will be the most damaged by climate change (Nardone *et al.*, 2010) which support India's 40% of the human and 60% of the livestock population. The increased heat stress will reduce feed intake (Mader and David, 2004) and high temperatures in the tropics may reduce the dairy milk yield by half to one-third of the potential of modern cow breeds (Parsons *et al.*, 2001). Studies have revealed that milk yield of crossbred cows in India are negatively correlated with temperature-humidity index (Mandal *et al.*, 2002a). The influence of climatic conditions on milk production has been also observed for local cows which are more adapted to the tropical climate of India. The estimated annual loss at present due to heat stress among cattle and buffaloes at the all-India level is 1.8 million tonnes, that is nearly two per cent of the total milk production in the country, amounting to a whopping over Rs 2,661 crore (Upadhaya, 2010). According to Tailor and Nagda (2005) heat stress has detrimental effects on the reproduction of buffaloes, although buffaloes are well adapted morphologically and anatomically to hot and humid climate. Upadhya *et.al.* (2007) stated that thermal stress on Indian livestock particularly cattle and buffaloes has been reported to decrease oestrus expression and conception rate. Maurya (2010) concluded that the length of service period and dry period of all dairy animals was increased from normal during drought. Roy and Prakash (2007) reported that severe heat stress may cause acyclicity/infertility in buffaloes during the summer months due to hyper – prolactinaemia. The conception rats of *Bos tautus* cattle will also decline at high thermal heat index and temperatures above 23.4°C (Amundson *et al.*, 2005).

The outbreak of the disease was observed to be correlated with the mass movement of animals which in turn is dependent on the climatic factors (Sharma *et al.*, 1991). Singh *et al.* (1996) reported that higher incidence of clinical mastitis in dairy animals during hot and humid weather due to increased heat stress and greater fly population associated with hot–humid conditions. In addition, the hot–humid weather conditions were found to aggravate the infestation of cattle ticks (Basu and Bandhyopadhyay, 2004; Kumar *et al.*, 2004). Changes in rainfall and temperature regimes may affect both the distribution and the abundance of disease vectors (Thornton *et al.*, 2009; Dhama *et al.*, 2013). Climate changes especially the increase in global temperature has led to rise and emergence/re-emergence of various infectious disease in human and animal population particularly the vector borne diseases like Blue tongue, Japanese encephalitis, Dengue, Cholera, Chikungunya, Malaria, Filariasis, food poisoning, Trypanosomiasis, Cryptosporidioses etc. Global warming also takes its toll on genetic diversity, the fourth assessment report of Intergovernmental panel on Climatic hage (IPCC) estimates that a 2.5°C increase in global temperature above the preindustrial level may put 20-30% of all plant and animal species assessed so far at high risk of extinction. The cost of biodiversity loss is difficult to measure; the loss in terms of cultural value, ecological services, loss of genes for disease and pest resistance, for environmental adaptation and for other desirable traits in both plants and animals cannot be ignored.

In this background, a study on climate change impacts on livestock as perceived by farmers and adaptation strategies initiated by them to sustain livestock production in Warangal district of Telangana was conducted.

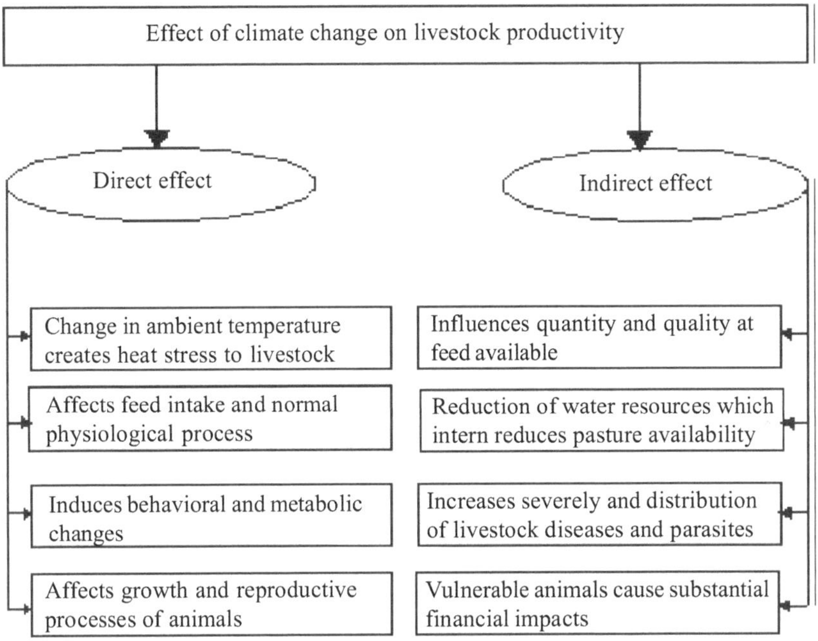

Fig. 1: Effect of climate change on livestock

Materials and Methods

The study was carried out in the operational area of KVK, Mamnoor, Warangal *i.e.*, in two revenue divisions in Warangal district of Telangana, and is located 148km northeast of the state capital of Hyderabad. Warangal district with a total geographical area of 12,846 sq km lies between the latitude 17-19' and 18-36' North and Longitudes of 78-49' and 80-43' East and is above the mid sea level by 870ft, bounded on the North by Karimnagar district, West by Medak district, South by Nalgonda district and by Khammam district on East and South-East. The district is mainly agrarian and agriculture is the main stay of the population.

Stratified randomly sampling technique was used for selection of mandals, villages and respondents. From each revenue division three mandals and six villages were selected. In this way 6 mandals and 12 villages from both the revenue divisions were taken into consideration, from each selected village ten livestock owners having at least two large animals were selected randomly to make 120 total respondents for the investigation. Semi structured interview schedule were used for the purpose of data collection. Besides primary data, some necessary secondary data were also collected from department of Meteorology. Frequency distribution, percentage, mean score and Cattle equivalent score etc. were computed by using Statistical Package for Social Sciences (SPSS) software and Microsoft excel.

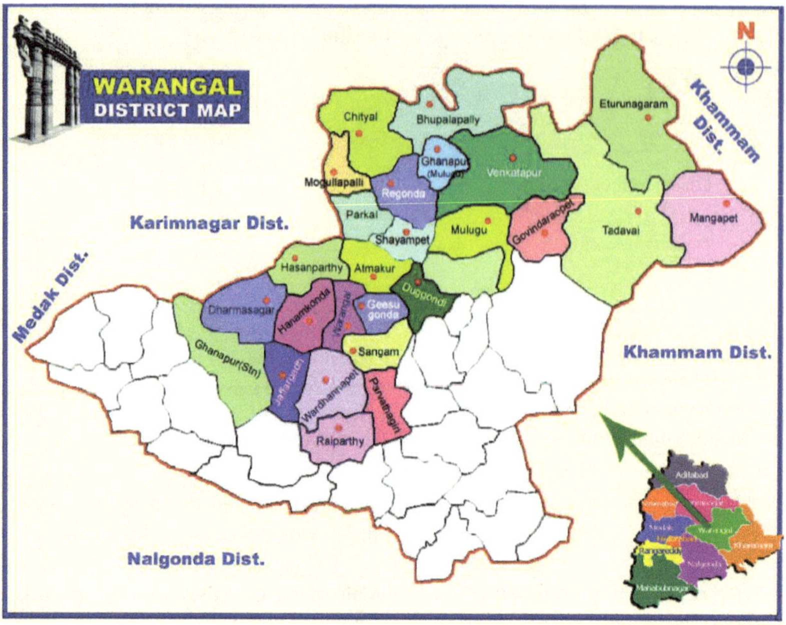

Results and Discussion

Farmers' perception on the impact of climate change Results presented in Table 1 indicated that majority of respondents (82%) perceived that climate change affects agriculture and animal husbandry, 62.5 percent felt negative impact of climate change on productive performance of livestock, 64 percent observed that climate change had negative impact on milk production and lactation length (57%). Majority of respondents (63.1%) perceived negative impacts of climate change on reproductive performance of livestock. Similarly (59.3%) respondents told that there were decreased length and intensity of oestrous period of their animals and (67.8%) respondents observed increased cases of repeat breeding in their animals. Majority of respondents (60%) replied that climate change have an effect on livestock disease incidences. Majority of respondents (76%) told that there were increased incidences of parasitic infestation in livestock. 82.5 percent respondents told that due to climate variability feed and fodder resources are decreasing and also (95%) respondent's perceived shortage of dry fodder. Large number of respondents (95.8%) replied that there was decreased quantity of self grown fodder/grasses and (96%) respondents replied that there were decreased water resources.

Table1: Perceived impact of climate change on livestock sector

S.N.	Perceived impact	Percentage
1.	Climate change affects agriculture and animal husbandry	82
2.	Negative impact on productive performance	62.5
3.	Negative impact on milk production	64
4.	Negative impacts on lactation length	57
5.	Negative impacts on reproductive performance	63.1
6.	Decreased length and intensity of oestrous period	59.3
7.	Increased rate of repeat breeding	67.8
8.	Effect on livestock disease incidences	60
9.	Increased incidences of parasitic infestation	76
10.	Feed and fodder resources are decreasing	82.5
11.	Shortage of dry fodder	95
12.	Decreased water resources	96

Strategies adopted by Farmers under Climatic Change scenario

For many years, livestock producers have traditionally adapted to climatic changes by building on their in-depth knowledge of the environment in which they live. Farmers own perception and local indigenous traditional knowledge help them in evolving measures and techniques to deal with situations arising due to climatic vagaries. A number of questions were asked to assess the strategies adopted by the local communities to cope with climate change. The respondents reported a diversity of coping strategies that included both modern and traditional methods (Table 2). Majority of the farmers (90%) preserve fodder crop in form of hay for adverse climatic condition; They also stored maize/sorghum straw, paddy straw, and crop residues to feed their animals in lean period; Majority of the respondents (85%) changed the planting dates; It was very interesting to note down that majority of farmers (81%) did traditional prayer and look foe government help (69%) to get rid of adverse climatic conditions; Majority of farmers (60%) told that they sown new crop varieties which required less water (draught resistant), less time to mature (early maturing), pest resistant and well adopted for water logging area (flood resistant); Plantation fodder tree lines around animal shed/ house to reduce effects of cold/ heat waves was an important coping strategy adopted by most of the farmers (55%); Migration along with livestock was one of the coping strategies of many of the framers (53%) during adverse climatic conditions; about 48 percent of the respondents kept more livestock and reduced reliance on crops; About 38 percent of the respondents told that they destock their livestock during adverse climatic condition; About 34 percent of the respondents told that they sold their animals to fulfil their daily requirement (food, clothes, school fee *etc.*); Only about 29 percent of the respondents insured their livestock; About 16 percent of the respondent did rain water harvesting.

Table 2: Perceived impact of climate change on livestock sector

Sl.No.	Strategy	Percentage	Rank
1.	Reduction in livestock number	38	IX
2.	Keeping more livestock and reducing reliance on crops	48	VIII
3.	Preservation of fodder	90	I
4.	Look for government help	69	IV
5.	New fodder crop variety/type	60	V
6.	Change planting dates	85	II
7.	Plantation fodder tree lines around animal shed/ house to reduce effects of cold/ heat waves	55	VI
8.	Selling of livestock	34	X
9.	Migration along with livestock during adverse climatic conditions	53	VII
10.	Livestock insurance	29	XI
11.	Rain water harvesting	16	XII
12.	Traditional Prayer	81	III

Other Coping Strategy adopted by Farmers

Providing cold water during hot and humid climate.

Provision of shade to reduce heat stress.

Provide fresh air/ fan/cooler during extreme hot condition.

Kept their animal outside during night during summer.

Loose housing system.

Conclusion

There is a serious threat of climatic changes (in the form of severe droughts, floods) undermining development programmes and millennium development goals aimed at reducing poverty. Climate induced disasters directly affect the livelihood of the farmers. Since livelihood of the farmers is based on agriculture and animal husbandry, all of the respondents said that decrease in the animal-agricultural production weakened the economic condition. As livestock is and will play very important role in rural economy, it is necessary to find suitable solution to reduce the ill effect of climate change on livestock production. Smallholder's intensification and linking them with corporate bodies and modern retail food supply chains needs urgent attention since they hold majority of livestock in the country and can pay a major role in food security and environmental stability. Techniques of remote sensing and Geographical Information System (GIS) need to be fully explored against various unpredicted outcomes because of fluctuating climate conditions. The recent advances in science and novel technologies /concepts need to be fully explored for their optimum potentials like genetic engineering, disease resistant varieties, embryo transfer technology, artificial insemination, superior genetics and breeding practices, cloning, nutrigenomics, immunomodulatory among others. If the crop livestock system has to respond in an ecologically sound and sustainable manner to the future demands, there is a need to tailor the research and extension for climatic resilience, sustainability and food security.

References

Amundson, J.L., Madar, T.L., Rasby, R.J. and Hu, Q.S. (2005). Temperature and temperature-humidity index effects on pregnancy rate in beef cattle. *Biometrology* 5: 9-14.

Basu, A.K., Bandhyopadhyay, P.K. (2004). The effect of season on the incidence of ticks, *Bull Anim Health Prod Afr* 52(1):39–42.

Dhama K., tiwari R., Chakrobarty, S., Kumar, A., Karikalan, M., Singh, R. and Rai, R.B. (2013). Global warming and emerging infectious diseases of animals and humans. *Int. J.Curr. Res.*, 5:1942-1958.

Kreft, S. and Eckstein, D. (2014). Global climate risk index (2014). Who suffers most from extreme weather events? http://germanwatch.org/en/download/8551/pdf.

Kumar, S., Prasad, K.D., Deb, A.R. (2004). Seasonal prevalence of different ectoparasites infecting cattle and buffaloes. *BAU J Res* 16(1):159–163.

Mader, T.L. and Davis, M.S. (2004). Effect of management strategies on reducing heat stress of feedlot cattle. *J.Anim.Sci.* 82:3077-3087.

Mandal, D.K., Rao, A.V.M.S., Singh, K. and Singh, S.P. (2002). Effects of macroclimatic factors on milk production in a Frieswal herd. *Indian J Dairy Sci* 55(3):166–170.

Maurya, R.K. 2010. Alternate dairy management practices in draught prone areas of Bundelkhand region of Uttar Pradesh, M.V.Sc. Thesis, IVRI, Izatnagar, India

Mendelsohn, R. (2014). The impact of climate change on agriculture in Asia. *J.Integrat.Agric.* 13: 660-665.

Nardone, A., Ronchi, B., Lactera, N., Ramieri, M.S. and Bernabucci, U. (2010). Effects of climate changes on animal production and sustainability of livestock systems. *Livestock Sci.* 130: 57-69.

Parsons, D.J., Armstrong, A.C., Turnpenny, J.R., Mathews, A.M., Cooper, K. and Clark, J.A. (2001). Integrated models of livestock systems for climate change studies. *Global Change Bio.* 7: 93-112.

Roy, K.S. and Prakash, B.J. (2007). Seasonal variation and circadian rhythmicity of the prolactin profile during the summer months in repeat breeding Murrah buffalo heifers. *Reprod.Fertil.Dev.* 19: 596-605.

Sharma, S.K., Singh, G.R. and Pathak, R.C. (1991). Seasonal contours of foot-and-mouth disease in India, *Indian J. of Anim. Sci.,* 61(12):1259–1261.

Singh, K.B., Nauriyal, D.C., Oberoi, M.S. and Baxi, K.K. (1996). Studies on occurrence of clinical mastitis in relation to climatic factors, *Indian Journal Dairy Science,* 49(8):534–536.

Tailor, S.P. and Nagda, R.K. (2005). Conception rate in buffaloes maintained under sub humid climate of Rajasthan. *Ind J Dairy Sci* 58(1):69–70.

Thronton, P.K., van de Steg, J., Notenbaert, A. and Herrero, M. (2009). The impacts of climate change on livestock and livestock systems in developing world. *Agric.Syst.* 101: 113-127.

Upadhayay, R.C. (2010). Annual milk production loss due to global warming, Animal Physiology, National Dairy Research Institute (NDRI) Press Trust of India/New Delhi September 26.

Upadhyay, R.C., Singh, S.V., Kumar, A., Gupta, S.K. and Ashutosh, A. (2007). Impact of Climate change on Milk production of Murrah buffaloes. *Italian J. of Anim. Sci.* 6(2s): 1329-1332.

Extension Management Strategies for Sustainable Agriculture, pp 245-248

New India Publishing Agency, New Delhi, India

26

Introspecting the Concept of Agricultural Development: A Multi Stakeholder Analysis in Thrissur District, Kerala State

Jaliya M.K.

India's agrarian culture and economy is undergoing drastic changes. In this progressive path of agricultural resurrection, a comprehensive multidimensional understanding of agricultural development is of great importance. In the present context of Kerala, where the local administrative bodies are vested with more powers to plan and execute need based agricultural development activities, a clear understanding of the concept of agricultural development among the major stakeholders shall serve as a firm base for development.

In this context, a study was undertaken to analyze the existing concept of agricultural development in Kerala as perceived by farmers, elected members of local administrative bodies, agricultural officers and agricultural scientists and map the common understanding as well as the deficiencies. The research was in Thrissur district of Kerala state and more of qualitative data collection procedures were followed for concept mapping, which was the basic tool of the research. Five blocks of Thrissur district, namely, Mullassery, Vadakkanchery, Chavakkad, Chowannur and Ollukkara were selected and 120 respondents were randomly drawn as respondents from the categories of progressive farmers, Panchayat Presidents, Agricultural Officers and Agricultural Scientists.

An open ended question was formulated to explore, explain and clarify the perceptions of these respondent groups on the concept of agricultural development. The focal question structured for this purpose was seeking the definition of agricultural development as perceived by each respondent based on his/her knowledge and experience.

The perception differences among the respondent groups were visually depicted by using the tool of concept mapping. It is a multi-step hybrid method that uses original respondent statements as units of analysis, aggregate qualitatively across individual conceptual schemes and enables the data to emerge through the use of multidimensional scaling and cluster analysis of the aggregated individual coding data. Concept mapping involves the following six steps:

- Preparation (including selection of participants and development of focus for the conceptualization).
- Generation of statements.
- Structuring of statements.
- Representation of Statements in the form of a concept map (using multidimensional scaling and cluster analysis).
- Interpretation of maps.
- Utilization of maps.

Ranking schedule with four sub-sections namely, physical dimension, economic dimension, ecological dimension and socio-psychological dimension was prepared after discussing with agricultural experts and incorporating productive suggestions from the different respondent groups.

From the study, comparison of percentage mean scores between various respondent groups conducted after the analysis of the responses of open ended questions revealed that Agricultural Scientists scored the most (75 %) and farmers scored the least (36.3). The scores of Agricultural Officers and Panchayat Presidents were respectively 50 % and 41 %. The extent of perception differences among the four respondent groups regarding the concept of agricultural development was significant when it was compared with a standard definition chosen for the purpose of the present study. The concept map of progressive farmers displayed that major emphasis was on production related activities of agricultural development covering cluster area of 45 % on the map. Map area covered by 'marketing and finance assistance' was 25 % in the farmers concept map and 20 % under eco-concern and 10 % under socio-concern.

The extent of emphasis on production related activities of agricultural development was less in the concept map of Panchayat Presidents (40 %). It was revealed that 30 % of cluster area was covered by socio-concern, 23 % under marketing and finance assistance and the remaining 7 % under eco-concern. From the concept map of agricultural officers, it was understood that major emphasis was laid on socioconcern (40 %), followed by production related conceptual focal area (25 %).

The concept map of agricultural officers was hinged on participatory approach and committed to social well-being of the society. About 40 % of total cluster area occupied fell under socio-concern. Area under production related conceptual focal area was 25 % of total area. It was followed by eco-concern (20 %) and marketing and finance assistance (15 %). The map of agricultural scientists stood out distinctly from the rest of the concept maps. The major emphasis was on marketing and finance assistance covering an area of 35 %. The second most concern was on the conceptual focal area of eco-concern (30 %) followed by production related activities (22.5 %). Progressive farmers and Panchayat Presidents concentrated more on production related activities on agricultural development. The divergence in the pattern emerged due to the increased concern of Panchayat Presidents on socioconcern when compared with that of progressive farmers.

The salient feature that differentiated the concept of agricultural officers from Panchayat Presidents was the increased emphasis of socio-concern by agricultural officers. It was examined to be 10 % greater than that of Panchayat Presidents. Panchayat Presidents found it more important to have production related activities under the concept of agricultural development. The concept map of Agricultural Officers was focused mostly on marketing and financial assistance (35 %) and ecoconcern (30 %).

Among the activity statements enlisted under physical dimension of agricultural development, conservation of top soil was ranked first by progressive farmers and Panchayat Presidents. It was ranked as second by agricultural officers and as fifth in the list by scientists. Agricultural officers ranked the activity statements, self sufficiency in agricultural inputs like quality seeds, planting materials, fertilizers at local level, as first under physical dimension. Agricultural scientists rated construction of scientific and well planned markets as the most preferred one under this dimension.

Formulation of integrated network of markets at district- state levels ensuring reasonable prices for agricultural products was ranked first by three groups of respondent groups, namely, Panchayat Presidents, Agricultural Officers and Agricultural Scientists. This activity statement was ranked second by progressive farmers. Progressive farmers considered insurance for agricultural crops on simpler terms as the most preferred activity statement under economic dimension.

Both scientists and agricultural officers had prioritized awareness against excessive use of agro-chemicals in farming as the most important one under ecological dimension whereas farmers ranked integrated use of organic methods in farming as the most preferred activity statement. Panchayat President respondent group ranked yet another activity statement, practice of sustainable and eco-friendly farming practices, as the most important one.

Under socio-psychological dimension, farmers, Presidents and Agricultural Officers have ranked the same activity statement as the most preferred one. They emphasized on the need to encourage group farming. Agricultural scientists ranked this activity statement as third important activity in the list. Agricultural scientists preferred the activity statement 'to assure deserving social status and respect to farmers' as the most important one under this dimension. Based on the concept maps developed by different actors, a Venn Diagram was developed and presented as follows:

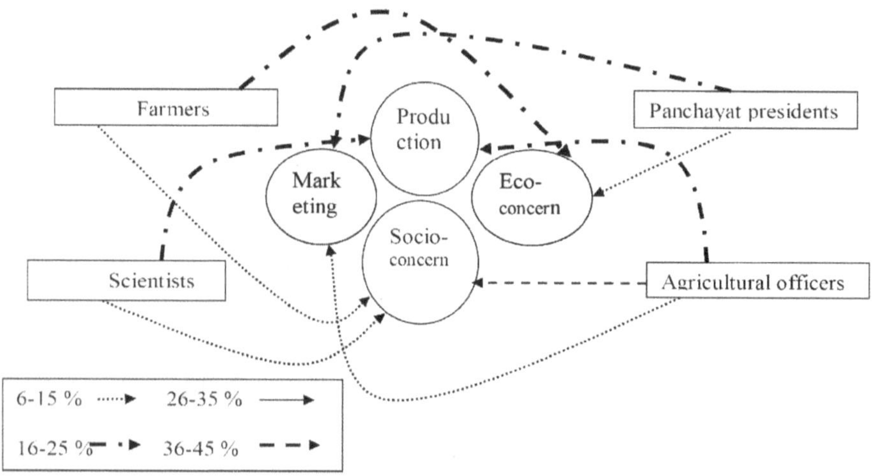

The study implies that any development planning in agricultural sector needs a multi stakeholder and participatory scrutiny of objectives, methodology and expected outputs to make the intervention realistic and meaningful.

Extension Management Strategies for Sustainable Agriculture, pp 249-253

New India Publishing Agency, New Delhi, India

27

Factors Influencing the Willingness to Pay for the Delivery of Agricultural Information and Communication Technologies in Rwanda

Nathan Kanuma Taremwa and Thimmaiyah Ranganathan

The digital revolution is radically shifting how people create, manage, share and publish information, as well as how they relate, collaborate, communicate and share resources. Information and Communication Technologies (ICTs), Social Media, Web 2.0 and mobile applications have also significantly changed the livelihoods of the local farmers especially in developing countries.

In Rwanda, Information and Communication Technology (ICT) is a central engine to driving economic transformation to a knowledge based economy, a fact Rwanda has acknowledged by allocating a budget to ICT as a percentage of its GDP that is at par with OECD countries. The Government of Rwanda (GoR) through its Vision 2020 is committed to the establishment of a globally competitive, diversified and balance economy that is driven by information, knowledge and skills an economy with an ICT-intensive modern industrial sector; a modern, efficient and competitive agricultural sector; and a vibrant ICT-driven, valued added service sector capable of serving as the engine for accelerated economic growth and development.

However, major challenges still exist and greatly influence the adoption and dissemination of agricultural technologies. In Rwanda, the major ICT constraint has been affordable, high-speed internet connectivity. Up until now, Rwanda has had to rely on satellite based data services, which are both slower and much more expensive than fiber based services.

Research Methodology

This study was carried out among the Coffee Growers' Cooperatives (referred to as ABAHUZAMUGAMBI) in Maraba Sector, Huye District in the Southern Province of Rwanda.

The objectives of the study are to (i). Establish the factors that influence the farmers' willingness to payfor the delivery of agricultural information and communication technologies and how these services if well adopted, would enhance the farmers' livelihoods (ii). To establish the technologies currently applied in Rwandan agricultural sector. Slovin's formula [($n = N / (1 + Ne^2)$ where n = Number of samples, N = Total population and e = Error tolerance], was applied to determine the sample size and the selection of respondents was done by judgmental and opportunity/convenience sampling techniques. The survey was divided into two main parts. The first part sought information on basic characteristics of households (age, education, dependents, occupation, expenditures, and membership in community organizations). The second part consisted of abiding game for alternative information delivery technologies.

Results and Discussion

Results indicate that most of the local farmers in Rwanda are willing to pay for the delivery of agricultural information and communication technologies. According to the education levels, 87.1% of the respondents from the coffee cooperative (ABAHUZAMUGAMBI) only attained primary level.

Table 1: Population by education level

Education level	Frequency	Percent
None	8	8.6
Primary	81	87.1
Secondary	4	4.3
Total	**93**	**100**

From table one above, it is concluded that the knowledge factor is very crucial to the farmers. Most of the farmers are not familiar with ICT developments especially those in rural areas and this greatly influences their willingness and ability for the delivery of agricultural and communication technologies.

Table 2: Willingness to pay versus sex of the respondent

		Willingness to pay		
		Yes	No	Total
Sex	Male	41	10	51
	Female	34	8	42
Total		**75**	**18**	**93**

Table 2 indicates that, 41 males are willing to pay against 34 females. Male farmers' willingness to pay is significantly higher compared to the females. An individual gender determines how he or she is willing to pay for agricultural information delivery technologies. This could be attributed to the fact that most decisions are taken by males about agricultural activities. In total, 75 respondents are those who are willing to pay, against 18 respondents who are not willing to pay.

Table 3: Methods of disseminating agricultural information

	Frequency	Percent
Extension agent	83	89.2
Radio	1	1.1
Television	1	1.1
Newspapers	2	2.2
Telephones	6	6.5
Total	**93**	**100**

Table 3 shows that farmers receive agricultural information disseminated through different methods with majority of the farmers obtaining information through extension agents (89.2%). Local farmers largely depend on extension agents to obtain the necessary agricultural information and these agents thus provide a pivotal role in enlightening them, raising their level of knowledge and eventually help the farmers to make appropriate decisions regarding farming activities. Extension workers are doing this noble work more effectively than the use of mobile phones, television, radio, and newspapers.

Table 4: Constraints in finding agricultural information

	Frequency	Percent
Information is costly	5	5.4
Information is often not updated	76	81.7
None	12	12.9
Total	**93**	**100**

Table 4 above highlights the constraints/challenge with obtaining agricultural information in Rwanda. Although a large percentage of farmers do obtain agricultural information (87.1%), the information obtained is often costly and somewhat none update. It is essential therefore that farmers be facilitated to get updated information regarding such services regarding products, and markets and especially on how the farmers can obtain the markets for their agricultural commodities at fair prices.

Table 5: Amount of money farmers are willing to pay for the required technologies

		Level at which they are willing to pay		Total
		1-70%	71-100%	
Kind of methods	Extension agent	6	77	83
	Radio	0	1	1
	Television	0	1	1
	Newspapers	0	2	2
	Telephones	1	5	6
	Total	**7**	**86**	**93**

Table 5 indicate that farmers are willing to pay more money on extension services than other technologies (77%) and more than 70 percent of the farmers are willing to pay through various communication technologies. Most farmers would be willing to pay for the services offered by extension agents than other forms.

Conclusion and Recommendations

Most Rwandan farmers are willing to pay for the delivery of agricultural information and communication technologies. The farmers perceive these services as very essential to enhance their levels in agricultural productivity and to transform the nature of agriculture from subsistence level to market oriented agriculture. Linking the farmers to the markets has also been identified as a potential factor that would not only improve the livelihoods of local farmers especially through income improvement, food security and poverty reduction among others. However, the cost of obtaining agricultural information is still very high and this greatly impedes the farmers from obtaining information from other sources besides extension agents. Major factors influencing the willingness to pay for agricultural information and communication technologies include: transaction costs, price for the delivery of ICT services, knowledge of the farmers and the associated risks and uncertainties.

References

Andrea, B. (2011). A comprehensive Assessment and Recommendations on strengthening the Rwandan Agricultural Extension System. *Journal on MEAS Assessment of Extension in Rwanda*, 4(4): 212-44.

Galin K. and Mohamed, K. (2010). Rural and Agricultural Development Communication Network (RADCON). *Journal of the Application of Information and Communication Technologies in Agricultural and Rural Development in Egypt.* Food and Agriculture Organization of the United Nations, ROME, 4(3): 13-28.

Ministry of Agriculture and Animal Resources (2009). National Agricultural Extension Strategy, Kigali-Rwanda, 54P.

Ministry of Agriculture and Animal Resources (2010). The Technical Centre for Agricultural and Rural Cooperative (CTA). ICT4Ag,Kigali-Rwanda.

Nicholas, O., Chris, J., Michael, C. (2013). Farmers' willingness to pay for Agricultural Extension Service, Lagos- Nigeria, *Journal of International Development,* 25(3):382-392.

Rivera, W. (2000). The changing nature of agricultural information and the conflictive global developments shaping extension. *Journal of Agricultural Education Extension* 7(1):31-41.

The Republic of Rwanda (2013). Economic Development and Poverty Reduction Strategy 2013- 2018, Kigali, Rwanda, 141P.

Extension Management Strategies for Sustainable Agriculture, pp 255-281

New India Publishing Agency, New Delhi, India

28

Deterioration of Tank Irrigation System in Tamil Nadu, India: Socio Economic Factors and Necessary Remedies

Muniandi Jegadeesan and Hidetoshi Miyazaki

Water is one of the most important production assets and securing access to and control and management of water is key to enhance livelihood base of any country like India where most people depends on agriculture. India is endowed with rich and vast diversity of natural resources, water being one among them. But water resources in India are unevenly distributed over the country. The more than 50 per cent of the total resources distributed with five states namely Uttar Pradesh, Odisha, Andhra Pradesh, Karnataka and West Bengal (MWR, 2008) and left other states much prone to water scarcity. Tamil Nadu is one such state in India has 13 Mha geographical area with 4 per cent of cultivable land and 3 per cent of water resources to cater the need of 8 per cent of its population. The average annual rainfall for the state is only 943 mm compare to all India average of 1170 mm. It varies from 1200 mm in the coastal areas to 550 mm in inland and per capita availability of the water in the state is only 900 cubic meters compared with 2200 cubic meter for all India. As more than half of the population (56 per cent) of this state is still depended on agriculture for their livelihood, the rain fall occur around October and November every year through Northeast Monsoon will decide their fate. This is crucial because Southwest monsoon which will give enormous rain fall in other states is blocked by Western Ghats. Thus, the tank irrigation system has been developed since ancient time to utilize scarce water for agricultural production and for all related water need for the villagers.

Mosse (2003:vi) described that the tank irrigation system is one of the vast network of thousands of water bodies that constitute a distinctive landscape which is medieval in origin but still the basis of livelihood in the dry southern plains. It is true that tank irrigation significantly contributes to agricultural production in India in general and in particular Andhra Pradesh, Tamil Nadu, Karnataka, and parts of West Bengal, Rajasthan and Uttar Pradesh as these states account for 62 per cent of the total area under tank irrigation (Minor Irrigation Census, 2001). The small earthen bunds are formed across the small streams and rivulets wherever feasible to form tanks in which water is collected and stored during the monsoon period and let down for irrigating small command areas through sluices and distributive channels. Over the years more modern form of irrigation such as canal irrigation building big dams across the rivers and energized well irrigation have pushed back the tank from their place of prominence. Even the sixties and early seventies tank irrigation constituted the largest single mode of irrigation in the states (Table 1) and even now after deliberate disregard shown to tanks by the administrators and people, the deteriorated tank system still provide lifeline for state's 20 percent net irrigated area. Growing population urged to increase food production but food grain production is almost stabilized, the Government now deciding to bring additional area under irrigation to improve its food production. But the scope for bringing additional area under irrigation has become not only limited as ,in Tamil Nadu already 92 per cent of surface water and 70 percent of ground water has been already tapped[1] (Sakthivadivel *et al.*, 1982) but also prohibitive in terms of cost.

Hence, the available accepted alternative lies in tank sources, it could be possible to bring additional area under irrigation or turning the fallow command area again into the farming by proper rehabilitating of existing tank irrigation system and economizing of the water usage by better water management techniques. It is estimated that using 39,202 tanks in Tamil Nadu, the potential area to be irrigated will be around 1,100,000 ha, (Palanisami, 2000) estimated loss due to defunct tanks will be around 50,000 ha. Hence, area that could be irrigated is around 1,050,000 ha but actual area irrigated is around 531,000 ha (PWD, 2008). Not even single year has shown tank irrigated area over 900,000 ha since 1997. Therefore the estimated area lost due to the mismanagement will be about 519,000 ha. This drop occurred in spite of the addition of hundreds of new tanks during the same period, which suggest that there has been wide spread mismanagement of existing tanks. Hence, by improving the performance of tanks in Tamil Nadu seems to be viable and best available option. The tanks have wider geographical distribution and hence any investment to improve and restore them would provide way for farmers to regain their lost livelihood. Moreover tank irrigation is the most desirable system of water harvesting from the ecological and sustainable point of view (Palanisami *et al.*, 2008). In present situation, the tank irrigation system is at a critical juncture and still undergoing profound changes. These changes have a potential to create unprecedented eco systemic and socio economic imbalances among the rural societies in Tamil Nadu. The Table 1 clearly indicates that deterioration of tank irrigation system over the period of time. Both all India and state level area irrigated by the tanks drastically decline. These statistics illustrate the enormity of the problems facing the state with respect to its water resources and startling disparities that exist in its utilization. Almost more than 50 per cent of the area under tank system becomes fallow. It is worth to notice that the fallow lands have increased from 1.76 Mha in the 1950s to 2.44 Mha in the 2001-02 and feared that, it is hovering around 2.5 Mha in the recent years, and estimated loss would be around Rs 500 million (State Planning Commission Annual Report, 2008: 25). It is needless to say among these; more than 50 percent of the area would come under tank command. The average net irrigated area by the tank has decreased from 19.2 ha from 1981-82 to 15.1ha in 1999-2000. Tamil Nadu human development report 2003 also cautioned that the shortage of water result on an average about 12 to 16 per cent of gross cropped area remain fallow every year (p:6).

Table 1: Trends in Irrigation in India and Tamil Nadu (area in Per cent)

	Source	1960-61	1970-71	1980-81	1990-91	2006-07
India	Canals	42.05	41.08	39.40	35.63	35.40
	Tanks	18.50	13.22	8.24	6.84	5.75
	Wells	29.56	38.22	45.70	51.04	55.10
	Others	9.89	7.28	6.66	6.49	6.21
	Total	100	100	100	100	100
Tamil Nadu	Canals	35.80	33.90	32.70	32.40	27.08
	Tanks	38.00	34.50	32.10	22.38	18.39
	Wells	24.20	29.80	33.80	44.61	54.22
	Others	2.00	1.80	1.40	0.61	0.31
	Total	100	100	100	100	100

Source: Season and Crop report of Tamil Nadu various issues, PWD (WRO) Citizen charter, 2008

The reason for this deterioration was studied by number of scholars in the different period of the time and they opined that the possible causes are numerous and often complicated and classified them as physical, socio economic and policy factors. The physical factors such as silt accumulation, encroachment and innovation of wells in tank command. The socio economic factors like disintegration of cohesiveness of village community, termination of common financial resources and deterioration of traditional irrigation institutes and policy factors including policy of government to support major and medium irrigation and policy to provide loan and subsidized electricity for ground water development as possible reasons. (Abel and Easter, 1971, Wade and Chambers, 1980, Easter and Welch, 1983, Janagarajan, 1993, Sakthivadivel *et al.*, 2001 and Palanisami *et al.*, 2008).Government of Tamil Nadu accepted the fact that these time honored assets had become dysfunctional. It was sorely disappointed about rapidly deteriorating status of tank irrigation and its shabby performance and planned and implemented various programmes to rehabilitate it in a phased manner. The attempts to modernize tank irrigation system are not new. Since after the independence 1947, the slew of efforts has been underway to develop and utilize its potential fully. The following Table 2 illustrates the important programmes implemented by state government with assistance of various international funding agencies and national level organizations in order to reinstate and reinvigorate the tank irrigation systems.

Table 2: Programme implemented by state government

S.No	Programmes	Year
1.	Special Minor Irrigation Programme (SMIP)	1951
2.	Desilting and Reclamation Programme (DRCP)	1961-62
3.	Drought prone Area Programme (DPAP)	1970
4.	Accelerated Minor Irrigation Programme (AMIP)	1974-75
5.	Tank Restoration Schemes (TRS)	1961
6.	European Economic Committee (EEC) Tank Modernization Programme	1984-1999
7.	State Tank Irrigation Project	1995-96
8.	Irrigated Agriculture Modernization and Water Bodies Restoration and Management (IAMWARM)	2007-2013

Source: PWD (WRO) Policy Note various years

All these programme intended primarily (i) to provide major repair and improve tank performance, (ii) to increase irrigation potential by constructing new tanks and modernizing existing tanks (iii) to improve existing tank structures (iv) to improve irrigation efficiency by reducing conveyance loss through promoting farmers participated water resources management strategies. In spite of all these attempts, there is big gap between potential existed and potential under use as indicated earlier. Over all progress of this entire programmes on tank water management has been neither sufficient nor comprehensive enough to reduce general trends of increasing water shortages and deteriorating tank eco system. Irrigation water security is vital for both for livelihood and food security. While the state has made large investment in the augmentation of water supply, there have been no commensurate efforts in the management of demand and in the promotion of efficient and economic water use. The money poured into improve minor irrigation especially tank irrigation in the state is enormous but outcome is not very impressive. The following Table 3 showed that the money spent on various tank programme between 1980s to 2005.

Table 3: Money Spent on restoring tanks in the state from 1980 to 2005

Funding Agency	No. of tanks covered	Funds (Rs in Million)
European Economic Community (EEC)	649	12.9
World Bank	620	10.44
NABARD	572	14.55
State Govt.	290	6.25
HUDCO	42	2.14
Union Government	46	1.59
World Bank	On going	157
Total	**2136**	**204.32**

Source: PWD policy notes and citizen charter various years.

The researchers are critic about fund utilized for the tank rehabilitation. Shanmugam (2006) concluded that programme implementing agencies only concentrating on repairing the bunds, sluices and surplus weirs as it is easy to make money and main problems silting and choking of feeder channels remain unaddressed. Again all these attempts made in smaller number of tanks and repeatedly the same tanks and widely ignored other tanks which for many years starving for government attention. This is especially true in the case of panchayat union tanks (52 per cent of tanks in the state comes under union management regime). Hence, no big attempt has been made until European Economic Community (EEC) came forward to modernize 649 tanks in Tamil Nadu in three phases during 1984-1999. It is believed that EEC tank modernization programme with huge investment has improved water use efficiency in the tank irrigation system. But no study has been conducted to assess the impact of tank modernization on tank performance. Hence it is important to study the impact of EEC tank modernization project on tank performance which will give good insight to policy makers, researchers and field workers. This study gain importance as several proposals from state government and other institution have been sent to various funding agencies indicating importance of tank modernization in revitalizing the tank irrigation including ongoing World Bank assisted Irrigated Agriculture Modernization and Water Bodies Restoration and Management (IAMWARM) at a total outlay of Rs 157 million within the period of six years from 2007 – 2013. While the above argument valid for the all tank irrigated states, in this paper we tested following hypothesis with help of date collected from 50 tanks and 500 tank farmers

1. EEC tank Project was effective for the improvement of tank performance.
2. Tank performance is better where participatory approach in tank management was adopted.

3. Water users association is active where socio-economic conditions are favorable.
4. Stable water supply is a detrimental factor for better tank performance.

With this the main focus of this paper is

1. To study the impact of modernized tanks with relation to socio-economic condition of the farmers.
2. To compare the effectiveness of modernization tank with non-modernized tank.
3. To compare production of paddy yield between modernized and non-modernized tanks.

Methodology

EEC has modernized 649 tanks spread over 17 districts of Tamil Nadu. It is decided to conduct the study in tank intensive districts. In Tamil Nadu two regions are considered as tank intensive viz., North region, comprising of Vellore, Tiruvannamalai, Kanjeepuram and Villupuram districts and Southern region, comprising of Madurai, Virdhunagar, Sivagangai, Ramanathapuram and Pudukottai districts. Southern districts had 40% share in total modernized tanks (267 out of 649 tanks). Hence, it is decided to study southern district of Tamil Nadu *viz.*, Madurai, Ramanathapuram, Virudhunagar and Sivagangai. Commonly, the tanks were constructed in cascade, so that surplus water from one tank will go to the next tank in the cascade or chain. Hence, the tanks located in the head of the chain has more favorable water supply than other tanks in the chain. So it is decided to identify all the modernized tanks with its chain and its location in the chain in selected districts. Materials and maps were collected from PWD records and block-level administrative offices. Personal visit was also made to have discussions with responsible PWD engineers. Using this material; we plotted all the modernized tanks in chain, in order to know the exact location of modernized tanks in the chain. Then, we selected tanks which were located in the head of the chain in order to eliminate factor of water supply, the crucial factor to determine tank performance.

Considering short period of time for the study, six tanks from Madurai, Virudhunagar and Ramanathapuram district were selected, and in Sivagangai alone 7 tanks as it has more number of tanks under modernization than other districts. Totally 25 EEC modernized tanks were selected for the study. In order to compare EEC tank performance, with the non modernized tanks, tanks located close to the selected EEC tanks with similar characteristics had been selected (6 each from 3 districts and 7 from 1 district). Thus, the total sample size for the study was 50 tanks (25 EEC and 25 Non-EEC tanks).This study was conducted during 2006 – 2007. Both primary and secondary data were collected for this study. Secondary data like total ayacut area and other hydrological details were collected from PWD official memoirs. Revenue related details were also collected from village administrative officer. Primary data was collected through pre-tested interview schedule, paying personal visits to the tank villages. All the collected data were entered in excel work book. Percentage analysis, average, correlation, t test, regression and equity analysis were used to analyze the collected data. Analysed data are discussed in the following sections.

Characteristics of EEC Modernized tanks

Tanks have been influenced by three important factors *viz*., hydrological factor, socio-economic behavior of farmers and administrative factors. Thus, data related to these factors collected and presented in Table 4. The table 4 shows that average *ayacut* area of these tanks was 120 ha and it has been distributed among 357 farmers. Thus, one farmer would have merely 0.34 ha of land to cultivate and earn for his family. Average family size of sample village was 5.2. It is conceivable that feeding 5 or more person from the meager of less than 1 ha of land with all its uncertainty is difficult task. These tanks had only one filling per year, but they had 1.36 fillings10 year before. Farmers of these tanks felt that 1.5 filling is needed to have successful crop harvest. One filling could supply water for 55 days. If they have 1.5 filling per season, available water supply will goes to 75 to 85 days, which will manage to get some good harvest. Here, again 36.2% of tank storage capacity has been reduced due to siltation. So, total dependence of paddy cultivation on tank water currently leads to miserable condition. The tanks get less than two month storage in 5 out of 10 years, commonly causing crop failure and yield reduction. In order to manage or escape from this strange situation, farmers need to provide supplemental irrigation using well water. Density of the well was also very less viz., one well will cater to the needs of 33 farmers or one well needs to give supplemental irrigation for 11 ha which is very difficult under present situations. Since, wells depends on tank storage for recharge, poor tank supply also affects the well recharge. Normally, to irrigate one hectare, about 20 hours of pumping is required and wells will not have sufficient water to pump continuously. The well water could not completely supplement for tankwater because the availability of electricity is only around 6 to 8 hours/day and also farmers complained about fluctuation of voltage may damage pump frequently. Even if all these factors eliminated Rs. 700-1200 is needed to pay for the water to irrigate one ha of land. It was estimated that about 38 percent of crop income of the non well owners is paid as water charge to the well owners in Tamil Nadu (Palanisami and Easter,2000:15). As most of the farmers are belonging to small and marginal category, they could not afford this cost even if water is available. The tanks usually can receive water from three sources like its free catchment, intercepted catchment and through supply channels from streams. It is assumed that only 20 percent of runoff from the intercepted catchment flows down to lower tank if the tank located in middle or tail end of the chain. So, it is imperative that runoff from the free catchment and through supply channels are important sources to augment required water. The runoff from catchment is again around 10-15 percent depending upon the soil type and the runoff occurs when the rain fall exceed 17mm/day in (AED, 1995). The siltation in the supply channel and wide spread encroachment in the catchment and tank foreshore will sizably reduces the water flow to the tanks.

Table 4: Characteristics of EEC modernized tank villages **(n=25)**

Villages	Tank performance (%)	No. of Households in Ayacut	Filling pattern (No)		Change in Water availability (days)		Siltation (%)		No of wells		
			10 yr before	Present	10 yr before	Present	10 yr before	Present	70's	80's	90's
Sowdarpatti	80.23	800	1.75	1	120	90	20	40	20	20	25
Saptur	82.12	700	2	1	90	50	10	40	20	30	30
Kudisery	79.45	1500	1.75	1	120	90	20	30	10	15	25
Thambipatti	84.50	150	1.50		60	40	10	25	15	25	25
Kottaiyur	82.21	400	2	1	65	40	10	40	25	25	25
Sundrarajapuram	83.12	300	2.5	1	60	40	10	25	25	25	25
Srivilliputtur	90.23	800	2.5	1.5	65	40	15	25	30	50	50
Vellur	86.15	300	2	1.5	50	30	15	40	0	0	5
Kunnathur	82.90	250	1	1	90	70	25	40	7	7	11
Sambakulam	92.15	200	2	1	150	35	20	50	13	15	15
Adanoor	81.52	100	1	0.75	50	30	30	60	5	5	5
Chitrakudi	81.30	250	2	1	190	70	20	30	4	4	4
Arasani	84.25	80	2	1	120	70	20	30	0	0	0
Theriruveli	80.50	650	1.5	1	40	60	35	25	0	3	4
Oriyur	72.25	500	1	1	100	75	35	30	5	5	4
Nallakulam	79.74	70	1	1	90	70	25	40	4	4	4
Aralikottai	80.13	150	2.5	1	60	40	10	25	5	5	5
Taraikudi	83.58	260	2	1.5	80	50	30	50	2	2	2
Mosukudi	84.10	150	1	1	30	40	40	30	0	0	0
Velankudi	85.25	220	2	1	90	60	25	45	3	3	3
Melrangiyam	84.23	40	1	1	30	50	50	60	0	0	0
Algan sirukudi	88.50	350	1	1	90	70	30	50	0	0	0
Nerkuppai	77.75	350	1	1	90	70	20	40	0	0	0
Padamathur	72.00	360	1	1	120	90	25	35	0	0	0
Sathamangalm	70.28	300	1	1	100	80	24	40	0	0	0

Mosse (1997) has noticed that due to the possibility of rain fed farming the encroachment in between the tanks is comparatively high in the black soil areas. Unfortunately the EEC modernization program ignored to address this problem. In some of the villages studied the farmers are opined that what is the point of spending so much on tank structure in which water has not been collected past 10 years. The main problem is that catchment area has been encroached upon and nothing has been done about it. Most of the farmers are looking for alternatives to alienate their deprived conditions. Livestock rearing was a viable option as most of the farmers have started practicing it. If the farmers could adopt some of the water management techniques like alternative wetting and drying, avoidance of field to field irrigation and cultivation of drought tolerant short duration varieties it is possible to get successful crop with available tank and well water. Sakthivadivel *et al.* (1982) observed that water use efficiency in the south Indian tanks declined to as low as 25 to 35 percent in many areas. The reason for the low water use efficiency were inadequate maintenance, lack of control over water release and excessive use of water at the farm level. In most of the tank where water user association is not in force water is generally allowed to flow continuously in the main canal from all the sluices until the tank water supply exhausted. The research showed that simple procedure of closing the sluices during the rainy days, when there is no demand for water has been shown to increase the net irrigated area by more than 20 percent and minimizing the risk of crop failure at a 17 per cent (Palanisami and Easter, 2000). But in the EEC project not concentrated much on On Farm Development (OFD) except lining the water distribution channels.

Table 5: Characteristics of Non EEC tank villages **(n=25)**

Villages	Tank performance (%)	No. of Household in Ayacut	Change in Filling pattern (No)		Water availability (days)		Siltation (%)		No of the wells		
			10 yr before	Present	10 yr before	Present	10 yr before	Present	70's	80's	90's
Poovanthi	81..12	1000	2	1.5	120	20	30	40	30	35	75
Nilaiyur	80.25	200	2	1	120	55	25	50	5	10	11
Eriyur	77.56	800	1	1	90	60	20	30	15	15	15
Mayaleri	73.50	150	1.5	1.0	120	60	20	40	4	6	6
Kambikudi	79.20	570	2	1	140	40	20	50	20	20	30
Sikkial	81.25	2000	2	1	90	50	10	40	15	10	3
Salvarpatti	75.23	500	1	1	45	30	25	50	10	15	30
A. Pudupatti	79.74	400	2	1	50	30	10	25	10	25	25
Ayardharmam	82.00	150	1.5	2.5	60	35	20	50	10	10	15
Piramanur	83.25	970	1.5	2.5	80	35	20	50	50	70	93
Theni	82.30	170	1	1	50	50	40	50	-	-	-
Koovalapuram	79.00	250	3	1	180	50	15	50	10	10	20
Kilankulam	84.52	127	1.5	1	120	60	20	50	0	0	0
Ponnampatti	79.36	200	1	1	120	60	220	50	0	5	8
Watrap	78.35	500	2	1.5	120	50	30	40	100	110	140
Silaimalkipatti	87.12	312	3	1	150	80	20	50	10	30	35
Jariusilampatti	78.58	150	2.5	1	100	60	10	60	-	5	20
Madapuram	83.88	700	1	1	90	60	20	50	-	15	20
Thuthai	75.23	60	1	1	100	80	10	20	-	-	10
Thevathanagudi	75.14	175	2	1	90	60	40	50	-	4	8
Kalangapuli	90.00	15	2	1	90	60	30	40	-	-	-
T. Kallikulum	83.11	300	3	1	150	30	25	40	-	-	2
Marathai	81.55	500	3	2	120	80	60	40	-	-	2
Vikaramapandiyapuram	85.23	110	1	0.5	70	40	50	75	-	-	2
Sekkanthidal	80.12	60	3	1.1	180	60	15	30	-	-	-

Characteristics of Non EEC tanks

It could be interpreted from Table 5 that average *ayacut* area of Non-EEC tanks is 207 ha and 414 farmers shared these lands. Thus, one farmer would have 0.5 ha of land. Average land holding of the state in 1980-81 is 1.07 ha and declined to 0.95 ha in 1995-96 as the numbers of farmers have increased (SPC, 2008). Thus in the study area both in EEC and Non EEC tank villages average land holding is less than state average. These farms are too small to operate and their credit worthiness is rated very poor. That sets limit for investment for intensive farming. One farmer and his family depended for their livelihood mostly less than 1 ha of land made their life miserable. Tank water supply was 109 days with thc average filling of 1.86 times per year10 years before, but decreased to 63 days, resulting in crop failure or poor harvest. Non-EEC tanks could supply water for 63 days and EEC tanks could supply for 55 days, but the difference is statistically not significant. Location of the tanks in the chain might have influenced the tank filling and supply. Number of days supplied by extra one week by Non-EEC tanks will not have any significant effects if it occur in the middle of the crop but made significant differences in end of the crop season, as these tanks were located in the same rainfall domain.This is mainly based on the calculation of number of stress days encountered by the farmers. The stress days for rice are defined as absence of standing water on the field is more than three days. Stress days are normally classified as early stress which occurs 31 to 61 days after planting and late or often consider as severe stress that occur 60 days after planting. The field study showed that here it occurs mostly after 60 days. In non system tanks[4]the average number of stress days rangesfrom 8 to 12 days. Thus the extra one week supply of water is not significant as the stress days are almost two weeks or more in the study area. Decreases in the duration of tank water supply during the last 10 years are, in average, 27 days in the EEC tanks and 46 days in the Non-EEC tanks. This is mainly because most of the EEC tanks have been provided with modernized sluices and modification in the surplus weir. But in Non EEC tanks continued with damaged and dilapidated sluices and surplus weirs. Also, in the case of siltation to total capacity, Non-EEC tanks had 20%, but the EEC tank has only 13%. Thus, it could be concluded that EEC works might have influenced positively but not significantly to stabilize water availability in the tanks over the period of time. Even though EEC works were not for full scale desilting, some of the onfarm treatment and sluice modernization might have influenced for this outcome. Farmers from these villages told that government rarely allot the fund for desilting, but the contractors is not doing desilting properly and just increase the height of the tank bund and surplus weirs. So, the problem of siltation still persists. Many field observations indicate that raising tank surplus weir is more easily achieved than desilting the tank bed (Mukundan, 1998 and Dikshit et al, 1993). Approximately 23 wells were to cater to the needs of 414 farmers in the *ayacut*, i.e. 1 well for every 9.09 ha of land. So supplementing irrigation through wells is almost negligible.Most of the farmer wanted to have wells of their own or community well in the tank area and government also promote well irrigation. Even though the development of well irrigation in the tank ayacut help the farmers to reap better harvest through supplemental irrigation and even increasing cropping intensity but it is lopsided development affects community participation both for maintenance of irrigation structure and management of the tank system (Sivasubramaniyan, 2006:8). Further development of well irrigation in Tamil Nadu is difficult. Central Ground Water Board (2003) noticed that out of 385 blocks in Tamil Nadu, 240 blocks has been declared as over exploited (135), critical (35) and semi critical (70) and 8 blocks were saline for ground water exploitation. Over the last 5 years the percentage of safe blocks has declined from 35.6 to 25.2 percent. The other research evidence also caution that in Tamil Nadu decline in tank performance has been observed when the number of wells/ha is increased (Palanisami and Balasubramani 1998 and Janagarajan 1993). Earlier (20 years back) all these tanks cultivated two rice crops per year, now they cultivated only one rice crop. If tank has some surplus water or utilizing available soil moisture in the paddy fields after harvest, farmers used to cultivate blackgram or chillies or groundnut. It could be observed from the villages there is not significant differences between EEC tank villages and Non EEC tank villages. Because, the possible alternatives like proper rehabilitation of tanks, strengthening the local institutions and creating awareness about water use efficiency and management among the farmers were remain unaddressed in both tank villages.

Roles of water user association in the tank management

Many studies confirmed that farmers' collective action will leads to better performance of common resources like tanks (Sakurai and Palanisami, 2001). With all its odd and forced enclosure movements of British and post independence the collective action declined although did not disappear. Earlier, *Kudimaramathu*, tank maintenance by people themselves, was sincerely practiced. The community had almost complete control and access over the water resources within their jurisdiction. The system has functioning well and there existed well laid out rules and regulations to undertake all the critical function of water management such as system maintenance, water sharing during the scarcity, conflict resolution, collective of penalty from defaulters and so. There existed hierarchy of functionaries to undertake all these activities. The caste structure played a crucial role in preserving and allocating responsibilities among various functionaries. The position of leadership always held by farmer from the upper caste and lower position which need hard labour provided to lower caste people. The remarkable feature about these informal institution is the fact that they have been able to conserve and manage natural resources much better than the usually externally imposed, technocratic and resource intensive management system. These local systems are resource efficient and effective (Agarwal, 2008). This is the reason why the international agencies and governmental programmes increasingly showed to renew traditional informal water user association or to develop such association in new. Traditional water use associations have been criticized on account of being invented and manipulated as gender biased, caste oriented institution for abuse of power. The power of decision making vested with some upper caste group and it is hereditary and little or no chance were given to other caste people in power sharing. Hence, it might fail to keep face with the changing socio economic and cultural values and government promoted social justice and equity have lead to the possibilities of losing its importance and become less relevant to present situation. Although these institutions have proved to be stable form of resource management in traditional societies, the combination of population growth, technological change, climate and political forces has destabilized many existing institution (Runge,1986:623).

Table 6: Role of water user association in the study villages (Multiple responses)

S.No.	Criteria	EEC Tanks **(n=16)** Normal year		Non-EEC Tanks **(n=13)** Normal year	
		No	%	No	%
1.	Supply channel cleaning	12	75	13	100
2.	Tank bed management	14	87.5	13	100
3.	Removing weeds in water spread area	6	37.5	4	30.75
4.	Equal water distribution	10	62.5	10	77
5.	Sluice repair and maintenance	8	50	5	38.45
6.	Approach government to dig community well	2	12.5	3	23
7.	Regular fish production and leasing	0	0	0	0
8.	Collect common maintenance fund	2	12.5	1	7.70
9.	Appoint common Neerkatti and Neerpachi	16	100	13	100
10.	Control well water price	1	6.25	2	15.38
11.	Encourage crop diversification and others	1	6.25	3	23.07
12.	Catchment area cleaning	0	0	0	0
13.	Approach government for funding	8	50	1	7.70
14.	Solve dispute between farmers	12	75	13	100
15.	Mobilize farmers for common work	14	87.5	12	92.30

Source: Author's Field work, 2007

Our field experience showed while it is true that in some villages the irrigation organization is less organized than others but it is inconceivable that in any village it is entirely absent. Hence, it is necessary to explore contemporary situation of water user association and their role perception and performance in the midst of water crisis. We have identified 15 different types of roles or work expected to be carried out by these association through discussion with traditional leaders, village irrigation functionaries and literatures. Among the 25 EEC tanks studied, water user association (WUA) is existed in 16 villages and in case of Non EEC tanks, out of 25 tank villages 13 villages has association. But in the case of presence of irrigation functionaries like Neerkatti[6], their presence observed in the 20 EEC tank and Non EEC tank villages. Table 6 indicated that role executed by the association in both type of the tanks. Slightly higher number of WUA noticed in EEC tanks. It is because EEC has advocated forming WUA during third phase of the programme. It could be observed from the table 6 that supply channel cleaning, tank bed management, appointing irrigation functionaries and management, mobilize farmers for common work and ensuring equal water distribution are considered important and executed by WUA in both type of tanks.

Maintenance of supply channels is very important to augment water supply from streams and rivulets. Every year before onset of the monsoon, around September the farmers organized themselves in a common place after being informed by Neerkatti through leaders. They usually cleaned the bushes and thorns and repair preaching in the supply channels. The every Household having land in Ayacut is requested to contribute labour for common work. The household is not participating in the common work without valid reason has to pay penalty money to the association. Irrigation functionaries are responsible for watching and guarding tank bed, surplus weir and sluices during the tank season. It is suspected that farmers, who encroached tank foreshore area, surreptitiously damage the surplus weir, breach the tank bunds and open the sluices in the night times in order to avoid submergence of their cultivated field. *Neerkatties* also watch the tank bunds carefully during the heavy rain and if he suspects the possibility of breach he will inform the villagers and mobilizes the labour to safeguard it. They also rested with responsibility of operating sluices. They are usually from the scheduled caste called "*Pallar or Devendrar*" and from particular family and widely hereditary. They called as *Neerkatti, Madayan, Thotti, Madayar, Neerpachi* etc in the different villages. These traditional institutions still play major role particularly when the tank receives normal or just below normal rainfall. *Neerkattis* are almost working as water managers in the tank system. Even though they are not technically qualified, they could judge by experience available tank water, and timing and amount of water release from the sluice very well. In normal rainfall year *Neerkatti* did not irrigate farmer's fields. During the water shortage periods, they are asked to irrigate all the farmer's field in order to assure equal water distribution.

The predominant role of *Neerkatti* was therefore to ensure equal distribution of available tank water to all. The *Neerkatties* were regularizing water supply in two ways. i)'*Murai Pasanam'* i.e. to irrigate the particular field at regular intervals where the *ayacut* area was divided in to 10 *pungus* (share). One *pungu* was about 63 acres. Water was rotated among these *pungus*. ii) '*Oru madai pachal*' i.e., to keep the lowest sluice open and to close the remaining sluices if the tank has more than one sluice. Using the water released from one sluice, *Neerkatti* could equally distribute water to the all the farmers. These arrangements avoided the conflicts among the farmers. It has been argued that traditionally *Neerkatti* have been Scheduled Caste (SC) and since SC had no lands with low social status, they could be trusted not to act arbitrately. However this argument is not convincing as *Neerkatti* seem to have considerable *maniyam* and were also land owners and cultivators. Traditionally *Neerkattis* were always the key witnesses that the community relied upon when the land and water disputes needed to be arbitrated. This was the status accorded to them traditionally by the community and does not necessarily have any relation to their social and economic status (Mukundan,1998:19). This is the reason the entire water user's association felt that the role of *Neerkatti* in tank is inevitable. Arputharaj (1982) have also found that the main problem in the tank system was uneven water distribution due to the lack an adequate water man to allocate water. Sometime, the WUA made request to the well owning farmers to not use tank waters whenever tank water is very scarce. WUA sometimes advocate in the midst of water scarcity that confined to the part of *ayacut* closest to the bund or only the head reach, but if a tail reach or faraway farmer has to cultivate if he is poor, either he gets a special allotment of water or he is given a plot in the head reach through exchange of plots with another cultivator (Sengupta, 1985). The Catchment and water spread area treatment, regularizing fish production, approach government for funds and seeking for community wells are not given due importance. This is because the villagers are felt that removing encroachment and cleaning water spread need huge money and authority. Approaching government official for common cause is not an easy task. They often poorly respond and not at all given importance to local villager's request. Presently WUA is not supported by the government officials and felt that they are not answerable to farmers rather than higher official. So, tank improvement which needed investment has to depend upon the mercy of government. Earlier benefit from the tank resources rest with traditional institution and acted as resource base, were utilized for tank repair and maintenance. Now it has no resource back up to execute the responsibility it has earlier. Baker (1984) summarized this situation as many irrigation tanks in India before British period were endowed with certain tank lands and usufruct rights which were given to the individual who has responsible tank maintenance. After the independence the same trend is continuing.

The government often views these traditional organizations simply do not have any status and unscientific and that methods of modern technology have to be used to break the old ways. The technocratic solution proposed also basically aimed at attaining the goal of modernization that shifting the control of management from the village to government. Agarwal (2008) cautioned that the level of recognition given to these institutions by the local people is crucial to secure sustainable natural resource management and it is beyond the doubt that institutions which were invented for the political gains of a few can ever be effective in this regard. Consider the proposal for the pilot projects of EEC tank modernization drawn up by the Public Works Department (PWD) responsible for maintenance of tanks with more than 100 acre of *ayacut* in the state which reveals that they hardly recognize existing institution as a institution. It states "presently there is no defined and recognized authority for water regulation from the tanks as well as within the command area. However, in order to ensure smooth and efficient functioning of this regulation, it is proposed to constitute two committees comprising official and non official members". After completion of first two phase of modernization only, they came to understand about the traditional institution and their social embeddedness. Hence, the committee recommended to form water user association in remaining phase three and extension. So in the most of the villages WUA has been formed for namesake as it was stipulated by international donor agencies (Down to Earth, 2006).

It does little good to create local institution for their own sake indeed according to the conception here of "Institution" as distinguished from "organization". One can create the latter. To the extent possible that they effectively meet the people need and expectation, they become former. So outside agencies must be careful that they promote only organizations that are likely to serve real needs. Otherwise the chances of these becoming instutionalised and self sustaining are slim (Uphoff, 1992:12). So all these reason together make no difference between EEC promoted water user association's role execution and Non EEC tank water association.The officials often argued that caste differences can cripple the efforts of water user association they created. They told that farmers from diverse socio cultural back grounds will have more difficult time in self organizing to govern tank resources and farmers assumed to be having problems of distrust and lack of mutual understanding. But researchers proved that heterogeneity is not strong predictor of collective activity. Rather heterogeneity is a challenge that can be overcome by good institutional design when the interest of those controlling collective choice mechanism are benefitted by investing time and effort to craft better rules (Varughese and Ostrom, 2001:747). But it is possible that for groups of farmers jointly using the same tank system to organize themselves so as to achieve relatively efficient outcomes from the continued use of resources. So it is understandable that water user association in the study area both in EEC and Non EEC tank villages execute their roles as they continued their belief, ideas and practices and custom are welded in to the system. EEC promoted tank institution failed to create credibility among the farmers that consideration such as equity and efficiency in the sense in which it seemed important from the PWD point of view, did not seem significant at all from the farmer's point of view. It is concluded that there is no significant difference between the role execution of WUA among the EEC and Non-EEC tank villages.

Impact of tank modernization on tank performance

Tank performance operationalised as the percentage of total area cultivated to total command area of the tank in a particular year. However the total performance redefined taking into the account of intensity of the wells in the tank command. But in this study, well density was relatively low (1 well/11 ha in EEC tanks and 1 well/ 9.09 ha) and well below the state average (1.21 well/ha) it does not influence tank performance significantly. Table 7 indicated that EEC tanks insignificant edge over Non EEC tanks in terms of tank performance with respect to area under cultivation in particular year. Tank has three aspects to consider viz., Physical, social and economic and technical aspects. Most importantly all these factors do not act in isolation and there is strong inter - connection among them which complicate each other for smooth functioning of the system. EEC tank modernization were attempted to administer remedies for physical improvement of tanks and partially provide technological solution but largely ignore the socio economic and institutional aspects of tank irrigation. In tank irrigation most of the problems are related to following aspects like maintenance, water distribution, siltation and encroachment. So many studies reiterated that together with siltation the encroachment will reduce the tank capacity by 30 to 35 percent. But EEC programme is not considered about this factor and neglect of this, we could observe wide spread siltation (36 to 47 %) and encroachment (36 to 45%) problems in the entire study area and farmers are much concern about it. In the case of maintenance, government is directly responsible for repair and maintenance. However this may not be feasible for most tanks because of their small scale and high cost of providing government services to each tank. This means that the farmer's cooperation and participation is one of the key to the whole problem. But table 7 shows that both participation and cooperation is around 40 per cent. EEC program did not give much attention to these institutional aspects. If the farmers can organize to maintain the tank system and distribute the water evenly among the farmers themselves then the tank irrigation is very effective. When the farmers do not organize, encroachment occurs, the tank silt up and water is wasted even when many farmers are short of water (Palanisami and Easter, 2000).

Table 7: Mean value of variables influencing tank performance among study villages

S.No	Parameters	EEC Tanks	Non-EEC Tanks
1.	Tank Performance (%)	82	78
2.	Filling pattern (No. of times)	1.36	1.28
3.	Water availability (No. of days)	56.5	52.2
4.	Siltation (%)	36	47
5.	Presence of WUA (%)	64	52
6.	Farmer's participation (%)	40	42
7.	Presence of Neerkatti (%)	84	76
8.	Maintenance of tanks (%)	44	36
9.	Farm income (Rs/acre)	6240	5975
10.	Water management (%)	12	12
11.	Equal water distribution (%)	40	38
12.	Employment opportunity (Mandays)	40	40
13.	Cooperation among farmers (%)	44	40
14.	Encroachment %)	36	45
15.	Well density (No. of wells/ ha.)	9	11

Source: Author's Field Work, 2007

It could be argued that where the government holds the rights to tank irrigation infrastructure and water, it may be in the public interest to maintain the system, but PWD staff charged with doing the work may often lack incentives to keep the assets in good condition. This is because rights to infrastructure and water do not provide a benefit directly to the PWD. Water and irrigation clearly valuable resources, but it does not yield benefits to the PWD. It is mainly because, irrigation fee is much low in Tamil Nadu and it is often merged with land tax and directly go to the revenue department rather than to the agency responsible for its management (Meinzen Dick, 2000:17). This is true that Palanisami and Meinzen dick (1999) have found that revenue from the tanks may go to the forestry department, fisheries department, village government panchayat, revenue department and even to the department of mines. None is directly paid to PWD which is responsible for tank maintenance and even what is paid to the state revenue department is only 7 per cent of the total revenue mobilized by fees, taxes, penalties and other tank resources. But it is the PWD's responsibility to fight with government to get proper share from the tank resources. They have well established organization structure and separate ministry for it. So far it could not take any initiatives to get what it deserves of it. It could be possible for the PWD as they can sizably reduce repair and maintenance cost if they created institution or acknowledged existing institution by providing legal status and resource base for after care. But it seems that they are not really interested in establishing water user association for sustainable management tank resources. For example the EEC rehabilitation projects aimed to "increase food production and rural income by achieving higher cropping intensity through improved water management and reduced loss". Their focus is clearly conserving storage and improving command area water use. Physical works like reconstruction of weirs and sluices and lining the channels accounted for 71 percent of the investment. About 27 percent went to administrative cost to meet government expenditure (ADB 2006). Only meager 2 percent was spent to address much crucial institutional aspects of tank irrigation. This trend continued first two phase of tank modernization.

As the project progressed, the views of farmers were taken to account to ensure sustainability after rehabilitation works. Hence EEC stipulated a tank user's association for each tank should be a precondition to take up rehabilitation. So the later period government forced to create such institution on order to satisfy funding agencies. Their intension towards such institution were described earlier in this paper needed mention here. Water user association existed only 64 percent of the study villages. Because of EEC regulation, government took step to initiate building institutions. So they have established WUA in all the rehabilitated tank villages. But our field study shows that only 64 per cent of the village had institution. It is understandable that WUA they created was active only during tank rehabilitation programme upto 1998. After completion of modernization officials responsible for these institutions could not follow up and failed to maintain the momentum (Jegadeesan and Fujita, 2008). It could be possible to suspect that existing institution in the EEC tank villages might be traditional institution, as the Non EEC tanks also have almost equal number of association and competently perform with EEC tank villages. Regarding water management and equal water distribution little more than one third of farmers (40%) perceived as canal lining foster equal availability of water to all the farmers in the command area.

One of the important objectives of the EEC programme is reducing transmission loss. It believed that it could be achieved by lining of the field channels. In tanks studied, the channel lining increased the per acre yield from 1300 kg to 1456 kg. While it is true that lining of the channel improves water use efficiency there has been wide spread frustration that farmers are not consulted about the lining in the planning and execution stage. The consent of the farmers is sought at the beginning of the project, subsequently designing, construction and implementation of the works are carried out without any discussion with farmers (Mukundan, 1998). As a result in many tanks the cement lining has been broken saying that it hinders field operation like ploughing etc. Easter and Welch (1983:12) commented that probably the cheapest method of reducing transmission losses, where the labour costs are low, is by proper and timely maintenance of the channels. Lowdermilk, *et al.* (1977) also believed that this problem can be alleviated by effectively organizing the users to maintain and improve their water cources. It could be concluded that farmers do welcome efforts on the part of government to augment water supply through desilting and repair the bunds, sluices and surplus weirs. The much of onfarm development did not seem necessary, even if it is necessary without active participation from the farmers who are ultimate use and maintain it will leads to no fruitful outcome. Often it led to resentment and silent non cooperation from the farmers or even active sabotaging of the OFD works as we have seen in the many tank villages.

Factors influencing tank performance in EEC and Non-EEC tanks

The factors such as filling pattern of the tanks, water availability in the tank, presence of water user's association, presence of *Neerkatti* (waterman), co operation among the farmers, dispute among the farmers in sharing the tank water, the employment opportunity after modernization and condition of tank physical structures might have influenced the performance between EEC and Non-EEC tanks. In order to test this, 't' test has been carried out (Table 8).

Table 8: Comparative t test between parameters influencing tank performance in EEC and Non-EEC tanks

S. No.	Variables	t value
1.	Filling pattern	1.290
2.	Water availability	1.010
3.	Tank structures	2.102*
4.	Water user's association	-0.596
5.	Waterman	0.293
6.	Cooperation among farmers	-0.278
7.	Employment opportunity	0.770
8.	Dispute among the farmers while sharing the water	2.028*

*Significant at 10% level

It could be interpreted from the table 8 that two out of eight variables selected for the test, showed significant relationship. Condition of the tank structures and dispute among the farmers while sharing the water were considered as crucial variables which differentiated the performance between the EEC and Non-EEC tanks. EEC programmes modernized the tank structures like sluices, surplus weir and canal below the outlet. Hence, it was easy to operate and maintain these structures by the waterman in the EEC tanks than in Non-EEC tanks. Outlet canals were also lined by EEC programme, hence, tank water released for irrigation reached tail enders relatively quicker than in Non-EEC tanks. However, canal lining was not done for entire command area of the tank. Maximum 50 % of the area was lined and average area under lined canal was much less. Yet current technologies for lining, particularly concrete lining tend to higher cost. Studies of concrete lining do not give clear cut conclusion concerning profitability (Taylor, 1980). Hence, it was concluded that even though EEC work has improved the tank performance than non modernized tanks but it was not up to the expectation of the farmers as there was no significant difference between EEC and Non-EEC tanks in terms of other parameters such as filling pattern, water availability etc.,

Equity in Water Availability and Paddy Yield

Equity ratio is defined asyield in kg per acre in head reach / yield in kg per acre in tail reach. If the ratio is greater than one, it indicates that head reach is benefited much from irrigation.

The ratio is close to one in modernized tanks compared to non-modernized tanks indicating that the modernization has improved the water supply and crop yield uniformly. Higher the ratio, wider the inequity in irrigation water distribution between the head and tail regions. The equity ratios for rice yield and water availability in EEC and Non-EEC tanks are presented in Table 9. It is inferred that in EEC tanks, the equity ratio for the rice yield is 1.06 which indicates that when the yield of paddy was one ton/acre in tail, it was 1.06 ton/acre in head reach. In Non-EEC tanks the ratio was 1.13 which indicates there was a subtle difference between head and tail reach. Hence, it is concluded that distribution of available water between head and tail reach is comparatively better in EEC tanks. This was due to the lining of the canal below the outlet which minimizes seepage losses considerably in EEC tanks. For Non-EEC tanks lining was not available hence the difference. Availability of water also marginally higher in EEC tanks than Non-EEC tanks. It might be due to modernized sluices and shutters provided in modernization program. However, the difference between EEC and Non-EEC tanks is not significant in terms of equity ratio.

Table 9: Equity ratio of paddy yield and water availability in EEC and Non-EEC tanks.

Particulars	EEC Tanks	Non-EEC Tanks
Yield (kg/acre)		
Head reach	2520	2448
Tail reach	2376	2160
Equity ratio	1.06	1.13
Water Availability (No.of Irrigation/ acre)		
Head	27	25
Tail	27	23
Equity ratio	1.00	1.08

Source: field work, 2007

Conclusion

The basic problem in most of the tank modernization programme including EEC tank modernization programme is failure to reach expected level of outcomes. In spite of criticism EEC programme have addressed some problems of tank irrigation which made subtle difference over Non EEC tanks. First,it has reduced conveyance loss and there by ensure equal water to all the farmers in the tank command. It is common problem in the tank command that tail end farmers not enjoy the water what his counterpart in head of the command would enjoy. The EEC minimized this disparity by lining the channel not significantly but noticeably. Ensuring equal water to all is important because it would create assurance over the water to all the farmers. If the farmers assured about their water supply, they start believe the agency or the institution. The relation of trust will lubricate cooperation and so reduces transaction cost. Instead of having to invest in monitoring and management of the system, better to create enable environment for farmer's participation. The agencies also trust farmers as they trusted them. The potential benefits realized by the farmer have incentive for collective action. But there is long way go as the trust takes time to build and aware that it need no time to break Secondly, better late than never, the EEC have recognized the importance of institution in the third phase of their programme implementation. They have requested government and other agencies involving in the tank modernization programme to form water user association for successful implementation of the programme. Even though they created the institution just for the political gain and it not serves real purpose, it is good that they started realizing the importance of such institution. Thirdly, it has initiated the incentives to the participation and there by attempt to create cordial relationship between government and villagers. Some could argue that it is passive participation in the name of consultation with farmers providing no role or power in decision making often said to be a means of indoctrinating the public values and priorities of the planner to ensure that they obtain public endorsement of their decision, rather than understanding of local needs and priorities (Smith, 1998). Lastly they have reduced discouragingly high water leakage by modernized dilapidated sluices and weirs. Tank irrigation system responded relatively well to the treatment given by EEC but then suffered a relapse. In total, rehabilitation of physical structures such as lining the field channels and modernizing sluices and weirs has been overemphasized and it does not appear that serious efforts were made to ensure the active participation of farmers and other institutional innovations to mobilize the people in removal and prevention of encroachmnets. It has misread the situation and misplaced the priorities hence misrespond to the occasion. As we argued earlier, it fostered tokenist, passive participation. They are very keen to utilize the funds but not ready to use resource existing in the tank villages for managing the systems. Another thing is that they have failed to conceptualize to include other user of the tank such as fisher man, agricultural labours, livestock grazer and other common villager. The Indian constitution gives supreme importance to the local elected body the village panchayat which is by constitution empowered with authority to manage minor irrigation, water management and watershed development did not come to picture of modernization. Government have recognized inexorable rise of encroachment in tank system and passed the Eviction of encroachment act recently, but not implemented as situation demand. The participatory strategy promoted by the government should seek active participation rather than passive and “also” participation.

They could understand that tank water flows within the ayacut and water itself make intricate connection between farmers. It espouse culture social network among them. It has some inbuilt mechanism, which allows flexibility to accommodate situation specific requirement. Hence, the government or agency would continue further in the assumption that farmers have much to contribute other than labours. They have ideas, management skills, and technical solution to the age old problems it should be harnessed through proper institutional back up. They have to be considered as collaborator and active partner of the programme rather than potential user or beneficiary. It is better to make need based investment. The efforts and expectation of the government and local people should not be elided to get good outcome. Consideration such as equity and efficiency in the sense in which it seemed important from the implementation agency point of view, did not seem significant at all from the farmer’s point of view. So it is recommended that priority goes to farmers rather than funding agencies or government. The resource base for tank system should be created utilizing tank resources such as water, fish, trees and silt etc., bring them under single management authority. There is wide spread mutual mistrust, unwanted disillusionment and regrets between line departments such as revenue, fisheries, forestry, public works, agriculture, agricultural engineering, and local administration and so on. This could be avoided by viewing the tanks as whole system rather than single entity. Government should have provided more authority to local electoral body as they empowered by the constitution to look after minor irrigation under its jurisdiction. It is necessary that before attempting to advocate new irrigation institution through legislative measures, it is important to reexamine the reason for disintegration of such institution and probe possibilities of reinstate it.

Notes:

1. Different opinion exists about state’s harnessed irrigation potential. Ministry of water resource says in its report on “More crop and income per drop

of water" as Tamil Nadu already achieved 100 percent irrigation potential. But,The state planning commission said 95 percent of surface water already tapped.

2. Tanks are classified as PWD tanks and Panchayat Union tanks. Tank with command area more than 100 acre comes under PWD management. Tanks with less than 100 acres of command area come under Panchayat Union Management.
3. The districts where more than one third of the area irrigated comes under tank irrigation is considered as tank intensive districts.
4. The tank command area is locally called as Ayacut. Ayacut area is usually larger than official estimate of PWD. The near by area of the tank command also use tank water whenever tank has surplus water. But it does not entered in official records as very rarely tank has surplus water.
5. Tanks which are connected to the near by reservoir or canal and there by get assure water supply are called as system tanks. The rain fed tanks called as non system tanks.
6. Kudimaramath is management of the water resource by villager themselves by providing common labour and other necessary maintenance. It has been deteriorated since long time. For details please see Mosse (2003), (Mukundan, 1998; Sengupta, 1985).
7. Neerkattis are the water distributor and sluice operator in the tanks. They are called in different names in north Tamil Nadu as kambu katti, kaladi *etc.*
8. Maniyam is grand given to irrigation functionaries for their services in common work.

References

Abel, Martin and K.William Easter (1971). Agricultural Planning: Focusing on regional constraints. *Economic and Political Weekly,* 6(30-32):1577-1596.

Agarwal, Asish (2008). Indigenous Institution for Natural Resource Management: Potential and threats. *Economic and Political Review,* June 7:21-24.

Agricultural Engineering Department (1995). Hand book on Run off Harvesting Structures. Govt. of Tamil Nadu.

Arputharaj,C. (1982). Problems and prospectus of Tank Irrigation in Tamil Nadu. Workshop on modernization of tank irrigation: Problems and issues, Center for Water Resources, Madras, India.

Asian Deveopment Bank (2006). The rehabilitation and management of tanks in India: A Study of Selected States. Philippines.

Baker, C.J (1984). An Indian rural economy, 1880-1995: The Tamil Nadu countryside, *Oxford University Press*, New Delhi.

Central Ground Water Board (2003). Dynamic ground water resources of Tamil Nadu, Govt. of India.

Dikshit,G.S., G.R.Kuppuswamy and S.K.Mohan (1993). Tank irrigation in Karnataka: A Historical Survey, Bangalore, Gandhi Sahitya Sangha.

Down to Earth (2006). Tamil Nadu's tank rehabilitation scheme does not hold water. *Down to Earth*, Vol. 15, No. 3, June 20.

Easter,K.W and Delane E.Welsch (1983). Socio economic issues in irrigation: Development and Distribution. Economic Report: ER 83-5, Department of Agricultural and Applied Economics, University of Minnesota, St.Paul, Minnesota.

Human Development Report (2003) Tamil Nadu: Human development report, Govt. of Tamil Nadu. *Social Science Press.* New Delhi.

Jangarajan,S. (1993). In search of Tanks: Some Hidden facts, *Economic and Political Weekly*, June 26, pp. A53-A60.

Jegadeesan,M. and K. Fujita (2008). The role of institution in tank water management in Tamil Nadu,India. *Forth coming.*

Lowdermilk,M.K., T.M. Freeman, A.C.Early, G.Radosevich and D.Kemper (1977). Organizing farmers to improve irrigation water delivery: The problems and prospects for solution in Pakistan. *Pakistan Economic and Social Review,* 15 (3/4), pp. 152-173.

Meinzen-Dick, Ruth (2000). Property rights and maintenance of irrigation sytem. Thematic paper No:6, Deutche Gesellschafur, Eschborn, Germany.

Ministry of water resources (2008). Water resources of India: A Hand Book.

Minor Irrigation Census (2001). Ministry of Water Resources, India

Mosse, David (1997). The symbolic making of common property resources: History, Ecology and Locality in a Tank irrigated landscape in South India. *Development and Change,* Vol. 28. Pp. 467-504.

Mosse, David (2003), the rule of water: statecraft,ecology and collective action in south India. *Oxford University Press,* New Delhi.

Mukundan,T.M (1998). The Ery system of South India. *PPST Bulletin,* 16: 1-33.

Palanisami and Ruth S.Meinzen Dick (1999). The endangered tanks of south India: Who will pay for the survival? Forth coming, *Economic and Political Weekly.*

Palanisami,K and K.William Easter (2000). Tank irrigation in the 21st century – What next? *Discovery Publishing House*, New Delhi.

Palanisami,K and R. Balasubramanian (1998). Common property and private prosperity: Tanks vs private wells in Tamil Nadu. *Indian Journal of Agricultural Economics.* 53, 600-613.

Palanisami,K. (2000). Tank irrigation: Revival for prosperity, *Asian Publication Services*, New Delhi.

Palanisami, K., Jegadeesan, M., Fuijta, K. and Kono, Y. (2008). Impacts of tank modernization programme on the tank performance in Tamil Nadu, India. *Kyoto Working Papers on Area Studies* No:5 (G-COE Series 3)

PWD (2008). Public Works Department (Water Resource Organisation) *Citizen Charter*, Govt. of Tamil Nadu.

Runge, C.F. (1986). Common property and collective action in economic development. *World Development*, 14(5):623-635.

Sakthivadivel, R., Savamuthu, S., Shanmugam, C.R., Balakrishnan, P.K. and Arputharaj, C. (1982). A pilot study of modernization in Tamil Nadu. Workshop on Modernization of Tank Irrigation: Problems and Issues, Center for Water Resources, Madras, India.

Saktivadivel, R., Gomathinayagam, P. and Tushar Shah (2004). Rejuvenating irrigation tanks through local institutions. *Economic and Political Weekly*, July 31:521-3526.

Sengupta, Nirmal (1985). Irrigation: Traditional vs Modern, *Economic and Political Weekly*, Special Issue. November, Vol. XX, Nos 45,46 and 47.pp.1919-1938.

Shanmugam, C.R. (2006). Tamil Nadu's tank rehabilitation scheme does not hold water. *Down to Earth*, 15(3)' June 20.

Sivasubramaniyan, K. (2006). Sustainable development of small water bodies in Tamil Nadu. *Economic and Political Weekly*, June 30, pp.2854-2863.

Smith, B.C. (1998). Participation without power: Subterfuge or Development. *Community Development Journal*, 33(3):197-204

State Planning Commission (2008). Annual Report on Balancing competing water uses and management – Need for an Ecological water resources policy, Govt. of Tamil Nadu.

Taylor, D.C. (1980). Economic analysis to support irrigation investment and management decision. Irrigation Water Management, IRRI, Manila.

Uphoff, Norman (1992). Local institution and participation for sustainable development. *Gate keeper Series, No:31*, International Institute for Environment and Development, London.

Varughese, G. and Elinor Ostrom (2001). The contested role of heterogeneity in collective action: Some evidence from community forestry in Nepal. *World Development*, 29(5):747-765.

Wade, Robert and Robert Chambers (1980). Managing main system: Canal irrigation's Blind Spot. *Economic and Political Weekly*, 15(37):A107-A112.

Extension Management Strategies for Sustainable Agriculture, pp 283-291

New India Publishing Agency, New Delhi, India

29

Indicators for Effective Extension Service for Extension Organisations

P. P. Murugan

Technology adoption is mounting every day in the field of agriculture. The greatest need of the day is to effectively communicate the information related to farm technologies to the millions of farmers to educate and to adopt them in order to modernise their agriculture. Inspite of our best efforts, only 25 to 30 per cent of the modern agricultural technologies have reached farmers. This modern technology however has been restricted to favourable farming situation only.

Hence proper mechanism for effective dissemination of technologies to its ultimate users is the need of the hour. Keeping this view in mind, this investigation entitled "Identification of Indicators for Effective Extension Services in Public, Private and Non-Governmental Organisations" was taken up to identify the indicators for effective extension services in public, private and non-governmental extension organisations

Indicators were defined as measures to verify the extent to which outputs or development objectives are achieved. In this study, indicators can be any extension function, activities, programmes, events, decision, condition or any thing, which helps of increasing the effectiveness of extension organizations.

In order to identify the most contributing indicators for public, private and non-governmental organization, data were collected from the beneficiaries and analysed using principal component analysis on a symmetric matrix type SPSS (Statistical Package for Social Sciences) and results were obtained.

For the above purpose, the data collected from all the respondents of public, private and non-governmental organisations were pooled together. Principal component analysis was carried out for the pooled data and the results were obtained. The analysis identified the existence of 10 groups of indicators. From each group indicators having the variance value of 0.4 and above were considered as important indicators.

The groups of contributing indicators found through analysis can be used as guideline for increasing the effectiveness of any extension organisation involved in developmental activities. The group of indicators found in the analysis is presented here under.

Group-I

The name of the indicators grouped under this category is presented in Table 1.

Table 1: Group-I. Indicators contributing for effective extension services **(n=180)**

Sl. No.	Indicators	Variance
1.	Having counselling centers	0.647
2.	Supply of quality inputs in time	0.631
3.	Popularising cost effective technologies	0.526
4.	Having model farms	0.506
5.	Following participatory technology dissemination	0.481
6.	Conducting demonstrations	0.462

The data from Table 1 reveals that indicators such as having counselling centres, supply of quality inputs in time, popularising cost effective technologies, having model farms, following participatory technology dissemination and conducting demonstrations were observed as contributing indicators for increasing the effectiveness of extension services under one group. The extent of variance level of these indicators were observed as 0.647, 0.631, 0.526, 0.506, 0.481 and 0.462 respectively.

To get farm advisory services regularly, respondents need the presence of permanent counselling centers within the village in a distance, which is easily accessible by villagers. Similarly for getting required inputs, respondents expect the assistance from the extension organization.

In the present day farming, cost of output is abnormally high due to use of chemicals. It is not matching with the cost of the price realised from selling the farm produce. This condition is mainly due to the costly nature of inputs. To overcome this problem, respondents seeking cost effective technologies. Scientists have generated many technologies, which are cost effective. The respondents would have expected the demonstration of these technologies through model farms. This requirement would have made the respondents to prefer the popularisation of cost effective technologies, having model farm and conducting demonstrations as the expected actions from the extension organisations.

In adequate manpower in the extension organization increases a gap between the extension worker and beneficiaries. In this situation, involvement of progressive and leader farmers in the dissemination of technologies will complement the extension work. Hence, the respondents of all the selected extension organization considered participatory technology dissemination as one of the contributing indicator for effective extension service.

The above mentioned advantages might be the reasons for the respondents of all the three selected organisations agreed these indicators as contributing indicators for effective extension service. In general it was observed that participatory oriented and input related indicators were found as most contributing indicators for effective extension services of any organisation involved in extension. As all the indicators found in this group reflects on extension activities, this group may be named as extension activity related indicators.

2. Group – II

The analysis further showed a group of indicators influencing for effective extension service. They are presented in Table 2.

Table 2: Group-II. Indicators contributing for effective extension services **(n=180)**

Sl.No.	Indicators	Variance
1.	Having programmes for popularising irrigation management technologies	0.576
2.	Promoting value addition	0.507
3.	Programmes for use of natural resources	0.488
4.	Promoting indigenous technologies	0.411

Table 2 reveals that the indicators of having programme for popularising irrigation management technologies, promoting value addition, programme for use of natural resources and promoting indigenous technologies have been grouped under one group. The variance levels of these indicators were seen with 0.576, 0.507, 0.488 and 0.411 respectively.

The respondents of pubic, private non-governmental extension organization felt the pinch of adequate rainfall received during the past few years. Hence they need programmes for effective management of available irrigation water. Predominant crop of the study area are commercial crops. Value addition of their farm produce of these commercial crops would fetches more money for farmers. Thus the importance of value addition has been widely felt.

There is lot of indigenous and natural resources available in the villages. Respondents are having inadequate knowledge on its importance. Proper identification and making use of these resources and integration with modern technologies would help the respondents to reduce the cost of cultivation and protect the environment.

The ultimate aim of any respondents is to increase the income and standard of living. As the above mentioned indicators were useful for increasing the income level of the respondents, they would have chosen as contributing indicators for effective extension service. The findings of this group indicated the importance of having programmes on organic farming, value addition and water management and this group may be named as programme related indicators.

3. Group-III

The details of the group III indicators responsible for effective extension service are presented in Table 3.

Table 3: Group-III. Indicators contributing for effective extension services **(n=180)**

Sl.No.	Indicators	Variance
1.	Use of modern ITs like internet, e-mail, etc	0.594
2.	Use of extension literatures	0.476
3.	Screen film or video shows	0.462
4.	Use of television	0.403

It could be observed from the Table 3 that the indicators viz., use of modern IT's like internet, e-mail etc., use of extension literatures, screening of film or video shows and use of television were the other effective indicator found in the analysis. The variance value was 0.594, 0.476, 0.462 and 0.403 respectively.

The above mentioned indicators were useful for transfer of technologies to the large number of people with in a short period of time. The cost incurred for transfer of technology through these facilities is comparatively low. Similarly the accessibility of these facilities is possible by the respondents especially in the present day of information and globalisation of this era. So the beneficiaries from the selected three organisations have chosen these indicators as contributing indicators for effective extension service. As all the above indicators were connected with communication, this group may be named as communication facilities related indicators.

4. Group-IV

The effective indicators found in the group IV are presented in Table 4.

Table 4: Group-IV. Indicators contributing for effective extension services **(n=180)**

Sl. No.	Indicators	Variance
1.	Having a committee for planning and monitoring	0.659
2.	Conducting regular village meetings	0.499
3.	Extension workers having regular contact with research institutions	0.468
4.	Co-ordination with sister organization	0.455

For effective transfer of technology, proper administration is necessary. Formation of a committee for planning and monitoring is needed for facilitating the effective extension services. Conducting regular village meetings is also important. These meetings make available the latest technologies to the respondents similarly extension workers gets feed back from the respondents.

Research institutions are developing number of new technologies. These technologies have to reach the beneficiaries through the extension workers. Hence extension workers have to get new information regularly from research institutions and transfer to the respondents. To obtain new technologies, the extension workers should have regular contact with research institutions. Similarly for carrying out development activities, the extension organizations should have co-ordinative with the sister organizations for effective services. For this purpose co-ordination with sister organisation is essential for effective extension service. Due to the explained reasons the respondents would have selected the above indicators. As all the above indicators concerned with administration, this group may be named as administration related indicators.

5. Group-V

The group-V indicators grouped in the analysis are enlisted in Table 5.

Table 5: Group-V. Indicators contributing for effective extension services **(n=180)**

Sl. No.	Indicators	Variance
1.	Group formation and extension workers work with group	0.487
2.	Having programmes for women development	0.480
3.	Involvement of rural youth in extension activities	0.434
4.	Guiding for credit	0.407

It could be seen from the table that group formation and extension workers working with group, having programmes for women development, involvement of rural youth in extension activities and guiding for credit were observed as the indicators contributing for effective extension service under one group, with the variance level of 0.487, 0.480, 0.434 and 0.407 respectively.

In the transfer of technology process, group approach is performing well. The synergetic effect of group also influence the members on changing their mind set. Hence the extension workers working with the group is one of the effective indicators perceived by the respondents of selected public, private and non-governmental extension organization. The respondents felt that the extension services available for women and youth development are not adequate for their development. The respondents required special attention for women development and involvement of rural youth in extension activities. This helps the women and rural youth to develop and contribute for the increase in the standard of living of the family.

Guiding for credit was considered as one of the important indicator for effective extension. It is widely felt that present procedure for getting the credit from credit institutions is comparsom some and common men from rural area could not easily approach these institutions. They seek guidance for getting credit. Since these indicators deals with human resources, this group may be named as human resource development related indicators.

6. Group-VI

The principal component analysis further showed a group of indicators influencing the effectiveness of extension services. The name of the indicators and their variance levels are given in Table 6.

Table 6: Group-VI. Indicators contributing for effective extension services **(n=180)**

Sl. No.	Indicators	Variance
1.	Implementing location specific programmes	0.434
2.	Popularising diversified farming	0.433

It could be observed from the Table 6 that two indicators namely implementing location specific programme and popularising diversified farming were grouped under one group. The extent of contribution of these indicators was 0.434 and 0.433.

All new and improved technologies may not be suitable for adoption in all places. Some time it may not be profitable one in particular area due to some reasons. The respondents felt that implementation of location specific programme is essential.

In the context of changing agricultural scenario, effective utilisation of available resources for maximum output is also important in farming. Considering prevailing climatic condition and changing needs of the people, the beneficiaries of selected organisations need popularising of diversified farming. This might be the possible reason for selecting the implementing location specific programmes and popularising diversified farming indicators as contributing indicators for effective extension. The above two indicators may be named as Approach related indicators.

7. Group-VII

The effective indicators found under this group are enlisted in the Table 7.

Table 7: Group-VII. Indicators contributing for effective extension services **(n=180)**

Sl. No.	Indicators	Variance
1.	Popularisation of seed production	0.541
2.	Employing qualified field extension workers	0.474
3.	Have village level plans	0.402

Table 7 highlights that the indicators like popularising seed production, employing qualified field extension workers and having village level plans were also found as responsible indicators for effective extension service. The level of varience of the indicators was shown as 0.541,0.474 and 0.402 respectively.

Among the various inputs used, respondents would have felt that seed is the important inputs in deciding the performance of the crops. To avoid high cost involved for seed purchase and non-availability of adequate quantity of seed in time, respondents felt that production of seeds in the farmers field is useful for them.

Similarly employing qualified field extension workers would have been felt that it will increase the quality of extension service. Planning the programme at village level pave way for meeting the actual requirement of respondents. These might be the possible reasons for the respondents expressed these indicators as effective indicators. Since these indicators describe seed production and services, this group may be named as seed production and quality services related indicators.

8.Group-VIII

It is further observed from the analysis that two indicators related to extension methods were also contributing for effective extension service and they are listed in Table 8.

Table 8: Group-VIII. Indicators contributing for effective extension services **(n=180)**

Sl. No.	Indicators	Variance
1.	Conducting field days	0.522
2.	Use of radio	0.485

It could be concluded from the table that the respondents of selected extension organization for this study also considered conducting field days and use of radio as the effective indicators. The level of varience of these indicators was observed with 0.522 and 0.485 respectively., These two indicators were extension method oriented indicators, it may be named as extension methods related indicators

9. Group-IX

The analysis on the responses of public, private and non-governmental extension organisation beneficiaries were also shown a set of two contributing indicators for effective extension and they are detailed in Table 9.

Table 9: Group-IX. Indicators contributing for effective extension services **(n=180)**

Sl. No.	Indicators	Variance
1.	Establishing community computer centers	0.439
2.	Providing required infra-structural facilities	0.407

It is understood from the Table 9 that the indicators related such as establishing community computer centers and providing required infrastructure facilities to extension service for effective communication were observed as contributing indicators for effective extension service. These indicators were having the varience value of 0.439 and 0.407 respectively. Hence these infrastructures related indicators prefered by the respondents as effective indicators and it may be named as infrastructure related indicators.

10. Group-X

Interactive meeting and environment protection related indicators. The indicators contributing for effective extension under group X are presented in Table 10.

Table 10: Group-X. Indicators contributing for effective extension services **(n=180)**

Sl. No.	Indicators	Variance
1	Arranging scientist - farmers interactive meetings	0.432
2.	Having programmes for environmental protection	0.411

It could be observed from the Table 10 that viz., arranging scientist and farmers interactive meetings and having programmes for environmental protection with the varience level of 0.432 and 0.411 were also found as contributing indicators for effective extension service.

To reduce the time gap in technology adoption, the farmer's expects direct meet with the scientists. For this purpose regular scientist and farmers meetings is needed by the respondents.

Due to over use of chemicals and fertilizers, soil, water and environment got polluted and the yield is having a decreasing trend. Now there is growing awareness among the beneficiaries to protect environment. This might be the possible reason for the respondents agreed these two indicators as contributing indicators for effective extension service.

The analysis on the collective responses from the respondents of public, private and non-governmental organization gave 10 group of contributing indicators for effective extension services of any common organization involved in extension.

Extension Management Strategies for Sustainable Agriculture, pp 293-296

New India Publishing Agency, New Delhi, India

30

Mobile Service – An Effective Tool for Agricultural Development

Akoijam Christina and Shantha Govind

Today, in many countries mobile phones are being used by farmers, not only as a person voice communication medium but also to provide access to information through sending message service (SMS), multimedia messages service (MMS) and access to the internet. Across the developing world, there are programmes that give farmers access to research and best practices, weather information and market prices via SMS, Interactive Voice Response (IVR) or call centers.

The mobile phone has reduced the gap among traders and farmers and at the same time farmers directly communicate with buyers and customers to find the good price of the products. The uses of mobile phones among farmers have played positive impact in their income and productivity because before sale of produce they communicate with buyers and sell their product in good price.

Employment of mobile technology in the agriculture will improve support to farmers, improve compliance with sector standards, increase access to high value markets and in the long run revolutionize the agriculture sector.

However, deriving the full benefits of mobile service in agriculture will depend on factors such as the effectiveness of public-private sector collaboration in agriculture extension services, effective analysis of farmers needs and proper focusing of mobile agriculture services to meet the identified needs. Content for mobile agriculture services must be fit for purpose- timely, localized and of high quality to be relevant for farmers.

Mobile Phone and Farmers

The mobile phone has been a great help among traders and farmers and at the same time farmers directly communicate with buyers and customers to find the good price of their product. Farmers before going to market simply contact one of the buyers who purchase production in good price. In the rural areas of developing countries mobile phones spread day by day and different organizations have launched many projects for the agricultural development and increase the production by using different technologies in agriculture. Mobile phones have created a new business opportunities for the poor farmers and has given the access to information about market, health, and weather services in remote areas.

The uses of mobile phones among farmers have played positive impact in their income and productivity. Mobile phones significantly reduce communication and information costs. Poor communication facilities lead to limited access to information and this can lead to loss f income. Farmers have another advantage of mobile phones, it save time, money and energy at the same time. The use of mobile phones can help farmers in decision making by making the right information available at the right time

The one of the most important role of mobile phones in developing countries showed that farmers use the mobile phones for getting the information from different markets and weather information while other communicate with agriculture experts to obtain the information about the use of pesticides in their farm.

There are no doubt the mobile phones technologies were used in different sectors of the society such as health, education, rural development and in agriculture for the economic growth in developing countries.

Mobile Phone and Indian Agriculture

In countries like India, services such as IFFCO Kisan Sanchar Limited (IKSL) and Reuters Market Light (RML) have already reached hundreds of thousands of farmers in rural areas.

Today, the mobile application is one of India's largest market information services, serving hundreds of thousands of paying customers in tens of thousands of villages. Via SMS, it delivers highly personalized, professional information to India's farming community, covering more than 250 crops, 1000 markets, and 3000 weather locations across 13 states in 8 local languages (Mehra 2010).

The impact is likely even larger than Reuters can count due to the widespread sharing of information that takes place within informal farmers. These mobile services mainly provide information regarding weather, crop/animal husbandry advisory, market prices, fertilizer availability, insects, pests etc. Farmers are taking great advantages of these services and making the most use of it.

The proliferation of mobile phones across the nation impinged on agriculture in various ways. Mobile phones used to help raise farmers' incomes, making agricultural marketing more efficient, lowering information and providing platform to deliver services and innovate. Every aspects of the technology is changing rapidly; the public sector; private sector, and the private citizens are constantly experimenting with applications for it; and governments are grappling with any number of strategies to ease the digital divide.

Mobile Benifits

Mobile confers distinct advantages as a communication link in isolated circumstances. Mobile users or farmers can determine when and where they can communicate and access information. Farmers reported benefits from mobile phones as a means of two-way communication as well as a means of access to the information service while at fields.

Farmers reported that they are able to make and received calls while working on the farm. This included the ability to describe plant diseases from the field to experts and to co-ordinate better with their hired labour. For most of the farmers it is time and trained savings, more yield, higher productivity and improvement in their income. Traders and commission agents reported improvements from their ability to deal with truck breakdowns and the ability to shift en route in response to changing market conditions.

Donner (2009) distinguishes between different livelihood functions of mobile phones, including mediated agricultural extension, market information systems, virtual markets and financial services.

As a telephonic device, the mobile enables access to information sources that may not otherwise be reachable. As an information platform to receive sms, menu or voice – message information, mobile provides the ability to get connected to new knowledge and information sources not previously available with the possibility of real time, highly tailored information delivery.

Even at this early stage, mobile phones are being used in Indian agriculture and are starting to deliver agricultural productivity improvements, an impact that is enhanced by the new mobile enabled information services.

In India, realizing the full potential of mobile phones is limited, however, by a set of constraints that prevent farmers from fully leveraging the information they receive. The barriers apply more to small than to larger farmers; large farmers are more able to leverage the benefits of the communication and information they can access.

The constraints include shortcomings in physical infrastructure affecting access to markets, storage and irrigation. Issues regarding the availability of critical products and services including seeds, fertilizers, medicines and credit to small farmers also exist.

Conclusion

Mobile has emerged as the technology of choice of the masses - be it in urban areas or remote areas. Cutting across the entire spectrum of society, mobiles is being used for a variety of purposes – for information, education as well as entertainment.

Although, the use of mobile phones is still very much limited, it has huge potential in meeting the information needs of the farmers. To enhance the use of mobile phones by the farmers, especially those belonging to low income groups, there is need to subsidize the cost aspects related to mobile services. Further, there is need to develop a comprehensive information base catering to the specific and diverse needs of the farmers. The information provided through mobile phones must be in local languages.

References

Adamides, G. and Stylianou, A. (2013). ICT and mobile phone use for agricultural knowledge sharing by Cypriot farmers, *Agris on-line Papers in Economics and Informatics*, (5) 3-10.

Ansari, M. A. and Pandey, N. (2013). Assessing the potential need and use of mobile phones in agriculture, *Karnataka Journal of Agriculture Science*, 26(3):388-392.

Chhachha, A. R. (2014). Impact of information and communication technologies in agriculture development, *Journal of Basic and Applied Scientific Research*, 4(1):281-288.

Donner, J. (2009). Mobile based livelihood services in Africa: Pilots and Early Deployments. In Communication Technologies in Latin America and Africa: A multidisciplinary Perspective, Barcelona: IN3 37-58.

Kameswari, V. (2011). ICTs for agriculture extension: A study in the Indian Himalayan Region, *The Electronic Journal on Information Systems in Developing Countries*, 48(3):1-12.

Mehra, A. (2010). Small technologies fuel big results in the developing world, The Huffington post, September 13, 2010.

Mittal, S. (2010). Socio-economic impact of mobile phones on Indian agriculture, Working paper no. 246, Indian Council for Research on 'International Economic Relations'.

Extension Management Strategies for Sustainable Agriculture, pp 297-303

New India Publishing Agency, New Delhi, India

31

Effectiveness Index of Mobile SMS – An ICT Platform for Information Dissemination

Hinduja, N. A., N. Kishore Kumar and R. Prakash

The ICT enabled extension systems act as a key agent for changing current agrarian situation by improving access to information, sharing knowledge, providing incredible opportunities and enabling the empowerment of farming communities. In order to extend the benefits of ICT's to the disorganized and disadvantaged farming sector, one of the road maps to achieve them are access of ICT's to villagers through Mobile telephony. The Government of India and Kerala have given importance for Mobile messages to provide information whenever required enabling farming community to take appropriate decisions in their farming practices there by acting as a decision support system. Every Tuesdays and Fridays a voice message in Malayalam language followed by text message in English language are being pushed to the farmers and extension personnel about various agriculture related information such as package of practices, departmental services etc., through the project, "Content development for Agri Kiosk and networking through Mobile SMS". This project is initiated by Department of Agricultural Extension, College of Agriculture, Vellayani and Indian Institute of Information Technology and Management, Kerala, Trivandrum. Hence, the present study was taken up with an objective of developing an index for measuring the effectiveness of these Mobile SMS to improve its perception and utility to the farmers.

Research Methodology

The study was conducted in the Thiruvananthapuram district of South Kerala region during 2013-2014. Ex-post facto research design is employed. Sixty farmers were selected through random sampling method from Thiruvananthapuram district of Kerala as respondents.

In this study Effectiveness of Mobile SMS is operationalised as the degree to which Mobile SMS is successful in achieving the objectives and producing the desired result. Effectiveness index of Mobile SMS is developed taking into consideration the conversational ability, quick availability, ability to exploit a considerable amount of knowledge, reliability, scalability, preservation and improvement of knowledge. Scale used by Varghese (2012) with some modifications was used. Nine items related to effectiveness of Mobile SMS were administered to the subjects in a five point continuum namely most satisfied, more satisfied, satisfied, least satisfied and not satisfied.Summation of the score of all items gave the score of each respondent. Based on the scores the effectiveness index of Mobile SMS for each respondent was calculated using the formula.

$$\text{Effectiveness Index (EI)} = \frac{\text{Individual subject's score}}{\text{Total score}} \times 100$$

Utility of Mobile SMS is operationalized as the extent of utility of Mobile SMS as perceived by the farmers. The method used by Hassan (2008) was used for measuring this variable with some modifications. It is measured in terms of characteristics of messages of an ICT which are timeliness, reliability, relevancy, retrievability, adequacy, format clarity, practicability and simplicity in four point continuum for all information, most information, some information and no information. Summation of the score of all items gave the score of the respondents. Based on the scores the Utility Index was calculated using the formula.

$$\text{Utility Index (UI)} = \frac{\text{Individual subject's score}}{\text{Total score}} \times 100$$

Perception in this case was operationalized as the farmer's opinion towards the features of the messages of Mobile. The method used by Balasubramani (2010) with necessary modifications was used. Eight positive statements and two negative statements were administered to the respondents in a three point continuum namely, most satisfied, satisfied and not satisfied. Summation of the score of all items gave the score of the respondents. Based on the scores the Perception Index was calculated using the formula.

$$\text{Perception Index (PI)} = \frac{\text{Individual subject's score}}{\text{Total score}} \times 100$$

Findings and Discussion

The results are presented along with the inferences drawn in the light of the objective set forth for the study.

Utility of the Mobile SMS

The respondents are categorized according to the utility of Mobile SMS into low, medium and high utility. Distribution of farmers based on extent of Utility of messages is represented in Table 1 and Figure 1. Majority of the farmers reported medium Utility (50%) for the Mobile SMS followed by both high Utility (25%) and less Utility (25%) based on quartiles.

Table 1: Distribution of farmers based on extent of utility of messages

S. No	Utility of the message	Number of respondents	Percentage
1.	Low	15	25
2.	Medium	30	50
3.	High	15	25
	Quartile-1	166	
	Quartile-2	201.8	
	Quartile-3	237.6	
	Range	65-312.5	

The Utility of the messages was medium for Mobile SMS because the content provided was not specific to the location and crops grown by the farmers and the Mobile SMS were pushed in English language which was a major constraint in understanding the content of the message.

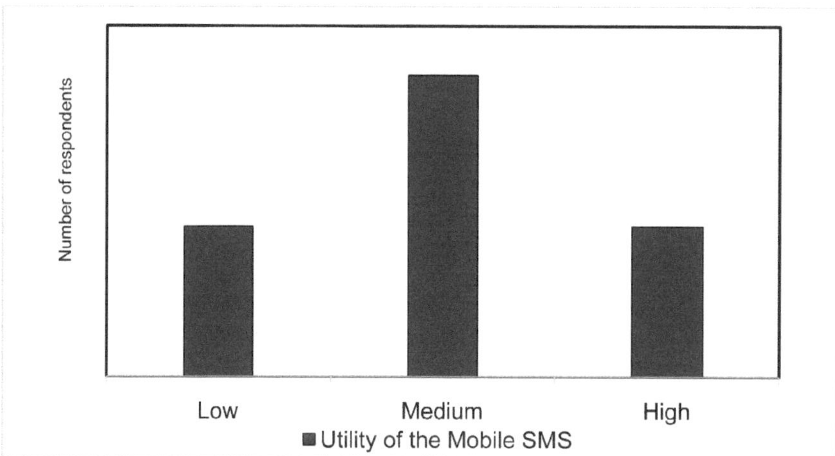

Fig. 1: Distribution of farmers based on extent of Utility of Mobile SMS

Perception towards Mobile SMS

Perception is the process of understanding sensation or attaching meaning based on experience to signs (Taneja, 1989). The respondents are categorized according to their perception towards Mobile SMS into low, medium and high perception category. The distribution of farmers for perception towards Mobile SMS is presented in Table 2 and Figure 2.

Table 2: Distribution of farmers according to their perception towards Mobile SMS

S. No	Extent of perception	Number of respondents	Percentage
1.	Low	14	40
2.	Medium	35	58.33
3.	High	11	18.33
	Quartile-1	150	
	Quartile-2	200	
	Quartile-3	233.3	
	Range	112.2-462.9	

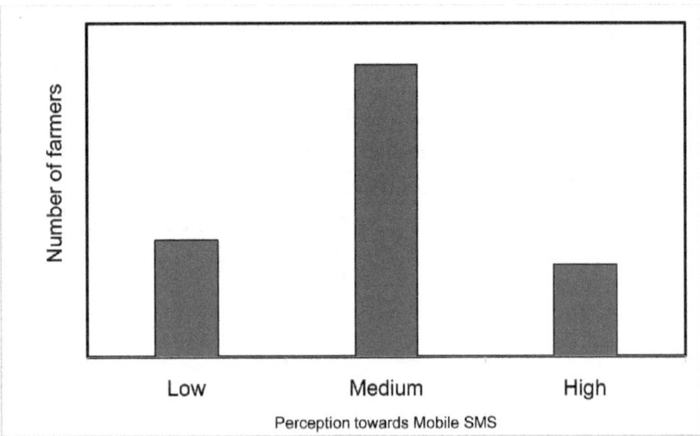

Fig. 2: Distribution of farmers based on extent of Perception towards Mobile SMS

It is clear from the Table 2 that majority of the farmers receiving Mobile SMS reported medium perception (58.33%) of the messages followed by low perception (40%) and high perception (18.33%) based on quartiles. Majority of the farmers in opined medium level of perception because majority of the attributes of the Mobile SMS were only opined to be satisfied by the farmers. Very few opined as most satisfied. This is due to the fact that many farmers contact extension personnel, Department officials in case of emergency and contingencies and age, education status also influenced the perception of the farmers.

Effectiveness Index of Mobile SMS

The respondents are categorized according to their perceived effectiveness towards Mobile SMS into low effectiveness, medium effectiveness and high effectiveness category. The frequency distribution of farmers for effectiveness of Mobile SMS is presented in Table 3 and Figure 3.

Table 3: Distribution of farmers according to their effectiveness of Mobile SMS

S. No	Effectiveness	Number of respondents	Percentage
1.	Low effectiveness	13	21.6
2.	Medium effectiveness	32	53.3
3.	High effectiveness	15	25
	Quartile-1	52.08	
	Quartile-2	65.97	
	Quartile-3	84.20	
	Range	24.3-154.3	

It is evident from the Table 3 that majority of the farmer's opined medium effectiveness (53.3%) of the messages followed by high effectiveness (25%) and less effectiveness (21.6%) for Mobile SMS based on quartiles. Medium effectiveness of the Mobile SMS is because most of the items were opined to be satisfied by majority of the respondents. Distribution of items according to their effectiveness for Mobile SMS according to farmers is represented in Table 4 and Figure 4.

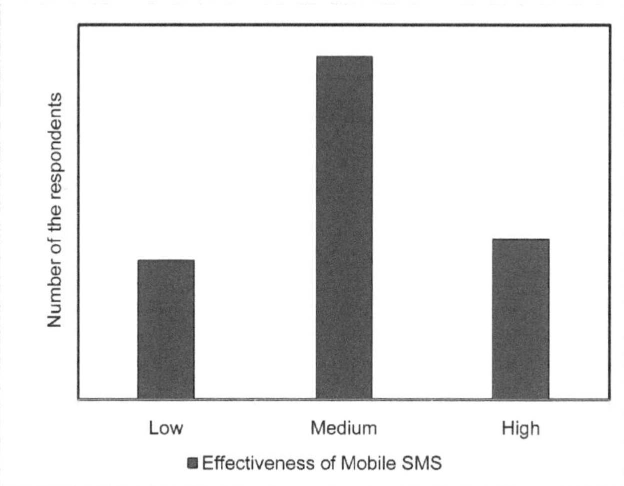

Fig. 3: Distribution of farmers based on Effectiveness of Mobile SMS

Table 4: Distribution of items according to their effectiveness for Mobile SMS

tems	Most satisfied (%)	More satisfied (%)	Satisfied (%)	Least satisfied (%)	Unsatisfied
Voice clarity	10	20	5.66	18.33	—
Reliability	15	31.66	51.66	1.66	—
Knowledge gain	20	28.33	48.33	3.33	—
Motivation to learn	23.33	33.33	41.66	1.66	—
Completeness of subjective matter	10	21.66	41.66	1.66	—
Readability of the message	30	21.66	53.33	13.33	3.33
Motivate to adopt the messages in farm	23.33	35	38.33	3.33	—
Information provided	15	30	40	15	—
Ability to exploit considerable amount of knowledge	18.33	25	41.66	13.33	1.66

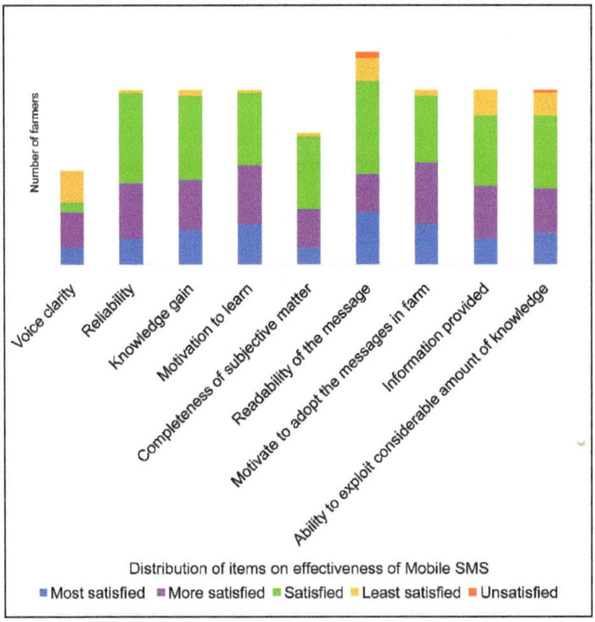

Fig. 4: Distribution of items according to their Effectiveness of Mobile SMS

A cursory glance through the pooled data for effectiveness of Mobile SMS depicts that almost all the items were observed to be satisfactory by the farmers. Voice clarity was opined by the farmers as only least satisfied due to speedy delivery of message. None of the items on variable effectiveness were opined as unsatisfied by the farmers which indicates that these Mobile SMS are effective in all aspects with some varying degree of effectiveness among the included components. Further improvement in more satisfied, satisfied, less satisfied categories increase the effectiveness of the messages given through the Mobile.

Correlation Between Effectiveness of Mobile SMS with Utility and Perception towards Mobile SMS

Correlation analysis was employed to assess relationbetween effectiveness of Mobile SMS with Utility and Perception towards Mobile SMS. The correlation coefficients were worked out and the significance was tested by comparing with the table values. It is represented in Table 5.

Table 5: Correlation between effectiveness of Mobile SMS with Utility and Perception towards Mobile SMS

Variable	Effectiveness index of Mobile SMS
Utility of Mobile SMS	0.500**
Perception towards Mobile SMS	0.514**

** Significant at 1% level

Effectiveness of Mobile SMS is positively correlated with Utility of Mobile SMS and Perception towards Mobile SMS at 1% significant level.Increase in Perception leads to more usage of the Mobile SMS and better understanding of the services which might increase the effectiveness of Mobile SMS. As Utility of the Mobile SMS increases, information requirements of the farmers are met which might influence the Effectiveness of this ICT tool.

Conclusion

In order to extend the benefits of this Mobile SMS and voice messages to the farming sector, effectiveness is an important dimension to be considered. Increase in Perception and Utility is the key principle to be focussed to improve the effectiveness of this Mobile SMS.

References

Balasubramani, R. (2010). An exploratory study on dissemination of dynamic market information through mobile telephony. M.Sc.(Ag.) thesis, TNAU, Coimbatore.

Hassan Nazreen, S. (2008). Effective agricultural information delivery system - As action research among farmers. Ph.D. Thesis, KAU, Thrissur.

Taneja, R.P. (1989). Dictionary of Education. Anmol publications, Ansari road, New Delhi.

Varghese, V. (2012). Effectiveness of selected ICT tools on cognitive and conative domain of pepper growers- An experimental study. M.Sc.(Ag.) thesis, TNAU, Coimbatore.

Extension Management Strategies for Sustainable Agriculture, pp 305-310

New India Publishing Agency, New Delhi, India

32

Role of ICT in Dissemination of Technologies in Agriculture

Gopika, M.H.

The transfer of technology keeps the wheels of agriculture development moving. The existing transfer of technology mechanisms and extension programs are either slow or ineffective in bridging the linkages between the research community and the farmers. Information Communication Technologies (ICT) in agriculture is an emerging field focusing on the enhancement of agricultural and rural development. The advancements in ICT can be utilized for providing accurate, timely, relevant information and services to the farmers, thereby facilitating an environment for more remunerative agriculture. However, all the ICT initiatives are not uniform with disparities between regions in the level and quality of telecommunications, information and the effort of individuals, public and private organizations and differentiated nature of demand of the farmers in different areas. As a result, there have been many successes, failures, lessons learned and experience gained, so far. While these initiatives are intended to address the needs of the farmers through ICT, their actual usage and their ability to bring significant impact on the farm productivity and socio-economic development of the intended beneficiaries is to be understood. ICT includes range of technologies like radio, television, telephone, multimedia, internet, video and tele-conferencing, lap-top computers etc. The introduction of Information Technologies as a strategic tool for agricultural development and welfare of rural farmers requires that the necessary IT infrastructure is in place. The rapid changes and downward trend in prices in various components of IT. Hence, it is relevant that the role of ICT in taking scientific knowledge/ technologies to the end users is very important. It involves application of innovative ways to use information and communication technologies in the rural development.

Meaning of ICT in Agriculture

ICT is defined as the gadgets and processes to access, retrieve, store, organize, manipulate, produce, present and exchange information by electronic and other automated means. It is also defined as the scientific, technological and engineering disciplines and the management technologies used in the handling of information, processing and application related to computers. ICT in agricultural extension and rural development is significant as it provide a medium to adequate access to agricultural information, despite the persisting. Specifically in extension, it is considered as a tool which enhances flow of information in application of agricultural extension services, by providing accurate, timely, relevant information and services to the farmers, thereby facilitating an environment for more remunerative agriculture.

Characteristic of ICT

1. Affordable
2. Scalable
3. Self-sustaining
4. Sensible
5. Participatory
6. Appropriate.

Important Roles of ICT

In the context of agriculture the potential of information technology (IT) can be assessed broadly as (a) Tool for direct contribution to agricultural productivity and (b) As an indirect tool for empowering farmers to take informed and quality decisions which will have positive impact on the way agriculture and allied activities are conducted.

The advantage in using new ICT is primarily in reducing time and distance in accessing information.

Enhancing Agricultural Production: By providing up to date information about pest, disease and erosion control, early warning systems, new varieties, new ways to optimize production and regulation for quality control.

Improving Market Access: By providing up to date information on market prices of commodities, inputs and consumer trends.

Capacity building and empowerment: Strengthen communities and farmer organizations their own capacities and better represent their constituencies when negotiating input and output prices, land claims, resource rights and infrastructure projects.

Application of ICT in Agriculture Development

The application of Information and Communication Technology (ICT) in agriculture is increasingly important. The main phases of the agriculture industry are crop cultivation, water management, fertilizer application, pest management, harvesting, post harvest handling, transporting of food products, packaging, food preservation, food processing, food quality management, food safety, food storage and food marketing. The following application of ICT in agriculture can be used for providing accurate, timely, relevant information and services to farmers on the above main activities of agriculture like providing online technical, market and weather related information, global information system, Expert systems, Question answer service, Tele education, E-governance, Creation of databases and Rural networking.

Problems in Adoption of ICT in Rural area

The common problems in adoption of ICT in rural segments are ICT illiteracy, availability of relevant and localized contents in their own languages, connectivity, inadequate community participation, poverty, poor infrastructure, easy and affordable accessibility and other issues such as awareness and willingness for adoption of new technologies among the rural peoples etc. One critical aspect in the usage of ICT's for farmers and their groups as seen in some of ICT driven initiatives is the involvement of human interface at the last mile indicating that there is a human dependency in transmission of Information/ Knowledge.

Some Successful Cases of ICT in Agriculture

ICT can bridge the gap between the farmers and researchers to make research more responsive to the actual needs. The potential of ICT to contribute to agriculture and rural development has been well recognized. The radio played a key role in technology transfer. The internet and associated computer technologies are expected to play a similar role in information and knowledge sharing in the ongoing transformation of agriculture. These new ICT will connect farming and rural communities to global information sources and markets. They will enable individuals, households and communities to learn and acquire new skills and technologies and also share innovations globally.

1. ITCs e-choupal

The Indian Tobacco Company (ITC) began e-choupal in June 2000 at Ujjain, Madhya Pradesh as a cost effective means of dealing directly with farmers to buy agricultural products for export. Choupal means village gathering place in Hindi. The system is now becoming a mega market for rural India. Farmers can strike orders with ITC directly through Internet Kiosks, the ITC and achieve savings by bypassing the intermediaries and the local entrepreneurs gain commissions on the transactions made through their Kiosks. The kiosks can also be used for companies to sell products and services directly to farmers and train the farmers on how to use them. The system is emerging as a one shop for selling and buying of range of products and services including government services in rural market. Within 2 years after its launch, e-chopal services reached 600,000 farmers in 6000 villages through 1000 kiosks. This project is under the following states such as Madhya Pradesh, Utter Pradesh, Andhra Pradesh, Karnataka, Haryana, Uttarakhand, Rajasthan, Maharastra, Kerala and Tamil Nadu.

Key factors for success

1. ITC has redefined the role of the local intermediaries and the procurement of facilitation and information gathering.
2. It has chosen to strengthen an existing situation in rural areas.
3. It is a participatory method.
4. It faces no significant competition from business perspective.

2. Expert System Service

An expert system is software that attempts to provide an answer to a problem, or clarify uncertainties where normally one or more human experts would need to be consulted.

Objectives of developing expert system

1. To enhance the performance of agricultural extension personnel and farmer.
2. To make farming more efficient and profitable.
3. To reduce the time required in solving the problems.
4. To help in performing the routine tasks thus leaving expert for other important task.
5. To maintain the expert system by continuously upgrading the database.

Potential advantages of expert system

1. Solves critical problems by making logical deductions without taking much time.
2. It combines experimental and conventional knowledge with the reasoning skills of specialists.
3. To enhance the performance of average worker to the level of an expert.

Application of expert system in agriculture

a. Crop production estimates.

b. Crop selection.

c. Soil management.

d. Plant diseases and pests management.

e. Weed management.

3. Kisan Call Center Service (KCCs)

Kisan call centre was launched on 21st January 2004 across the country. It aims at providing technical advice to the issues raised by farmers instantly in local language. The KCC is synthesis of two hither to separate technologies namely the Information and Communication Technology and the Agriculture Technology. To optimally utilize the strength of both these systems, it was proposed to take full advantages of professionally knowledge of agricultural scientists and extension officers, so as to facilitate its reach to the farming community.

Needs for KCCs

1. To overcome information asymmetry.
2. To overcome the constraints of distance and time in providing extension services to farmers.
3. To leverage extensive telecom infrastructure rolled out to deliver extension services to the farming community.

Objectives of KCCs

1. To disseminate the knowledge as well as to solve the everyday problems in the farmer's own language.

2. To create a database which collects the actual location specific need of farmers.

3. To provide an opportunity to students and agricultural scientists to play a dynamic role in the transformation of Indian agriculture to a sustainable one.

4. To establish a network of relationship among the scientists/experts, policy makers, extension workers, farmers and other stakeholders.

Conclusion

Information Technology is the major means of meeting the needs of the requirements of the present society. The efforts should be made to incorporate ICT in all endeavours related to Agricultural development. The organizations and departments concerned with agricultural development need to realize the potential of ICT for the speedy dissemination of information to farmers, timely and quality information inputs for decision making, in improving the efficiency and productivity of agriculture and allied activities.

References

Anonymous (2008). ICTs for Sustainable Food Production and Agriculture, www.infodev.org/en/Document.660.pdf

Anonymous (2008). Why do People use ICT in Agriculture? An Overview. Htt://www.e-agriculture.org

G. Perumal (2004). The Impact of Information Technology in Indian Agriculture, *Asian J. of Extn.Edn.* XXIII: 122-128.

http://www.e-choupal.com

http://www.kisancallcentre.net

Extension Management Strategies for Sustainable Agriculture, pp 311-316

New India Publishing Agency, New Delhi, India

33

Constituents of Contribution of Leaders in SHG Performance

T. Rathakrishnan and S.R. Padma

In India, self employment has been recognized as an essential force of development in rural areas. It has emerged as a strategy designed to improve the socio-economic life and mainly focuses on extending the benefits of development to the poorest in the rural areas improving their standard of living and self-realization. SHGs in India are integrating the low income segments with rest of the rural community by ensuring them a better contribution in a more equitable share in the benefit of developments. These Groups are not only speeding up economic growth, but also providing jobs and improving the quality of rural life towards self-reliance. Self-employment needs a very wide ranging and comprehensive set of performance, relevant to all aspects of rural economy and covering rural people including skilled, unskilled and landless labours and artisans of Rural India. The Rural Indians put their entrepreneurial skills in all the rural development performance but their economic status has not improved to the expected level. They have much potential, despite they are ignorant of converting their skills into reality.

The formation of common-interest groups consisting primarily of women has a substantial impact on their lives. The impact of SHGs on women's empowerment and social security has been invariably an improvement from the status quo but there is a need for support in several areas which are analyzed in this report. The status of women has generally improved as they have developed stronger confidence which has changed gender dynamics and their role in the household.

In south India, significant improvements in fertility rates, female literacy, contribution in development programmes and economic independence are evident. Women are able to fight for their rights and entitlements and have emerged as a force to be reckoned with. Further, SHGs are becoming more than just financial intermediaries, instead they have emerged into a more political and social unit of society. More importantly, the penetration of microfinance to the poorest of the poor is still weak and needs a wider reach.

Methodology

The study was conducted among the SHG leaders distributed throughout the Omalur taluk of Salem District. Omalur Taluk consists of three blocks namely, Omalur block, Kadayampatti block and Thirumangalam block. As more number of SHGs have been under operations in these three blocks, all the three blocks were selected. The selected respondents were registered and received their identity cards from the NGO, Salem.

To assess the determinants of contribution of leaders in SHG performance, the study was undertaken with One Twenty SHG leaders of the Omalur Taluk.

Findings and Discussions

a. Motivating factors for joining in SHGs

In order to unearth the possible factors that influenced the respondents to join in SHGs, a set of twelve factors were identified, studied and presented in the Table 1. The sample leaders were requested to rank the factors based on their perceived level of importance in influencing them to join in SHGs.

Table 1: Motivating Factors for Joining in SHGs

Motivating Factors	Percentage	Rank
Economic necessity	55.71%	I
For availing the financial assistance	45.68%	II
Improvement in the social status	44.52%	III
Unemployment	41.70%	IV
Motivation by friends	38.75%	V
Avail the trainings offered by training Institutes	38.12%	VI
To gain additional knowledge	37.77%	VII
To acquire more skills and specialisation	33.59%	VIII
For giving employment opportunities	27.49%	IX
For village development	20.63%	X
Compulsion of family members	16.76%	XI
To spend Leisure time	13.31%	XII

Table 1, revealed that Economic necessity (55.71%), availing the financial assistance (45.68 %) to improve the social status (44.52%) were the primary factors motivated them to involve in SHGs. The result has been in line with the result of Arularasan (1992) who indicated that economic necessity of availing financial assistance from the banks was the important motive behind the sample respondents who joined the SHGs.

Unemployment (41.70%), motivated by friends (38.75 %) and to avail the trainings offered by training Institutes (38.12%), to gain additional knowledge (37.77%) were also influenced the respondents to join in SHG. In the current social climate, women's survival is very difficult without adequate knowledge on institutions and their surroundings. Particularly the poor rural women were suffered a lot. On this areas as the respondents had low level of education and hence, they could not get adequate knowledge. Therefore the respondents would have been preferred to join in SHGs to get required knowledge and awareness for doing their jobs.

And also to acquire more skills and specialisation (33.59%), for giving employment opportunities to the villagers (27.49%) the farm women are willingly involve themselves in SHGs. The study results of Ponmani (1993) also supported that 30.00% and 25.00% of the respondents joined in the SHG due to their unemployment and or giving employment opportunities to others respectively.

About one-fifth also preferred to join in SHGs with the conscious of village development. They were found to have the desire of getting the government schemes to their villages concerned for the development of their villages.

b. Self confidence

The respondents self confidence level was assessed through five point continuum on a set of five statements and they were categorised as low, medium and high.

Table 2: Distribution of respondents according to their Self—Confidence

Sl.No	Category	Percent
1.	Low level	15.80%
2.	Moderate level	62.50%

It is indicated that 62.50% of the respondents possessed a moderate level of self confidence followed by high and low level of self confidence (21.70% &15.80% respectively).

Self confidence is essential for SHG leaders to perform their assigned roles viz.,to motivate the members towards collective thinking and action , convinced the members for conducting meeting at regular intervals and create awareness of present social position. This led to conclude that the SHG leaders possessed moderate to high level of self confidence.

c. Level of aspiration

The level of aspiration influences ones action in a group and his or her decision making pattern. Hence it was assessed and the results have been presented.

Table 3: Distribution of Respondents according to their level of aspiration

Sl .No	Category	Percent
1.	Low level	11.70%
2.	Moderate level	75.80%
3.	High level	12.50%

As observed from Table 3 that majority of the SHG leaders (75.80%) had moderate level of aspiration, followed by high level of aspiration (12.50%) and 11.70% of the respondents had low level of aspiration. SHG leaders were aspiring for government job for their children. The SHG leaders if they have aspiration naturally must perform better in their performance through which getting established with other sources of contact. Thus, their desire aroused to put their children in government jobs.

d. Decision making pattern

In any Institution /organisation decision making is an inevitable component required for a leader/manager, SHG is not an exception as the groups performance and performances are totally influenced by the leader. The decision making pattern of the sample was studied and the responses have been incorporated.

From the Table. 4 it could be observed that majority of the SHG leaders (58.33 per cent) had medium level of decision-making pattern, followed by low level of decision-making pattern and 10.00 per cent of the SHG leaders had high level of decision - making pattern.

Generally joint decision with husband (45.83 percent) was taken for performance such as accepting technologies ahead of others and contribution in training programmes and development performance (66.66 percent). Organizing community work in the village, organizing self help groups and arbitrating disputes involves group action by the members, NGO's and relatives, so the SHG leaders sought for joint decisions for above said performance. In general, majority of the SHG leaders preferred joint decisions.

This reason might be due to the performance in SHG involved group action and team spirit of SHG members. So majority of the SHG leaders took joint decision with their group members.

Table 4: Distribution of respondents according to their Decision making pattern

S.No	Performance	Type of Decision									
				Joint Decision							
		Independent Decision		Joint decision with their family members or husband		Joint decision with SHG members		Joint decision with NGO officials		Joint decision with friends and relatives.	
		No	%	No	%	No	%	No	%	No	%
1.	Accepting Technologies ahead of others	5	4.16	115	95.83	-	-	-	-	-	-
2.	Organizing Community work		_	20	16.66	60	50.00	30	25	10	8.33
3.	Organising SHGs	-	-	-	-	100	83.33	20	16.66	-	-
4.	Arbitrating disputes					110	91.66	10	8.33		
5.	Contribution in training programmes and development performance	5	4.16	80	66.66	-	-	30	25	5	4.16

Conclusion

Rural women in India need to be both economically and socially independent in the existing economic scenario. It is attainable through Group approach which is a viable setup to disburse credit to the rural women to empower them in rural areas. Self Help Groups provide the benefits of economies of scale, cost effective alternative for different financial services, collective learning, democratic and participatory culture, a firm base and plat form for dialogue and cooperation. In India, Self Help Groups represent a unique approach to financial intermediation. The approach combines access to low-cost financial services with a process of self management and development programmes for the women who are SHG members. SHGs are formed and supported usually by NGOs or by Government agencies. Linked not only to banks but also to wider development programmes. SHGs are seen to confer many benefits, both economic and social. SHGs enable women to grow their savings and to access the credit which banks are increasingly willing to lend. SHGs can also be community platforms from which women become active in village affairs, stand for local election to take action to address social or community issues.

In order to motivate the leaders and members to increase their self confidence, NGOs can play a effective role to explain the benefits obtained by other leaders through success stories. Majority of the SHG members preferred training on different subject areas. The SHG members were more interested in income generating enterprises like poultry, farming, animal husbandry, mushroom cultivation etc, The enterprise could generate self employment opportunities and provide incremental income. The leaders might have understood the rationality of each of the enterprises to boost the family income and thereby to upgrade the standard of living. The uninterested group can be motivated by NGOs using suitable extension techniques like projecting success stories, training cum demonstration etc.

References

Arularasan, G.S. (1992). Expost-facto analysis of self-employment training on vocational settlement pattern of rural youth. Unpub. M. Sc., (Ag.) Thesis, AC&RI, TNAU, Coimabtore.

Ponmani, M. (1993). Rural women's contribution in the training offered by government and voluntary agency, Unpub. M.Sc.,(Ag.) Thesis, AC&RI, TNAU, Madurai.

Extension Management Strategies for Sustainable Agriculture, pp 317-328
© 2016, Editors, T. Rathakrishnan, H. Philip, P.P. Murgan and Jagadeesan
New India Publishing Agency, New Delhi, India

34

e-SAP ICT Dissemination Tool in NEK Region of Karnataka – Empirical Analysis

G.N. Maraddi, A. Prabhuraj, G.M. Hiremath and Sathish

e-SAP is one of the important emerging and efficient dissemination tool for agricultural pest management. e-SAP refers to electronic solution for agricultural pest management. e-SAP is a ground – breaking application built on the backdrop of agricultural pest management. Its foundation lies on a unique platform that facilitates the storage access and transfer information between all stakeholders in agriculture. It enables rapid and delivery of farm data in various forms including multimedia to researchers, policymakers and other users in real time. e-SAP has been successfully built & tested and deployed by UAS Raichur in association with Tene Agriculture Solution Pvt.Ltd, Bangalore. e-SAP is novel ICT platform and dedicated field device capable of making two way information exchange in real time is available today. With huge potential for revolutionizing transfer of technology system, it has opened doorway for the future particularly in India & world in general. The potential is such that any new agricultural technology can be communicated in an extremely effective manner in real time directly to the field, for instance, if new pest management strategy has to be disseminated to many field workers spread across a vast geography, a press the button in the computer lab all it takes to deliver it instantaneously to all of them, such information can be assessed anytime, anywhere, irrespective of connectivity. Simultaneously, field situations across the space & time are instantaneously made known to the decision maker.

Materials and Methods

The research study was conducted in northeastern region of Karnataka (NEK) to assess the perception of the stake holders of e-SAP dissemination tool. The ex-post-facto research design was used for the study with 384 stakeholders during 2014. The random sampling procedure is being used in six northeastern districts to draw optimum sample size for the study. A scale was developed to measure the perception of respondents by following the standard procedure for both farmers & field facilitator. Around ten independent variables in addition to one dependent variable is being used for the research study. The standardized schedule used to investigate the information from the stakeholders of e-SAP. Sphersmans simple correlation test was used to know the relationship between selected variables.

Results and Discussion

Overall perception of farmers towards e-SAP in NEK region

It was observed from the Table 1 that,overall perception of farmers towards e-SAP. The trend is relatively more members of farmers belongs to high category in Raichur, Yadagir, Bellary and others except Gulbarga which has equal number in all categories of perception. Totally around 37 per cent (36.98 %) of the respondents belongs to medium perception (32.55 %) and only 30.47 percent of the respondents belong to low perception category. The results are in line with the results obtained by Sankri (2013), who has conducted a study on perception of the extension officials towards TNAU Agritech portal and reported that more than half (61.67 %) of the respondents have medium level of perception followed by nearly one forth (23.33 %) of the respondents in low perception category. Similar results were also reported by Hayrol and Bahaman (2009) where in overall, ICT were perceived moderate to high in term of their importance for the development of their village among the village development committee members.

Statement-Wise mean perception score of farmers towards e-SAP in NEK region

A cursory look at the Table 2 indicated that, statement of perception men score about e-sap in NEK region, that it makes pest disease identification and diagnosis easier, timely information injudicious use of pesticides by farmers, provides opportunities for interaction with experts, reduces the cost of pest and diseases management through relevant solution with the mean score of 2.48. Whereas e-SAP is credible in nature, easy to understand and adopt, increase yield raised with the mean score less than above mentioned statements. It seems maximum score 3, out of three, more than two shows the favourable perception towards e-SAP. A outset, majority stands are being used to the inputs and got encouraging results.

Personal, Socio-economic and Psychological characteristics of the respondents

From Table 3, it could be inferred that in total nearly half (48.70%) of the respondents belonged to be middle age category followed by young (35.94 %) and old (15.36 %) aged categories. The probable reasons for majority of the respondents to be in middle age categorymight be that, usually farmers of middle age are enthusiastic and having moderate experience in farming have more work efficiency than older and younger ones. The results are in conformity with the findings of Suresh (2004). It is evident from the Table 3, one third (30.99 %) of the respondents were illiterates. 24.48 per cent of them were educated up to high school followed by middle school (17.45 %) and PUC (13.54 %). The probable reasons for the majority of respondents to be illiterates might be poorer economic condition, non realization of importance of formal education and also ignorance and non availability of educational facilities at village level. As per the results of Table 3, majority (38.54 %) of the respondents were medium farmers. Nearly one third (30.21 %) of them were small farmers followed by big (25.52 %) and marginal (5.73 %) farmers category. The probable reasons may be transfer of ancestral lands from generation to generation. It was observed from the (Table 3) that, majority (45.31%) of the respondents were in high income group and 27.86 per cent of them were in low income group followed by medium (14.06 %) and semi medium (12.76 %) income categories respectively. The probable reasons, which could be attributed for varied income level of the respondents, might be due to the size of land holding, price fluctuations in the market and practicing other occupations. With regards to farming experience, 41.41 per cent of the respondents belonged to low farming experience category followed by medium (33.85 %) and high (24.74 %) farming experience categories respectively. The above results could not be attributed to the fact that majority of the respondents were middle aged with medium land holdings. The above findings are in line with the findings of Vijaykumar and Narayanagowda (1999). It was observed from the (Table 3) that, majority (45.57 %) of the respondents had medium achievement motivation followed by high (28.13 %) and low (26.30 %) achievement motivation. Achievement motivation is a psychological variable. It differs from individual to individual. It is assumed that achievement motivation forces the individual towards reaching some goals, which he set for himself. Higher the association with the individual higher will be his efforts. This can be attributed to the social status of a respondent feels to keep by achieving greater goals. The findings are the in accordance with the studies conducted by Vijaykumar (2001) and Nagesh (2005). With regard to risk orientation it was observed from (Table 3) that, majority of the respondents (43.23%) had medium risk orientation followed by high (29.17 %) and low

(27.60 %) risk orientation respectively. It depends up on personal, psychological and socio-economic characteristics. The individual with more experience, better risk taking ability. All these are the possible reasons because of which majority of them possess medium risk orientation. It was observed from the Table 3 that,majority (44.53 %) of the respondents belonged to medium management orientation followed by low (28.91 %) and high (26.56 %) management orientation category respectively. The probable reasons might be due medium level of achievement motivation and risk taking ability of the respondents.It was observed from the (Table 3) that, majority (47.14 %) of the respondents belonged to medium innovative proneness category followed by low (27.08 %) and high (25.78 %) categories respectivcly. The medium innovativeness of the respondents might be due to their marginal size of the land holding and majority were middle aged which had restricted them to go always for new things. Apart from this their low level of the extension contact, medium level of extension participation contributed to their medium level of innovativeness. Similar results were also reported by Nagesh *et al.* (2011). It was observed from the (Table 3) that, majority (46.88 %) of the respondents belonged to medium level of information management followed by high (28.39 %) and low (24.74 %) information management categories respectively.

Table 1: Overall perception of farmers towards e-SAP

Categories Districts	Low		Medium		High	
	Freq	%	Freq	%	Freq	%
Raichur (n1=97)	36	37.11	10	10.31	51	52.58
Gulbarga (n2=126)	42	33.33	42	33.33	42	33.33
Yadgir (n3=82)	30	36.59	13	15.85	39	47.56
Bellary & others (n4=79)	25	31.65	22	27.85	32	40.51
Total (N=384)	**117**	**30.47**	**142**	**36.98**	**125**	**32.55**

Table 2: Statement wise perception mean score about e-SAP

Sl. No	Statements	Mean score				
		Raichur	Gulbarga	Yadgir	Bellary & others	Total
1.	e-SAP makes the pest and disease identification and diagnosis easier	2.29	2.59	2.77	2.75	2.58
2.	e-SAP provides latest and updated information in pest management	2.04	2.50	2.00	2.39	2.26
3.	When new pest or disease is observed it is difficult to get timely solution through e-SAP	1.93	2.14	1.66	2.03	1.96
4.	e-SAP provides timely information for the pest or disease identified	2.28	2.56	2.68	2.59	2.52
5.	e-SAP provides reliable and specific information on pest and disease management	2.41	2.32	2.59	2.57	2.45
6.	e-SAP is user friendly as it contains multimedia content	2.23	2.26	2.61	2.63	2.40
7.	e-SAP reduces the injudicious use of pesticides by farmers	2.27	2.73	2.50	2.52	2.52
8.	e-SAP provides opportunities for interaction with experts	2.33	2.32	2.62	2.72	2.47
9.	e-SAP provides opportunity for expressing opinion of farmers	1.90	2.29	2.22	2.52	2.22
10.	e-SAP provides information in local language and recommendations from e-SAP are easy to understand and adopt	2.34	2.48	2.59	2.67	2.51
11.	It reduces the cost of pest and disease management through relevant solution	2.22	2.60	2.51	2.58	2.48
12.	e-SAP enhances the rapid dissemination of pest management technologies	2.31	2.40	2.70	2.63	2.49
13.	Information available from e-SAP is credible in nature	2.08	2.29	2.48	2.41	2.30
14.	Information on pest and disease can be accessed from anywhere and anytime	2.26	2.33	2.59	2.67	2.43
15.	e-SAP should be extended to all parts of the state	2.16	2.35	2.66	2.67	2.43
16.	e-SAP helps in increasing the yield level of identified crops	2.13	2.58	2.41	2.43	2.40
17.	e-SAP helps to reduce the extent of damage by the identified pest	2.22	2.59	2.59	2.65	2.51

Overall Perception of Farm facilitator towards e-SAP in NEK region

It was observed from the Table 4 that, overall perception of Farm Facilitator towards e-SAP in NEK region of Karnataka state. About half of Farm Facilitator belongs to high perception (>45.95–range interval) category followed by low perception category (31.58 %) with range of <42.04 and only 21.05 percent of Farm Facilitator were identified in medium category, range interval of 42.04 to 45.59. This trend is due to rapid and speed availability of information, accuracy in synthesis with multimedia data capturing capacity in addition to web based application, which ensures the data generation from field to disseminate through field devices back to the respondents. Further it will support decision support system of stake holders in agriculture system.The results are in line with the results obtained by Sankri (2013), who has conducted a study on perception of the extension officials towards TNAU agritech portal and reported that more than half (61.67 %) of the respondents have medium level of perception followed by nearly one forth (23.33 %) of the respondents in low perception category.

Statement-Wise Mean Perception score of Farm facilitator towards e-SAP in NEK region

A access look at Table-5 that statement wise perception score of farm facilitator revealed that, usage of e-SAP helps the farmers to reduce the cost of pest management with the mean score of 2.63 followed by access any information regarding pest and disease management at any time and any where, and e-SAP helps in identification of pest in addition timely decisions with regard to pest management with th mean score of 2.63 followed by e-SAP influences farmers to adopt, latest pest and disease management practices, provides opportunity to farmers for interaction with experts and it is easy to complete work in short time in addition to it act as ready reckoner for advising farmers with the mean score of 2.53. Whereas immediate solution to the identified pest (2.00), latest information of pest management, in addition e-SAP helps in effective and efficient job performance (mean score of 2.47). The results gets partial support from the results obtained by Singh *et al.* (2012) in their study reviled that majority of the e-farmers (91.82 %) perceived it as a better source of timely and quick information, followed by source of strong linkage with KVK (81.23 %), need based information (77.13 %), easy to understand the message (77.00 %).

Table 3: Personal, Socio-economic and Psychological characteristics of the Farmers

Sl. No	Particulars	Categories	Raichur (n1=97)		Gulbarga (n2=126)		Yadgir (n3=82)		Bellary & others (n4=79)		Total (N=384)	
			Freq	%	Freq	%	Freq	%	Freq	%	Freq	%
1	**Age (years)**											
	Young	18-30	36	37.11	41	32.54	35	42.68	26	32.91	138	35.94
	Middle	31-50	46	47.42	54	42.86	41	50.00	47	59.49	187	48.70
	Old	>50	15	15.46	31	24.60	6	7.32	6	7.59	59	15.36
	Mean	37.53										
	SD	11.05										
2	**Education**											
	Illiterate	0	24	24.74	54	42.86	23	28.05	18	22.78	119	30.99
	Primary	1 to 4th	3	3.09	5	3.97	4	4.88	7	8.86	19	4.95
	Middle school	5-7th	13	13.40	20	15.87	22	26.83	12	15.19	67	17.45
	High school	8-10th	34	35.05	22	17.46	13	15.85	19	24.05	94	24.48
	PUC	11 and 12	9	9.28	16	12.70	1	1.22	14	17.72	52	13.54
	Degree and above	Graduation and above	14	14.43	9	7.14	23	28.05	9	11.39	33	8.59
	Mean	6.55										
	SD	5.28										
3	**Land Holding (acres)**											
	Marginal	up to 2.50	1	1.03	8	6.35	6	7.32	7	8.86	22	5.73
	Small	2.51 to 5.00	28	28.87	39	30.95	29	35.37	20	25.32	116	30.21
	Medium	5.01 to 10.00	46	47.42	40	31.75	28	34.15	34	43.04	148	38.54
	Big	Above 10.00	22	22.68	39	30.95	19	23.17	18	22.78	98	25.52
	Mean	9.61										
	SD	7.62										

(Contd.)

(Contd. Table)

4 **Annual Income (Rs.)**											
High	Above 51,000	41	42.27	77	61.11	39	47.56	17	21.52	174	45.31
Medium	34,001 to 51,000	13	13.40	21	16.67	14	17.07	6	7.59	54	14.06
Semi medium	17,001 to 34,000	7	7.22	19	15.08	5	6.10	18	22.78	49	12.76
Low	Below 17,000	36	37.11	9	7.14	24	29.27	38	48.10	107	27.86
Mean	81796.88										
SD	109130.4										
5 **Farming experience (years)**											
Low	Mean-0.425*SD	34	35.05	59	46.83	35	42.68	28	35.44	159	41.41
Medium	Mean+0.425*SD	35	36.08	29	23.02	24	29.27	30	37.97	130	33.85
High	Mean+0.425*SD	28	28.87	38	30.16	23	28.05	21	26.58	95	24.74
Mean	16.17										
SD	10.39										
6 **Achievement motivation**											
Low	Mean-0.425*SD	29	29.90	21	16.67	24	29.27	27	34.18	101	26.30
Medium	Mean+0.425*SD	43	44.33	62	49.21	37	45.12	33	41.77	175	45.57
High	Mean+0.425*SD	25	25.77	43	34.13	21	25.61	19	24.05	108	28.13
7 **Risk orientation**											
Low	Mean-0.425*SD	23	23.71	28	22.22	27	32.93	28	35.44	106	27.60
Medium	Mean+0.425*SD	41	42.27	59	46.83	36	43.90	30	37.97	166	43.23
High	Mean+0.425*SD	33	34.02	39	30.95	19	23.17	21	26.58	112	29.17
Management orientation											
Low	Mean-0.425*SD	31	31.96	29	23.02	25	30.49	26	32.91	111	28.91
Medium	Mean+0.425*SD	39	40.21	61	48.41	38	46.34	33	41.77	171	44.53
High	Mean+0.425*SD	27	27.84	36	28.57	19	23.17	20	25.32	102	26.56

(Contd.)

(Contd. Table)

Innovative proneness											
Low	Mean-0.425*SD	29	29.90	26	20.63	25	30.49	24	30.38	104	27.08
Medium	Mean+0.425*SD	41	42.27	69	54.76	38	46.34	33	41.77	181	47.14
High	Mean+0.425*SD	27	27.84	31	24.60	19	23.17	22	27.85	99	25.78
Information management											
Low	Mean-0.425*SD	27	27.84	30	23.81	17	20.73	21	26.58	95	24.74
Medium	Mean+0.425*SD	41	42.27	61	48.41	41	50.00	37	46.84	180	46.88
High	Mean+0.425*SD	29	29.90	35	27.78	24	29.27	21	26.58	109	28.39

Table 4: Overall perception of farm facilitators towards e-SAP

Sl. No.	Categories	Frequency	Percentage
1.	Low	6	31.58
2.	Medium	4	21.05
3.	High	9	47.37

Table 5: Statement wise perception mean score of farm facilitators about e-SAP

Sl. No	Statements	Mean score
1.	e-SAP influences the farmers to adopt latest pest and disease management practices	2.53
2.	Using the e-SAP helps the farmers to reduce the cost of pest management	2.63
3.	e-SAP improves the productivity of various crops	2.37
4.	Operating e-SAP device is difficult	1.47
5.	It takes lot of time to get information on particular pest or disease	2.47
6.	e-SAP provides latest information on pest management to the farmers	2.32
7.	e-SAP provides opportunity to farmers for interaction with experts	2.53
8.	Farmers can access any information regarding pest and disease management at any time and from anywhere	2.63
9.	e-SAP provides information on pest and diseases and their management only	2.63
10.	e-SAP helps in rapid dissemination of latest technologies in pest management	2.42
11.	e-SAP helps in easy identification of pests	2.63
12.	e-SAP helps the farmers in making timely decisions regarding pest management	2.63
13.	e-SAP provides immediate solutions to the identified pests	2.00
14.	e-SAP provides information in local language which is clearly understandable by farmers	2.58
15.	High skills and knowledge is required to operate e-sap device	2.63
16.	It makes easy to complete a work in short time	2.53
17.	It act as ready reckonor for advising farmers	2.53
18.	It helps in effective and efficient job performance	2.47

Table 6: Correlation between selected independent variables and perception

Sl. No.	Variables	Correlation coefficient
1.	Age	-0.341**
2.	Education	0.260**
3.	Land Holding	0.054
4.	Farming experience	-0.279**
5.	Annual Income	0.049
6.	Achievement motivation	0.21
7.	Risk orientation	0.47**
8.	Management orientation	0.56*
9.	Innovative proneness	0.19
10.	Information management	0.59**

Zero order Correlation between Independent Variables Farmers with their Perception level

It was observed from the Table 6 that, age and farming experiences of the farmers had shown negative & significant relationship with their perception level at 1 percent level of probability. It seems as age and farming experiences increases the corresponding by favourable perception decreases towards e-sap dissemination tools. Since it requires exposure to recent ICTs tools, to use of tablet to get timely and relevant information, which is lacking among older people, hence age and farming experiences negatively significant with perception level of farmers.Whereas education of the farmers had shown positive and significant relationship with their perception towards e-SAP tool. As education widen the mental horizon of the individual, which helps empowerment of grasping capacity of recent ICT tool in excess of information. So use of tablet and utilization of e-SAP is more significant. Hence the education of respondents had significant relationship with perception level towards e-SAP. Hayrol and Bahaman (2009) reported that factor of age found to have negative and weak relationship with perception towards the importance of ICT.

Conclusion

It SAP- novel ICT tool which enables the rapid and delivery of farm data in various forms including multimedia to researchers, policymakers and other users in real time. e-SAP has been successfully built & tested and deployed by UAS Raichur in association with Tene agriculture solution pvt.ltd, Bangalore. e-SAP is ICT platform and dedicated field device capable of making two way information exchange in real time is available today. With huge potential for revolutionizing transfer of technology system, it has opened doorway for the future course of action particularly in India & world in general. Perception analyses of stake holders had skewed towards higher side. It shows that wider potentiality for application at farmers field level. In addition it empowers field facilitator to diagnose the pest incidence, disease incidence, Physiological disorder, weeds as well as scientific advising to ensure timely decision making system of farmers. Intern this will helps in management pest, disease, disorder in least cost, there by increases the marginal return of crop enterprises.

References

Hayrol, A.M.S. and Bahaman A.S. (2009). Perception towards importance of information and communication technology among rural administrators. *Journal of Agriculture & Social Sciences,* 5: 67-72.

Nagesh, (2005). Study on entrepreneurial behaviour of vegetable seed producing farmers in Haveri district of Karnataka. M. Sc. (Agri.) Thesis, University of Agricultural Sciences, Dharwad.

Nagesh, B., Halakatti, S.V. and Hanchinal, S.N. (2011). Study on entrepreneurial behaviour of pomegranate growers. *Agriculture Update*. 6(3&4):122-125.

Sankri, S. K., Anand R, N., and shibi S. (2013). Perception of the extension officials towards TNAU agritech portal. In: proceedings of international conference on extension educational strategies for sustainable agricultural development-a global perspective. University of agricultural sciences, Bangalore, India. pp 220-221.

Singh, S. K., Gautam, U. S., Anupam, M., and Tushar, A. (2012). In: proceedings of international conference on extension educational strategies for sustainable agricultural development-a global perspective. University of agricultural sciences, Bangalore, India. pp 220-221.

Suresh (2004). Entrepreneurial behaviour of milk producers in Chittoor district of Andhra Pradesh – A Critical study. M. V. Sc. Thesis, Acharya N. G. Ranga Agricultural University, Hyderabad.

Vijay Kumar, K. (2001). Entrepreneurship behaviour of floriculture farmers in Ranga Reddy district of Andhra Pradesh. M. Sc. (Agri.) Thesis, Acharya N. G. Ranga Agricultural University, Hyderabad.

Vijaykumar, A. C., and Narayanagowda, K. (1999). Adoption of improved cultivation practices among Rose growers. *Journal of Extension Education*, 10: 2331–2334.

Extension Management Strategies for Sustainable Agriculture, pp 329-336

New India Publishing Agency, New Delhi, India

35

Impact of Web-Education on Knowledge and Symbolic Adoption of Farmers – An Experimental Study

P. Mooventhan and H. Philip

"Access to Internet will soon be universal, and it can provide unrestricted low-cost access to information, as well as highly interactive distance learning. The Internet will not only facilitate interactions among researchers but also quickly improve their ability to communicate effectively with the potential users of their research knowledge".

Prof. M.S. Swaminathan

Web-education Technology

"Educating the people through web" this process is called web-education. The World Wide Web provides new opportunities and hopeful future for distance education over the Internet and the distance education is becoming a widespread used for quite a long time through Newspapers, Radio and TV. Now-a-days mostly the World Wide Web (WWW) getting vital role in distance education, e-learning and distance learning since the networking technologies are emerging very rapidly in the media dominant era. The connectivity of the internet to different kind of people around the world is increasing steadily, like TV and Radio. The World Wide Web and other Internet–based collaborative modern communication technology tools have significantly enhanced the ability to train and educate people electronically.

Information provided through new communication technologies such as modern information technology gadgets should be need based, demand driven, site specific and should be in local language. With this background the present experimental research study entitled, "Impact of Web-Education on Knowledge and Symbolic Adoption of Farmers – An Experimental Study" was planned.

Objectives of the study

1. To study the knowledge levels of agricultural web-education learners.
2. To study the symbolic adoption of agricultural web-education learners.
3. To find out the relationship between personal and socio-psychological factors of the learners with the knowledge gain and symbolic adoption.
4. To find out the problems faced by the agricultural web-education learners.
5. To find out the suggestions of the learners about agricultural web-education.

Research Methodology

This study was conducted in Coimbatore and Trichy districts of Tamil Nadu, because these two districts had close proximity with agricultural institutions such as Tamil Nadu Agricultural University, Coimbatore; Forest College and Research Institute, Mettuppalayam and Anbil Dharma Lingam Agricultural College and Research Institute, Trichy respectively. From these districts, the list of villages provided with computers and internet facilities were obtained from collectorates. Among the villages, six villages *viz.,* Odanthurai, Chikkarampalayam and Marudur from Coimbatore district and Ediyapatti, Anthanallur and Navallur Kuttapattu from Trichy district were selected at random. From each of these villages, thirty respondents are selected with the use of simple random sampling. The respondents are expected to have a minimum educational qualification of 5th standard and above. They will have to undergo pre and post exposure knowledge test.

Abstract of Sampling procedure

a. Number of districts : 2

b. Number of villages : 6 (@ 3 villages/district)

c. Number of respondents : 180 (@ 30 respondents/village)

Result and Discussion

Effectiveness of selected learning module in terms of knowledge gain

The knowledge aspects of the selected subject matter areas (TNAU website) were exposed to the six experimental groups (Group I, Group II, Group III, Group IV, Group V and Group VI) through single treatment. The knowledge levels of the subjects before and immediately after exposure were assessed to find out the knowledge gain. The maximum attainable knowledge score was 31. The result of the selected groups knowledge gain is presented in Table 1.

Table 1: Mean knowledge gain due to exposure to the treatment

S.No.	Groups	Name of the Villages	Mean knowledge gain		Mean knowledge gain	Percentage
			Pre exposure	Post exposure		
Coimbatore District						
1.	Group I	Maruthur	4.12	22.12	18.00	60.00
2.	Group II	Chikkarampalayam	3.52	21.27	17.75	59.16
3.	Group III	Odanthurai	8.92	21.72	12.80	42.66
Trichy District						
4.	Group IV	Anthanallur	4.12	21.07	16.95	56.53
5.	Group V	Ediyapatti	3.42	19.98	16.00	53.53
6.	Group VI	Navallur Kuttapattu	7.99	21.72	12.02	40.06

(n=30 for each group)

It could be seen from Table 1 that all the groups namely Group I, Group II, Group III, Group IV, Group V and Group VI had effective in terms of knowledge gain. From the result, in Coimbatore district the Group I (Maruthur) had highest (60.00 %) knowledge gain followed by Group II (Chikkarampalayam) 59.16 per cent and Group III (Odanthurai) 42.66 per cent and from Trichy district the Group IV (Anthanallur) had maximum (53.53 %) knowledge gain followed by Group V (Ediyapatti) 53.53 per cent and Group VI (Navallur Kuttapattu) 40.06 per cent. Among the six groups, four groups namely Group I, Group II, Group IV and Group V had no significant different in knowledge gain. But the Group III and Group VI had minimum knowledge gain (42.66%) and (40.06 %) respectively. Even through these groups got maximum score in the pre exposure knowledge test. It might be due to the fact of their close proximity with Forest College and Research Institute and Anbil Dharmalingam Agricultural College and Research Institute respectively.

Table 2: Effectiveness in terms of symbolic adoption

S.No.	Questions	Number*	Percentage
1.	Do you propose to use the alternative crops like sweet sorghum, atrophic, sugar beet recommended by TNAU website?	86	47.77
2.	Do you propose to use the recommended crop management practices recommended by TNAU website?	97	53.89
3.	Do you propose to use the agricultural produce processing technologies recommended by TNAU website?	98	54.44
4.	Do you propose to use the chemical herbicide recommended by TNAU website?	110	61.11
5.	Do you propose to use the post harvest technologies recommended by TNAU website?	120	66.66
6.	Do you propose to use the organic farming procedures recommended by TNAU website?	126	70.00
7.	Do you propose to use the chemical pesticides recommended by TNAU website?	130	72.22
8.	Do you propose to use the agricultural farm machineries recommended by TNAU website?	137	76.11
9.	Do you propose to use the course material related to agricultural technologies recommended by TNAU website?	146	81.11
10.	Do you propose to use the fertilizers dose recommended by TNAU website?	147	81.66
11.	Do you propose to use the web-education course material recommended by TNAU website?	154	85.55

(* Multiple response obtained) (n=180)

From the Table 2, it is inferred that about 47.77 per cent of respondents were ready to adopt alternative cropping such as sweet sorghum, sugar beet, and jatropha as per the recommendation of TNAU website followed by 53.89 per cent of respondents were ready to adopt crop management practices, 54.44 per cent of respondents were ready to adopt agricultural produce processing technologies, 61.11 per cent of respondents were ready to adopt chemical herbicides and 66.66 per cent of respondents were ready adopt post harvest technologies.

Further, nearly 70.00 per cent of respondents were ready to adopt organic farming procedure, 72.22 per cent of respondents were ready to adopt chemical pesticides, 76.11 per cent of respondents were ready to adopt farm machineries, 81.11 per cent of respondents were ready to use course material related to agricultural technologies, 81.66 per cent of respondents were ready to adopt recommended fertilizers and most of the respondents ready to adopt the technologies given in web-education course material for specific crops as per the recommendation of TNAU.

Problems in Utilizing Agricultural Web-education Technology

It is necessary to study the constraints in any social research. The constraint analysis helps the planners, administrators, development workers, scientists and others to frame policies and to implement schemes. Hence the constraints faced by the farmers in utilizing web-education technology were studied. The multiple responses of constraints were worked out. The results are presented in Table 3.

Table 3: Problems expressed by agricultural web-education learners

S.No	Problems	Number*	Percentage
1.	Lake of information in local language	180	100.00
2.	High cost for establishment	172	95.55
3.	Not suitable to illiterate people	169	93.88
4.	Difficult in handling of computer mouse	165	91.66
5.	Difficulty to type the URL to access the web page	160	88.88
6.	High cost for internet connection	157	87.22
7.	Lack of skill to use modern information technology gadgets	142	78.88
8.	Lack of familiarity towards modern information technology gadgets	140	77.77
9.	Lack of relevant information in the website	132	73.33
10.	Lack of prompt reply to online queries and clarification	97	53.88
11.	Inability to make or receive electronic payments	90	50.00
12.	Slow down loading and uploading of photos/files/ videos	82	45.55
13.	Lack of update information	75	41.66
14.	Difficulty in reading online information	62	34.44

(* Multiple response obtained) (n=180)

The result could be observed from Table 3. The major problem as expressed by all respondents (100%) was lack of information in local language and this was due to the fact that most of the web information published in English, high cost for establishment (95.55%), illiterate people cannot use (93.88%), difficulty in handling of computer mouse (91.66%) and this might be due to lack of practice and proper training on the use of mouse in computer and its accessories, difficulty to type the URL to access the web page (88.86%) was expressed because all the URL addresses are exclusively in English language, high cost to connecting internet (87.22%) and this was due to the reason of unavailability of internet providers at village level, lack of skill to use modern information technology gadgets (78.88%) as most of the respondents lacked training on the use of modern gadgets, lack of familiarity (77.77%), lack of relevant information (73.33%), lack of prompt reply to online queries and clarification (53.88%), inability to make or receive electronic payments (50.00%), slow downloading and uploading of photos/files (45.55%), lack of update information (41.66%)

and few per cent of the respondents expressed that they had difficulty in reading online information.

Suggestions for Effective Utilization of Agricultural Web-education Technology

Any study would be complete if it gives suggestions for development in the field of investigation. The suggestions rendered by the respondents for effective utilization of web-education technology were studied. The results are presented in Table 4.

Table 4: Suggestions for effective utilization of agricultural web-education technology

S.No	Suggestion	Number*	Percentage
1.	Information may be in local language	180	100
2.	Location specific information may be provided	172	95.55
3.	Training may be provided to learn computer in detail	170	94.44
4.	Regular update of website information	167	92.77
5.	Installation of software in tamil of word, excel, power point, e-mail can be installed in information kiosk	160	88.88
6.	Valarum vellanmai may be published in online	157	87.22
7.	Information kiosks may provide location specific weather data	152	84.44
8.	Linking of Village Information Centre with State Agricultural University, Krishi Vigyan Kendra and development departments	149	82.77
9.	On line marketing facilities for farm produce	137	76.11
10.	Keeping the agricultural website addresses on the desktop of computers in Village Information Centre	102	56.66
11.	Self Help Groups can run the information centre	91	50.55
12.	More number of information kiosk	71	39.44
13.	Use of video conferencing	51	28.33

(* Multiple response obtained) (n=180)

Suggestion rendered by the respondents could be seen from Table 4. Major suggestions were giving information in local language expressed by all the respondents (100.00%) for easy comprehension by all farmers, followed by location specific information (95.55%) to reach all farmers who are spread across the state and country with diversified climatic condition, training to operate computers (94.44%) due to digital divide among rural farmers on the usage of computer, regular update of website information (92.77%) to fulfill the latest and current agricultural information need of the farmers, software in Tamil of word, excel, power point, e-mail may be installed in information kiosk (88.88%), Valarum Vellanmai may be published in online (87.22%), information kiosk may provide location specific weather data (84.44%), linking of Village Information

Centre with State Agricultural University, Krishi Vigyan Kendra and Development departments (82.77%), on line marketing facilities for farm produce (76.11%), keeping the agricultural website addresses on the desktop of computers in Village Information Centre (56.66%), Self Help Groups may run the information centre (50.55%), more number of information kiosk (39.44%) and very few per cent of the respondents suggested that use of video conferencing is required for effective utilization of web education technology.

Conclusion and Recommendation

The study revealed that the non-computer owning respondents lack skill to operate computer and to know about technologies given through web-education. So information kiosks should be installed in villages and skilled trainers should give proper and regular trainings. The results clearly depict that majority of the respondents were unaware of website-education, despite the fact that they are more interested to know about it. So large scale awareness campaigns and mass media can be utilized in a big way to promote web-education technology.

The study revealed that respondents were satisfied with the tool web-site but not convinced with the technology existed on it. So the scientific community should take initiatives to evolve technologies, which would compromise farmers in terms of affordability and effectiveness. The study revealed that respondents are more satisfied with the web-education and transfer of technology through computers. Further, this finding has indicated that they are inclined towards accepting Cyber Extension. So the farm graduates who are unemployed can be manned in Information shops and can facilitate farmers to know timely, accurate information through World Wide Web.

The awareness level and utilization behaviour of websites was found to be low and this indicated that there is an urgent need for development and popularization of such websites. The variables such as educational status, mass media exposure, innovativeness, farm status and farming experience were found to act as critical variables. So, while preparing web-education in future, one should take care of above variables. Public and private computer training institutions and entrepreneurs may be entrusted with the task of opening information clubs on partnership with farm association to teach, train farmers and keep away their fear in using this new technology in rural areas. This may be done as a Government policy initiation.

References

Anandaraja, N. (2002). Developing Farmer Friendly Interactive Multimedia Compact Disc and Testing its Effectiveness in Transfer of Farm Technology. Unpub. Ph.D. Thesis, TNAU, Coimbatore.

Anonymous (2003). http://www.mssrf.org

Karunakaran, B. (2004). A Study on the Potential of Modern Information Technology Gadgets for Agricultural Development. Unpub. M.Sc. (Ag.) Thesis, TNAU, Coimbatore.

Senthilkumar, M. (2003). Field Testing Cyber Extension Techniques for Transfer of FarmTechnology-A Feasibility Study. Unpub. Ph.D. Thesis, TNAU, Coimbatore.

The Chronicle year book (2005). Information Technology. Chronicle Publication Private Ltd. New Delhi.

Extension Management Strategies for Sustainable Agriculture, pp 337-342

New India Publishing Agency, New Delhi, India

36

User Friendly Interactive Multimedia Module – An ICT Application

P. B. Pradeep Kumar, S.V. Prasad, B. Vijayabhinandana K. Dhanasree and D. Sekhar

Information Technology can play an important role in disseminating agricultural information to the farmers by using various multimedia sources. Farmers are more desirous and anxious to get quick, accurate and authentic information in the changing scenario of agriculture at local and at global levels. Dissemination of the required and recent agricultural information to the farmers in scattered villages at the variegated geographical situation in India is very difficult task. Transfer of technology to farmers is not a one time exercise because new farm technology is being constantly evolved. A continuous flow of technologies in an appropriate manner is vital to provide quick benefit for the development of the farmers. There has been a technological explosion in the field of agriculture. This demands that the farmer has to know all aspects of technology prior to its adoption. Hence Interactive Multimedia Module (IMMM) has potential to reach the unreached.

Interactive multimedia is a sub field of computer science which extends the hypertext concept of non-linear and non-sequential links of textual material that may be digitally encoded for storage and retrieval through computer based systems, including images, sound, graphics, moving video, animation, text and photographic quality stills. Using the multimedia data types such as video, graphics, images or animation any participant can introduce illustrative material into the meeting to facilitate better communication. It informs, educates, persuades and entertains with dazzling pictures and animations, engaging sounds, compelling video clips and raw textual information.

Materials and Methods

The present investigation is mainly intended to develop an Interactive multimedia module in cotton in IPM in cotton. With this developed module the The perception of the respondents about content of the IMMM and effectiveness in terms of knowledge gain of the respondents is tested. Keeping above backdrop in mind, the present study is modulated with the following objectives.

1. To study the perception of the respondents about the content of the multimedia module.
2. To test the effectiveness of the developed module in terms of knowledge gain by the respondents.

Keeping in view the above objectives, the relevant literature on various aspects of interactive multimedia module were reviewed. The study was conducted in Guntur and Adilabad districts of Andhra Pradesh. Experimental research design was adopted for the study. The sample of the study comprised a total of 180 cotton growing farmers. A comprehensive interview schedule and Interactive multimedia module covering all facets of the IPM in cotton was developed. The findings of the study were summarized and presented below.

Results and Discussion

Perception of Respondents about the content of the Interactive Multimedia Module

Information is treated as most important resource, in addition to the resources such as man, machine, money, material and time. The degree of effectiveness of the developed IMMM can be assessed through respondents perception. In this context, perception is the process by which one can able to see, hear and understand things. Hence, it is assumed that greater the perception of IMMM as perceived by the viewers in terms of satisfaction derived from various components of IMMM, greater would be the viewers exposure and post exposure activities.

On the basis of perception index score, the respondents were categorized into three groups viz., highly satisfied, satisfied and not satisfied. The details of frequency and percentage distribution under each category were furnished in the Table 1.

Table 1: Distribution of respondents according to their overall level of perception **(n = 180)**

Sl. No	Category	Frequency	Percentage
1.	Highly satisfied	128	71.11
2.	Satisfied	32	17.78
3.	Not satisfied	20	11.11
	Total	**180**	**100.00**

It could be explained from the above Table 1 about the overall perception of the IMMM that nearly 71.11 per cent of the respondents were highly satisfied followed by 17.78 per cent of respondents were satisfied and minimal of 11.11 per cent of respondents were not satisfied.

Table 2: Perception of respondents about various components of IMMM and suggestions for improvement **(n = 180)**

Sl. No.	Perception	Highly Satisfied (%)	Satisfied (%)	Not Satisfied (%)	Suggestions for improvement
1.	Adequacy of information	73	14	9	-
2.	Motivation to learn Subject matter	71	15	14	-
3.	Provides complete knowledge	50	32	18	-
4.	Motivate to adopt the message in farm	60	26	14	-
5.	Message is understandable	67	18	15	-
6.	Practical utility of message	56	34	10	-
7.	Use of simple words and sentences	46	34	20	-
8.	Readability of text matter	65	22	13	-
9.	Colour combination of text	100	0	0	-
10.	Clarity of voice	40	10	50	Increase the sound and clearity
11.	Quality of illustration	65	24	11	-
12.	Text Vs video clipping balance/ synchronization of voice	78	11	11	-
13.	Duration of the video	82	10	8	-
14.	Understandability of video clipping	86	14	0	-
15.	Attractiveness of video clippings	85	9	6	-
16.	Naturality of the module	100	-	-	-
17.	Pleasant background music	50	0	50	Decrease the volume of the background music

Nearly 73 per cent of the respondents were highly satisfied with the adequacy of information followed by 14 per cent were satisfied and a minimal of 9 per cent were not satisfied.

The interest shown by the respondents was very high for this learning module *i.e.* 71 per cent of the respondents were highly motivated towards the module and 15 per cent were satisfied followed by 14 per cent not satisfied. The simple words, sentences, sequential and logical presentation, adequacy of messages coupled with photographs, voice and good interactivity of IMMM might have increased their inquisitiveness to learn more about IPM practices in cotton.

About 50 per cent of the respondents were highly satisfied followed by 32 per cent satisfied and 18 per cent not satisfied with the IMMM providing complete knowledge. This might be due to the reason that the module covers all the IPM practices.

Majority of the respondents *i.e.* 60 per cent felt that they were highly satisfied followed by the 26 per cent satisfied and 14 per cent not satisfied with the motivation to adopt in the farm.

Majority of the respondents 67 per cent were highly satisfied with the understandability of the message followed by 18 per cent of the respondents were satisfied and 15.00 per cent were not satisfied. This reason for high satisfaction might be due to sequential arrangement of the module.

Around 56 per cent of the respondents said that they were highly satisfied with the practical utility of the content in the field, followed by 34 per cent satisfied and 10 per cent were not satisfied.

A section of around 46 per cent of the respondents was highly satisfied with the words and sentences used in the IMMM, followed by 34 per cent were satisfied and 20 per cent were not satisfied.

They were about 65 per cent respondents who were highly satisfied with the respect to readability of the text matter, followed by 22 per cent satisfied and a meager of 13 per cent were not satisfied. The reason for high satisfaction among the respondents might be that, the text was written in vernacular language i.e. *Telugu.*

Almost all the respondents were highly satisfied with the colour combination of the text as it was written with white colour on dark background.

Around 40 per cent of the respondents were highly satisfied with the clarity of voice followed by 10 per cent satisfied and 50 per cent were not satisfied. As half of the respondents were not satisfied with the clarity of voice, it was again corrected as suggested by the respondents.

A majority of 65 per cent were highly satisfied with the quality of the illustration followed by 24 per cent satisfied and 11 per cent not satisfied. This might be due the presence of good and colourful illustrations having pleasing effect on the farmers and it increased the curiosity and interest in viewing.

Majority of 78 per cent of the respondents were highly satisfied with the balance of text and video clipping, followed by 11 per cent satisfied and 11 per cent were not satisfied.

A vast majority of 82 per cent of the respondents were highly satisfied with the duration of the video followed by 10 per cent satisfied and 8 per cent were not satisfied.

Most of the respondents 86 per cent were highly satisfied with the understanding of the video clippings, closely followed by 14 per cent satisfied and their were no respondents in the degree of not satisfied. This might be due to the presence of live pictures of the pests and their control measures.

Around 85 per cent of the respondents were highly satisfied with the attractiveness of the video clippings followed by 9 per cent satisfied and minimal of 6 per cent not satisfied. The old adage a picture is worth a thousand words is crucially relevant. This might have increased their attractiveness of video clippings.

Almost all the respondents were highly satisfied with the nature of the module. As it was having real pictures and video of pests and IPM practices.

A majority of around 50 per cent of the respondents were highly satisfied with background music followed by 50 per cent were not satisfied. As half of the respondents were not satisfied with the background music, it was again corrected as suggested by the respondents. The music was pleasant as it reduces the boredom while viewing the IMMM.

Effectiveness of the Interactive Multimedia Module in terms of Knowledge gain

Knowledge is recognized as one of the most important components of human behaviour, which give impetus to adopt a technology. A proper understanding of improved practice is prerequisite for its adoption by a cotton farmer at his farm.

The knowledge in IPM practices in cotton which were exposed to the respondents with the developed IMMM was collected. The knowledge of the subject matter before and after the exposure of IMMM were assessed to find out the knowledge gain. The maximum attainable score was 25. Knowledge of farmers before the exposure of the IMM Module were categorized into low, medium and high levels was depicted in Table 3 and also after exposure of IMMM were also categorized into low, medium and high levels as depicted in Table 3.

Table 3: Distribution of the respondents according to level of knowledge before, after exposure of IMMM and Knowledge gain

Sl. No.	Level of Knowledge before and after exposure of IMMM	Frequency (Before)	Frequency (After)	Percentage (Before)	Percentage (After)	Knowledge gain	
						Frequency	Percentage
1.	Low	114	19	63.33	10.55	9	5.00
2.	Medium	45	21	25.00	11.67	12	6.67
3.	High	21	140	11.67	77.78	159	88.33
	Total	**180**	**180**	**100.00**	**100.00**	**180**	**100.00**

It was found from Table 3 that most of the respondents had low level of 63.33 per cent followed by medium level of 25.00 per cent and high level of 11.67 per cent before the exposure of IMMM to the respondents. But after the exposure of IMMM to the respondents, their was drastic increase in their knowledge level. High Knowledge level of cotton farmers of 77.78 per cent followed by medium level 11.67 per cent and low level 10.55 per cent. Also we can conclude that this Interactive Multimedia Module could disseminate farm information effectively to the clients and thereby increase one's knowledge.

After calculating the difference in knowledge gain between pre and post exposure of IMMM in IPM practices it was concluded that majority of (88.33%) of respondents belong to high level followed by medium level (6.67%) and low level (5.00%).

Conclusion

The extensive use of modern information technology needs to be promoted for communication between researchers, extension workers and farmers to transfer technologies and information in a cost effective manner. ICT has many potential applications in agricultural extension. It can bring new information services to rural areas where farmers, as users, will have much greater control than before over current information channels.The use of ICT is an important pillar of Agriculture extension and in the current scenario of a rapidly changing world, it has been recognized as an essential mechanism for delivering knowledge (information) and advice as an input for modern farming directly to the farmer himself.

Extension Management Strategies for Sustainable Agriculture, pp 343-346

New India Publishing Agency, New Delhi, India

37

Participatory Need Based Extension Through WhatsAPP for Effective and Efficient Sharing of Technology

M. Senthilkumar, M. Pandiyan, B.J. Pandian and A. Suganthi

Information and Communication Technology is ruling the world in all walks of life and access to mobile phones and internet facility is growing in India at a rapid rate in recent years. However, the access to internet based technologies have confined primarily to the urban areas. Rural communities have not been able to gain to the same extent from IT. As a means of agricultural technology transfer to farmers, information technology has had a limited impact.

Over the years, Radio and TV have vastly increased their reach, as also reception facilities and played a major role in communicating with an audience with low literacy skills. The traditional electronic media played a key role in rapid and effective dissemination of general information and advice to farming communities. This includes market information; market led production planning, on farm and posts harvest management/value addition, e-contracting, market networks and market intelligence. The World Wide Web especially portals, web sites and other pull technological services are also contributing to the technology transfer of general information / knowledge.

However, Harnessing Information Technology for agricultural extension is very essential in the wake of challenging environment. The Extensive use of modern information technology for communication between researchers, extension workers and their farmers to transfer technologies and information more cost effectively and efficiently is necessary. Information technology should be made available, particularly, to those with specific inquiries to guide them in adopting the more knowledge intensive forms of agriculture, which will expand in future. Extension mechanisms will have to be driven by farmers' needs, be location specific and address diversification demands. Technologies required to address total farming systems are knowledge intensive. The focus is on availability, accessibility knowledge-based technologies to upgrade and improve the skills of the Extension functionaries, farmers. Extension services must maintain close and effective relationships with sources of appropriate information, expertise, practical, specific problems from the farmers.

Effective communication, the information should flow freely, unrestricted, downward, and upward and across people in real time and the message should be complete and clear. Without affecting the initiative, motivation and morale of the farmers, scientist and extension functionaries, and the messaging service is able to control the activities without any deviation.

In the recent past and at present a number of successful ICT initiatives across globe, India and in Tamil nadu has successfully demonstrated the use of ICT for technology transfer through off line (CDs, IMCDs, Videos, Expert Systems) and Online (IVRS, Web sites/ portals, Social networking sites like Face Book, Twitter, mail networks) mode.

WhatsApp Messenger is a cross-platform mobile messaging app and it is a platform to share real-time information which allows user to exchange messages, audio, video, photographs. It allows users to create groups, broadcast/send unlimited images, video and audio media messages simultaneously to one person, or to a group. WhatsApp Messenger requires a smart mobile phone with internet data access.WhatsApp is the largest community of users of any IM client on any device, ever, has been on the rise since its inception. With over 600 million users it is the most popular application for Instant Messaging, and has found mainstream acceptance and popularity now world over. WhatsApp was purchased by Facebook for 22 Billion US Dollars, which is a huge amount.

This paper addresses a more complex role, but simultaneously requiring a flexible approach allowing specific information to be customized for different farmer-groups by using Instant messaging tool / application namely whatsApp for location specific sharing of technology/ information by the scientists, Extension workers and farmers as individuals and in group.

Steps in using WhatsApp as a tool for Effective and efficient sharing of Agricultural Technology to select farmers.

- Scientists and farmers have to be trained in the use of Instant Messaging application namely WhatsApp in creating, sharing, recording and retrieval of agricultural information/ technology.
- Access to be made to necessary hardware and connectivity namely availability of Smart Mobile devices and Internet connectivity to all the stakeholders (Farmers/ Scientists) involved in the project.

- The knowledge sharing process has to be documented by all the stake-holders.
- The reusability of shared knowledge retrieved from this extension model by for research and extension may be studied and the constrains faced by the users in using WhatsAPP can be documented.

I. Creating Virtual groups

Select farmers will be enrolled as Change Specialists from select villages. It may of FIGs/ CIGs/ SHGs or any farmer who is interested in sharing and learing.

II. Capacity building of Change Specialists and Scientists

Awareness training may be provided to all the enrolled Change specialists and Select Scientists on the need and importance and application of the use of IM tool namely WhatsAPP and on basic mobile computing skills. The training has to cover

a. Knowledge management and technology transfer process
b. Operation of Smart mobile phones
c. Enabling Internet connectivity
d. Installation of IM messaging application from the internet and its features
e. Basic operations / computing in smart phones and trouble shooting
f. Enrolling in WhatsApp as member
g. Creating and Joining group
h. Creating a broadcast group/ lists – Techniques of broadcasting
i. Initiating a messaging service / reading a incoming message
j. Recording pictures, videos, storing/ sending to individuals / groups.

III. Provision of Mobile Computing Devices for Instant Messing through WhatsAPP

The Smart mobile computing device has to be made available to all the selected change specialists.

IV. Studying the Reusability of WhatsAPP contents generated and constraints in using the IM

The contents thus generated over a period has to be analyzed for the nature of technology requirements, technologies transferred, problems, suggestions, feedback by the users, contents reused by the members and constraints faced by the farmers in using the IM.

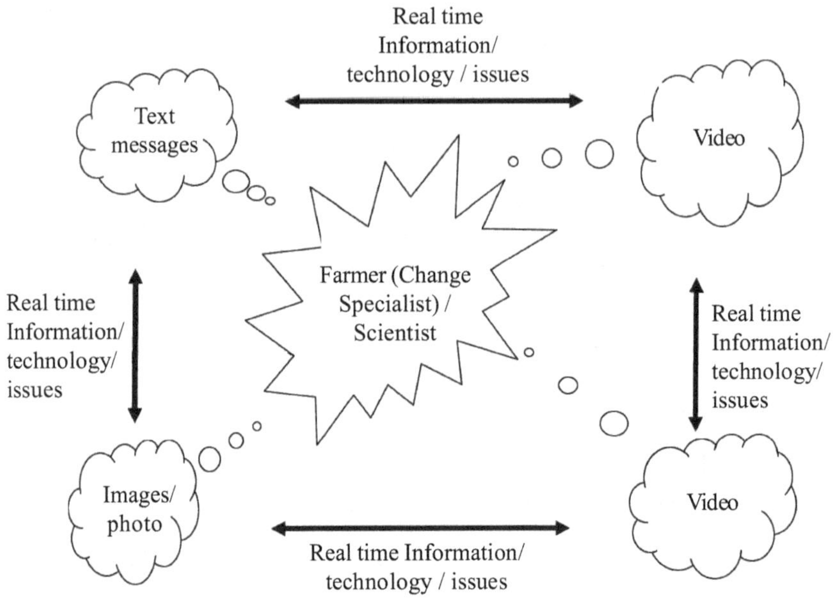

Model depicting the Technology sharing model through WhatsAPP

Extension Management Strategies for Sustainable Agriculture, pp 347-350

New India Publishing Agency, New Delhi, India

38

Facilitative Extension Strategies for Hi-Tech Neo Micro-Irrigation Cluster Farmers

B. Shanmugasundaram, M.C. Narayanan Kuttty and Chitra Parayil

Kerala State receives an average annual rainfall of about 3000 mm but with high variability. The demand for water in Kerala is mainly for drinking, agriculture, power generation and aquaculture as well as for prevention of salt-water intrusion. Currently 71 per cent is used for irrigation , 18 per cent for domestic use and the rest is for other purposes. However, Kerala is one of the States with lowest per capita water availability among the Indian states. Basak (1998) reported the per capita water availability in Kerala was only 69 per cent of that of the country in 1981 though the general perception is contrary to this. Moreover, the water availability shows a steady decline over the years since 1901. It was 59,206 L/day in 1901 which got reduced to 11,062 L/day in 2001. The decline is sharper in surface water than rain and ground water levels. Thus, there is a widening gap between population growth (demand) which is positive and supply of water which is negative. Hakkim (2013) reported that Kerala has land, water and environmental features that are unique in the country. Steep and undulating topography, highly uneven rainfall, high population density,one of the lowest per capita water availability and the fact that majority of the State's river hardly carry any water during the summer months illustrate the need for adopting proper water conservation techniques. Further Kerala is frequently facing severe drought followed by acute water scarcity for the last two decades. And the major consumer of water resources is Irrigation with about 70% of the total resources being used by this sector. Compared to National average, unit land of Kerala receives 2.5 times more rainfall but the same unit of land has to support 3.6 times more population, Hence, for self sufficiency, unit land of Kerala has to produce 3.6 times more food and biomass and also has to provide 3.6 times more water.

Micro-irrigation is used extensively in many water scare countries like Israel, Arabian states, Mexico *etc*. In India its development is slow as compared to other developing countries. Micro irrigated area in India grew slowly but steadily.While it covered just 1000 ha when this was launched in 1985 , it made progress in the following decades. Though India had large irrigated area of 69 Mha , yet it has only 4-5 Mha *i.e*., less than 7% under Micro irrigation Experiments and farm trials have been going in India from 1970 onwards. Progressive farmers in Andhra Pradesh, Karnataka, Kerala, Maharashtra and Tamil Nadu adopted this method in the late 70's. From a mere 1500 ha's in 1985 , the area under microirrigation has really gained momentum in recent years. A report of Central Water Commission, Ministry of Water resources (2012) says that among the top five states using Micro irrigation it was found that adoption is more seen in Maharashtra (35 %) followed by Andhra Pradesh (24%) Karnataka(13%), Gujarat (12 %) and Tamil Nadu (9%). With this background in mind the present study was undertaken to assess the fertigation demonstration in the farmers field and to suggest strategies for upscaling of Hi-tech micro-irrigation technology in the farmers field by following cluster approach.

Materials and Methods

The demonstration on Micro-irrigation were initiated in 2012 and continued till 2013. Farmers with some awareness were selected. The demonstration was conducted by involving the Agricultural Officers of State Department of Agriculture who played a facilitative role in identifying innovative farmers and monitoring of the fields along with Scientists of Regional Agricultural Research Station. The demonstrations were conducted in Palakkad districts of Kerala. About 20 farmers representing 2 villages namely Elavancherry and Kollengode were registered as a group and included for the demonstration. The principal crops grown by the farmer includes Bitter gourd and Snake gourd.The Technological interventions include the conduct of soil testing in the farmers field and recommendation of need based fertilizers, conduct of demonstration on drip fertigation in Bitter gourd and Snake gourd, Providing a part of water soluble fertilizer for the fertigation unit for small scale demonstration and the extension interventions included in the project were Conduct of training at different stage of the crop, Conduct of Exposure visit to Jain irrigation unit, at Udumalpet and visit to Precision farming project being implemented by Tamil Nadu Agricultural University, Coimbatore at Dharmapuri distirct of Tamil Nadu. Periodic review meeting to seek their problems and give need based solutions by inviting experts from the University were given.

Results and Discussion

Table 1: Yield performance of different vegetable under Micro Irrigation

Crop	Yield (t/ha) Traditional method	Drip fertigation method	Per cent yield increase (%)
Bittergourd	19.31	30.64	58.67
Snake gourd	26.32	41.64	58.21

Demonstrations on yield performance of vegetables like Bitter gourd and Snake gourd was laid out in farmers field at Elavancherry village of Palakkad district during 2012. Based on the demonstrations conducted the yield of Bitter gourd and Snake gourd was arrived in comparison with the traditional method of cultivation being practised by the farmers. The results presented in Table 1 clearly reveal that an yield increase of nearly 60.00 per cent can be achieved by adoption of drip fertigation in vegetables.

Table 2: Comparative cost of cultivation Bitter gourd and Snake gourd under drip fertigation method of cultivation (per ha per season)

S. No	Operation	Cost of cultivation (Rs)			
		Bitter gourd	Per cent	Snake gourd	Per cent
1.	Land preparation	22125	6.06	22125	6.66
2.	Planting and Trailing cost	85625	23.46	72625	21.87
3.	Fertilizer application cost	152360	41.74	137360	41.38
4.	Weeding cost	8750	2.40	8750	2.64
5.	Irrigation cost	13125	3.60	13125	3.95
6.	Plant protection cost	20000	5.48	15000	4.52
7.	Harvesting cost	63000	17.26	63000	18.98
	Total cost	**364985**	**100.00**	**331985**	**100.00**

The results presented in Table 2 on Comparative cost of cultivation of Bitter gourd and Snake gourd revealed that nearly half of the cost was spent on Fertilizers (41.74 per cent) followed by planting (23.46 per cent) and Harvesting (17.26 per cent) cost. The total cost of cultivation works out to Rs 3,64,985 per hectare. The results presented on Cost of cultivation of Snake gourd revealed that nearly half of the cost was spent on Fertilizers cost (41.38 per cent) followed by planting and trailing cost (21.87 per cent) and Harvesting cost (18.98 per cent). The total cost of cultivation works out to Rs 3,31,985-00 Dhawan (2000) reported that drip irrigation is more suited for horticultural crops in terms of cost and net return.

The reason for fertilizer cost accounting to nearly 42 per cent in Snake gourd and Bitter gourd is due to the adoption of more organic manure application like Farm yard manure, Groundnut cake, Poultry manure, Bone meal, Neem cake and water soluble fertilizers. The second rank occupied for planting and trailing cost is due to the high cost of Hybrid seed which costs around Rs 7500 per Kg. Further the Trailing cost which requires Bamboo poles, String and rope also makes the farmer to invest in such cost as it is absolutely necessary for the growth of crop.

Conclusion

It can be concluded from the demonstration that Drip fertigation in vegetables like Snake gourd and Bitter gourd is definitely promising. However the problem with respect to the technology is that many farmers cannot adopt the technology as majority of the farmers in Kerala are small and marginal farmers and many factors like high cost of the Micro-irrigation system, Complexity of technology and socio economic issues like lack of access to credit facilities, fragmented land holdings. As far as farmers are concerned a general opinion is that the installation charges, cost of pump set and overhead water tank is not included in the subsidy cost which thereby hinder the rate of adoption. Other factors such as approved drip irrigation agencies which are not present in adequate number in a particular district, lack of after sale service and lack of knowledge on use and maintenance turned out to be a demotivating factor among the farmers who adopt micro-irrigation. Further the cluster approach of formation of group was found to be helpful for the project implementation team and the farmers.

References

Basak, P. (1998). Water resources of Kerala-Myths and realities. In: Water Scenario of Kerala. Eds: E.J.James, K.N. Ramani and P.S. Harikumar. State Committee on Science, Technology and Environment, Government of Kerala, Thiruvananthapuram. pp1-6.

Dhawan B.D. (2004). Drip irrigation: Evaluating Returns. Economics and Political Weekly, pp: 3775-3780, Oct 14.

Hakkim. A. (2013). Role of Micro irrigation and Fertigation in Hi-Tech Farming, Proceeding of National Seminar on Hi-tech Farming- Marching towards Attaining Sustainability, Jan 21-22, 2013, Jointly organized by Precision Farming Develoment Centre, State Horticulture Mission, Kerala and National Committee on Plasticulture Applications in Horticulture, New Delhi: 100-109.

Report of Central Water Commission (2012). History of Irrigation Devlopment and Management in India. Ed: Rangachari, Kanjilia, Gupta A.C. and Baweja. M.L. New Delhi: pp162.

Extension Management Strategies for Sustainable Agriculture, pp 351-355

New India Publishing Agency, New Delhi, India

39

Awareness and Utilization of Kisan Call Centre by the Farmers of Kerala

Shely Mary Koshy and A. Sakeer Husain

The Indian agriculture is on the threshold of a second revolution. It is becoming increasingly clear that the next leap will come from the Information and Communication Technology (ICT). Even though ICT has developed very well, its penetration to the farming community seems to be very less. At the same time, it is difficult for the extension agent to personally attend the queries of every farmer individually. With the decreased number of extension workers, there is a need to use the latest technologies for delivering extension services.

The telecom network in India is very impressive and almost every village has got telephone services. This telephone network can be used for delivering information to the farmers. Thus Kisan Call Centres (KCCs), a facility to get agricultural information over telephone, play a significant role. The aim of KCC in India is to provide information to the farming community through a toll free number, 1551. This service, funded by the Central Government, started functioning from 21-01-2004. The main objective of KCC is to address the need of the farming community, by making use of increased tele-density and Information Technology, making professional help and information available to farmers at their doorstep, putting the farmers directly in contact with agricultural experts and making authentic field data available for agricultural policy decisions by the authorities. However there is a need to evaluate the performance of the service in order to understand how far the service could reach and create an impact in the farming community.

In this backdrop, the study was undertaken with the following objectives

1. To assess the awareness of farmers about Kisan Call Centre.
2. To analyze the extent of utilization of Kisan Call Centre.

3. To elicit constraints experienced by the farmers in using the services of the centre.

Review of Literature

According to Batchelor and Sugden (2003) simpler technology often produces better results. Telephone access and use can add considerable value to the communication systems of the poor in developing countries.

Sowmyashree (2007) reported that the highest number of calls made from Karnataka state to the kisan call centre was to avail information on plant protection (38.68%), followed by soil and nutrition (10.43%), veterinary (8.89%).

Lavanya *et al.* (2009)reported that half of the users of KCC suggested to reduce the speed delivery of messages delivered by the level I officials. Practical solutions suitable to the local situations were expected by farmers from the officials of level I Centre. Recommendation of low-cost technologies was the need of less than half of the users of level II KCC.

Anurag and Kumar (2012), found out thatmobile phone was the most popular ICT gadget among the farming community. A farmer needs information mainly on pest and disease (45%), field preparation (44.9%), nutrient management (41%), input availability (39.5%), market information (32%) and weather information (22%). Most of the farmers had faith in the information provided by their friends and neighbours (61.5%) followed by ICT providers (53%) and local input dealers (46.4%).

Savithramma *et al.* (2014) reported that 10 per cent of the respondents had correct knowledge about KCC telephone number, while remaining 90 per cent did not know the KCC telephone number.

Methodology

The study was carried out in Thrissur district of Kerala state. Hundred farmers were randomly selected from Thrissur District to study the awareness of farmers on the Kisan Call Centre (KCC) functioning in Kerala. To study the utilization and constraints experienced by the farmers, 30 farmers were selected randomly from the list of farmers available with the data base of the KCC who had used the services of the KCC during 2011-12. Data were collected using structured interview schedule.

Findings and Discussion

The study revealed that a vast majority (86%) of the farmers were not aware of the service of the Kisan Call Centre. This can be attributed to the lack of wide publicity among the farmers about the KCCs functioning across the country. Importantly, all the farmers who were aware of the KCC utilized its services. This calls for concerted propaganda among the farmers about the availability and services of KCC. Table 1 shows the sources of information to farmers about KCC.

Table 1: Source of information about Kisan Call Centre

Source of information	Farmers	
	No.	%
Agricultural Extension Worker	7	23.33
Friend	1	03.34
Newspaper	9	30.00
T.V.	6	20.00
Radio	7	23.33

Among the users (farmers) of the KCC, 73 per cent came to know about the Call Centre through different mass media such as newspaper (30%), radio (23%) and television (20%), while 23 per cent knew about the services of the Centre through the local agricultural extension worker. At times, the services of the KCC are advertised through mass media, and this has become the major source of information. Among the farmers who used the services, more than half of the farmers availed the service 2-3 times and 30 per cent contacted the Centre four or more times (Table 2). It is vivid from the table that the farmers are not regular users of KCC. This points to the need of further strengthening the services of the KCC.

Table 2: Utilization pattern of Kisan Call Centre by the farmers

Variable	Category	Farmers	
		No.	%
No of times called (frequency)	1	3	10.00
	2-3	18	60.00
	4-6	6	20.00
	>7	3	10.00
Purpose of call	Plant protection	62	34.40
	Crop cultivation	48	26.70
	Marketing	24	13.30
	Supply of input	28	15.60
	Schemes	18	10.00

Further, as evident from Table 2, most of the farmers contacted KCC to get information on plant protection (34.40%) followed by crop cultivation (26.70%). This is in line with Sowmyashree (2007) and Savithramma *et al.* (2014). It is quite expected that plant protection queries top the list, as this is the major technology dimension on which farmers are more concerned and seek advices. Crop cultivation aspects were also seen important for farmers. However, the number of queries on marketing, input supply and schemes were found to be comparatively less, may be due to presumption of the farmers that only agricultural technological information are available through KCC.

Table 3: Satisfaction of farmers by using the services of Kisan Call Centre

Level of Satisfaction	Farmers	
	No.	%
Highly satisfied	5	16.70
Satisfied	19	63.30
Neutral	4	13.30
Dissatisfied	2	06.70

Regarding the satisfaction of the farmers by using the services of KCC, majority of the farmers fell in the satisfied category. Among the callers only 6% expressed dissatisfaction about the service, while 13 percent expressed their neutral stand. When asked about the constraints faced, 10 percent of the farmers were of the opinion that the solution provided by the call centre was not an immediate solution. Similarly, 10 per cent farmers opined that they had difficulty in getting connected to the call centre.

The suggestions put forth by the farmers to improve the services of KCC were to provide field services along with advisory service, provide more effective recommendations and to recommend chemicals that are available in the market.

Conclusion

Our country has showcased an impressive growth in the field of information and communication technology. Kisan Call Centre is such a service that can solve almost all the problems faced by the farmers with regard to queries in agriculture. The only thing to be taken care of is to advertise the service more strongly and also the agricultural extension workers should take effort to make the farmers aware of the existence of the service. This is an era where almost all the individuals in India is having a mobile phone; therefore by using the service, farmers can save time and the solutions for his farm problem is just a call away. But quality has to be ensured in recommendations, which is the lynchpin in the success of KCC.

References

Anurag, T.S. and Kumar, S. (2012). Interactive Information Dissemination System: Architecture for Disseminating Information to Farmers. *Information Technology in Developing Countries,*22(2): 2-5.

Batchelor, S. and Sugden, S. (2003). *An Analysis of info Dev Case Studies: Lessons Learned.* Info Dev, Washington. Available: http:// www.sustainableicts. org/ infodev/ infodevreport. pdf [22 Sept. 2013].

Somyashree, G.T. (2007). A Critical Analysis on the Functioning of Kisan Call Centres in Karnataka. MSc. (Ag) thesis, UAS, GKVK, Bangalore 89p.

Lavanya, P., Karthikeyan, C. and Venkata, P. J. (2009). Needs and expectations of the farmers with the services of Kisan Call Centre. *J of Rural Development,* 28(9):1- 2.

Savithramma, Manjunath, M., Parvathamma, T. and Naika, K.V. R. (2014). Farmers Awareness of Kisan Call Centre and the Symbolic Adoption of Advices in Karnataka. *J. Ext. Educ.* 26 (2): 5236- 5240.

Extension Management Strategies for Sustainable Agriculture, pp 357-366

New India Publishing Agency, New Delhi, India

40

Attracting Rural Youth and Fostering Entrepreneurship in Agriculture: Lessons for Indian Agricultural Universities from Michigan State University (MSU), USA

R. Venkattakumar, Manoj P. Samuel and P. Venkatesan

Indian economy is basically an agrarian economy (Saxena, 2012) and the majority less literate and unskilled population living in rural areas does not fit into the employment market. Employing 55% of labour force, agriculture sector is a major employment provider (Chand, 2014). However, the seasonal nature of Indian agriculture leads to seasonal and cyclical unemployment. This situation can be altered by generating employment opportunities through nurturing entrepreneurship in agriculture (Rehman *et al.*, 2012).

In India, transfer of technology remains in the domain of the State Department of Agriculture (DOA) and State Agricultural Universities (SAUs) are mandated to serve a limited extension role (Sulaiman and van den Ban, 2000). Most of the SAUs continue to be dominated by top-down, monolithic structures that follow a limited extension mandate. The SAUs have conventionally involved in extension activities such as training, communication and information and services through Farm Science Centers. A single-window facility of the Agricultural Technology Information Centre (ATIC) is established in some SAUs for delivery of research products, information and other services. The emerging socio-economic scenario and change in knowledge structure of agriculture explicitly indicates that the traditional agricultural extension role of the SAUs cannot sufficiently address the challenges of the new trends in agricultural development (Bonny and Mathew, 2010).

Large-scale migration of rural youth from farming to urban areas has caused concern among the country's agricultural policy makers. To address this issue, the Indian Council of Agricultural Research (ICAR) has formulated a pilot project called Attracting Rural Youth in Agriculture (ARYA). The concern is mainly about the school and college dropouts, who migrate to cities in search of menial jobs. The plan is to address issues such as providing training on innovative and sustainable agricultural practices to help make agriculture a profitable venture. This would involve a series of activities, including creation of infrastructure facilities in various aspects of farming and providing marketing linkages and in exploring opportunities in secondary agricultural operations such as value addition of crops and hiring and servicing of mechanised farm implements. The SAUs have a major role to play in this mission (http://asia.ypard.net/news/icar-focus-retaining-rural-youth-agriculture/).

Further, there is a strong and immediate need to establish link between study lab of SAUs and farm must be established, so that innovative products developed are used in agricultural and food processing sector. The government has already taken steps to promote entrepreneurship in agriculture and attract entrepreneurs to do business in agriculture through National Agricultural Education and Entrepreneurship Project in 12th five year plan.

With the above-mentioned background, the role of SAUs in promoting entrepreneurship in agriculture and building the capacity of rural entrepreneurs through their mechanism becomes imperative. However, the Land Grant Universities of USA have their tradition and emerging innovations in promoting entrepreneurship in agriculture. This paper discusses about the Michigan State University (MSU), USA and its initiatives to foster entrepreneurship in agriculture and what lessons the Indian SAUs system can learn from MSU.

Case study of Michigan State University (MSU), a Land Grant University

Educational institutions such as MSU are committed to foster entrepreneurship to energize the rural economy of local, state and federal governments of the USA. Through its, education, research and outreach activities, MSU caters to the diverse needs of entrepreneurial community.

a. Creating Entrepreneurial Communities (CEC) Programme

The CEC programme is a unique long-term, comprehensive approach to economic development for local economies by helping rural communities nurture and recruit entrepreneurial talent. Communities takes responsibility and support entrepreneurs by removing barriers and by increasing assets, in each of the four CEC Hallmarks of entrepreneurial communities such as social networks, resources, policy and entrepreneurial culture (Figure 1). The projects that communities select to work on are influenced by their local and regional asset base, the specific needs of their entrepreneurial populations, the talents and community's readiness to develop projects (Fails, 2011).

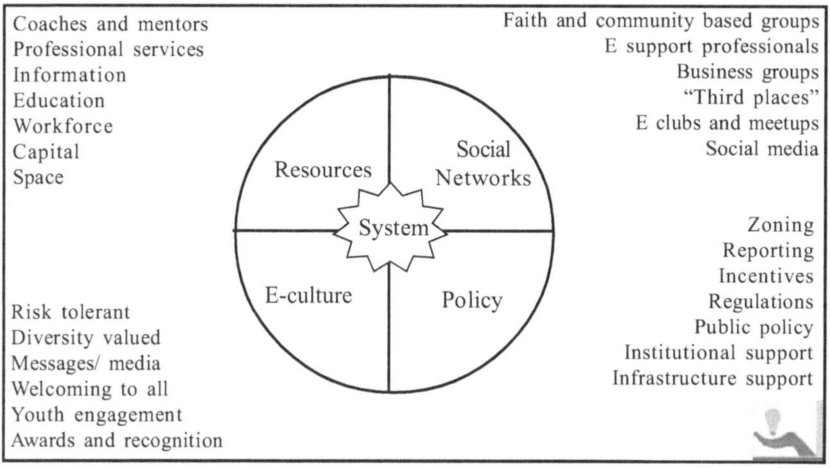

Fig. 1: Hallmarks of Creating Entrepreneurial Community (CEC) programme of Michigan State University (MSU) (Fails, 2011)

b. MSU Product Centre

The MSU Product Center established in 2003 focuses on assisting entrepreneurs who develop and commercialize high-value, consumer-responsive products and businesses. The Product Center's 'Innovation Counsellor Network' is strategically positioned around rural areas of Michigan. The Innovation Counsellors are certified to facilitate the clients to identify markets, innovate new products, make critical decisions and guide from concept to launch.

Product Center's 'Concept Definition' helps entrepreneurs decide if their concept is something they want to pursue. If they have already developed their concept, 'Venture Development' can help them create a business plan to effectively communicate with bankers, investors, and customers. If they have already developed their product but need assistance with nutritional facts labelling or processing authority to meet Federal Food and Drug Administration requirements, 'Venture Development' can provide assistance on a fee for service basis.

'MSU Client Services' pinpoints the entrepreneurs about the necessary resources at MSU as well as the resources of state and federal agencies and private companies to move projects forward. 'Client Services' works in conjunction with the Center's 'Strategic Marketing Institute'. The 'Marketing Institute' develops the information base needed to support the initial screening and evaluation of concepts, products and businesses. The center also has initiated a new line of programming directed towards existing Stage 2 businesses that have sustainable revenue and are looking to make major expansions in sales and production. This new initiative is named as the 'High Impact Venture Action Team' or 'HI-VAT'. Since 2004, the centre organized 24184 one-to-one counselling for aspiring entrepreneurs, facilitated 2125 entrepreneurs towards concept development, 1232 towards business expansion, 1200 specialized services and successfully launched 254 business ventures (http://productcenter.msu.edu/about).

c. MSU Innovation Centre

By integrating innovation, technology transfer, support for start-ups and partnerships with businesses and communities, the 'Innovation Centre' of MSU brings path-breaking ideas to the marketplace as rapidly as possible in order to improve the quality of life in rural communities near and far and to help Michigan and the United States compete in a global environment (http://innovation.msu.edu/about.html).

1. Spartan Innovations

'Sartan Innovations', a component of 'Innovation Centre'helps to turn MSU innovations, ideas, technologies, creative activities into successful businesses. Spartan Innovations'experts screen innovations for start-up potential as part of technology transfer, review technology and business strategy, engage seasoned entrepreneurs to guide early stage technology development and capital attraction. The centre uses grants mechanismto drive technology development, engages with various sources of venture capitalists, informs start-up strategies and gains access to state of art technology. Spartan Innovations' CEO-in-Residence programme provides technology/product development leadership and executive management for MSU start-up enterprises (http://www.spartaninnovations.org/).

2. MSU Business-CONNECT

MSU Business-CONNECT, another component of 'Innovation Centre' serves as MSU's front door for corporations, directing businesses to MSU resources starting from recruiting students to finding a researcher for a specific project and enabling two-way strategic communications and planning for business customers. Either the research or development team of business firms can work with MSU faculty researchers to develop new ideas or to refine or test their innovations (http://www.businessconnect.msu.edu/).

d. The Institute for Entrepreneurship (IE)

IE provides support for entrepreneurial start-ups with an emphasis on providing access to and refining technological innovations having market potential and established companies and organizations renew their "entrepreneurial spirit" through consultancies (http://ie.broad.msu.edu/).

e. International Business Centre of Broad College of Business

The International Business Centre of Business School runs Michigan Export Growth (MEG) Programme. The MEG Programme is providing professional consultancy services at no cost to Greater Lansing area companies to expand their business to international markets. Companies are matched with student teams led by the center's staff and faculty (http://global.broad.msu.edu/ibc/).

f. Small Business and Non-profit Clinic of MSU College of Law

The Small Business and Nonprofit Law Clinic, a public interest law clinic, provides an important rural community service to start-ups and existing small businesses and nonprofit organizations. The goal is to offer legal business counsel and services that enable the entrepreneur to run his or her business successfully. It also reviews, prepares and negotiates necessary legal documents (http://www.law.msu.edu/clinics/sbnp/).

g. Innovation Club for Entrepreneurs (ICE) of Land Policy Institute (LPI)

The ICE of LPI, MSU provides a new forum for young entrepreneurs of the Greater Lansing Area to share and explore ideas in a positive and supportive environment, to create or strengthen existing networks for collaboration and find resources to bring ideas to life (http://www.landpolicy.msu.edu/).

h. MSU Business Library Model

'Research Guides' of MSU Business library are the resources for entrepreneurs towards getting started in business, accounting, advertising and marketing, business ethics and corporate social responsibility, hospitality & tourism, human resources management, insurance, international business, international statistics, management information systems, mergers and acquisitions etc.

Michigan local resources, Mintel academic guide, business newspapers, personal finance resources, popular business collections, ProQuest statistical datasets, supply chain management resources, media resources, Wharton Research Data Services (WRDS) guide, Writing for Business series etc. are the other useful resources of MSU Business Library.

The library promotes an online 'Business Model Generator' (Stratgyzer Alpha) for the use of entrepreneurs. This supports the start-up entrepreneurs who are with business ideas to plan their business models through a business model canvas that has nine blocks such as key partners, key activities, key resources, value propositions, customer relationship, customer segments, channels, cost structure and revenue streams. The library also has set up a marketing blog to help the start-ups to find their initial markets. Another scheme on Inter-Library Loan (ILL) promises entrepreneurs that if MSU does not own the material the entrepreneurs need, the library may be able to get it from another library for free of cost for the entrepreneurs (http://www.lib.msu.edu/bus/).

i. Food Hubs promoted by Centre for Regional Food Systems (CRFS)

'Food hubs' are centrally located facilities with a business management structure facilitating the aggregation, storage, processing, distribution and/or marketing of locally/regionally produced food products. The 'Michigan Food Hub Learning and Innovation Network' is led by CRFS, MSU in collaboration with the Michigan Department of Agriculture and Rural Development (MDARD) and private sector players. This network facilitatescross-learning, access to innovation, financial and technical assistance, research and education and business-to-business collaboration across food hubs (http://foodsystems.msu.edu/). The conceptual framework on which the MSU offers its services to entrepreneurs is depicted in Figure 2.

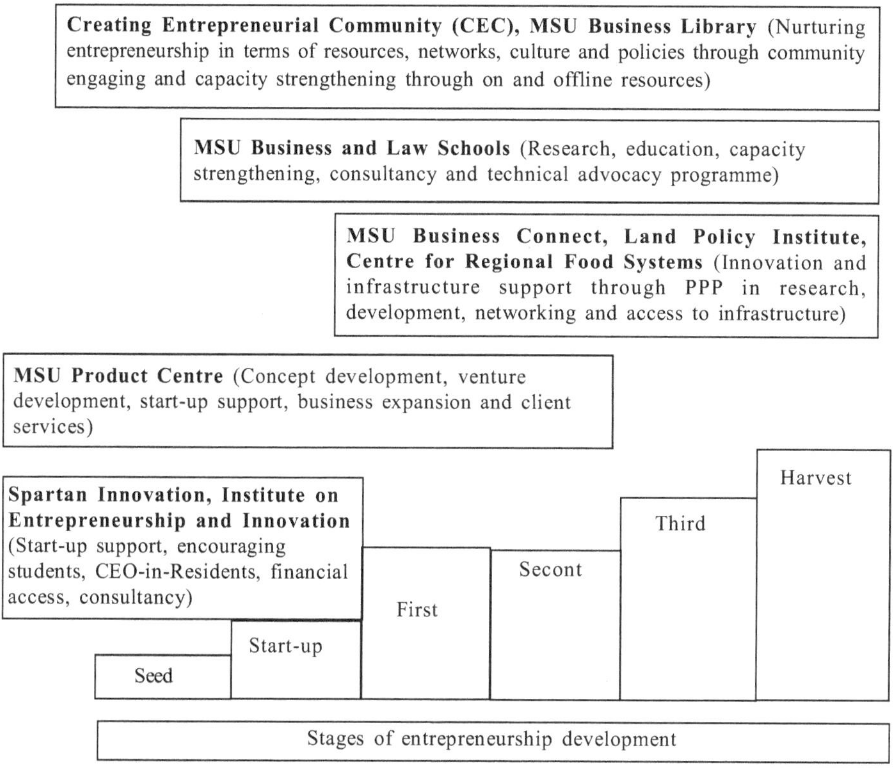

Fig. 2: Conceptual model depicting initiatives of Michigan State University (MSU) in promoting different stages of entrepreneurship (Jon-Gillespie-Brown, 2007) in agriculture

Potential Interventions and Implications for India

The following are some of the potential interventions (Policy, institutional and capacity building/ educational interventions) derived from MSU system that may be piloted in India through SAU system.

Institutional Interventions

- Many of the SAUs have the 'extension educator' model to transfer the university technologies to farming community. Similarly, Farm Science Centres (KVK) recruit subject matter specialists (SMS). This is the high time to transform these extension educators of SAU and SMS of KVK as 'Innovation Counsellors' who can link the entrepreneurs of rural areas to innovation, infrastructure and various resources. The 'network' of such innovation counsellors may be utilized for building the capacity of community leaders too.

- 'MSU Product centre' model can be piloted in selected SAUs of the country. The Directorate of Extension Education (DEE) may have this structure with innovation counsellors as the human resource. The 'concept development', 'Venture development', 'Client services' and 'HI-VAT' models of MSU product centre can be piloted to foster entrepreneurship in agriculture and rural development arenas.
- The lack of collaboration between public, private and non-governmental sectors in fostering entrepreneurship need to be addressed. The MSU 'Business CONNECT' model may be piloted in SAU system to foster public-private research partnership. SAU may amend policies that encourage public-private partnership in research, intellectual property (IP) management, technology commercialization and innovation-led entrepreneurship.
- Similar to 'Spartan Innovations' arrangement of MSU, the SAUs may plan business incubation and start-up support to entrepreneurs; encourage students and aspiring individuals to practicetheir business concepts(igniting innovations) and mentoring entrepreneurs towards business management.
- SAUs as well as Deemed-to-be universities of Indian Council of Agricultural Research (ICAR) may adopt the 'MSU Business Library' model to foster competency enhancement of entrepreneurs. Libraries of these organizations may develop online guides, resources, databases, information packages and courses etc. Awareness campaigns, webinars, teleconferences etc. may be organized for the local entrepreneurs. A network of these business libraries may be created for better service delivery.

Capacity Building/ Educational Interventions

- The role of community in engaging entrepreneurs and the potential rural youth is very vital and will have a long-term impact on enhancing rural livelihoods and overall socio-economic development. It may be possible only through sensitizing community leaders and local policy makers. The SAUs has a major role to play as that of Michigan State University's (MSU's) Creating Entrepreneurial Community (CEC) programme in collaboration with KVK, private and non-governmental institutions and provide pragmatic capacity building in this regard.
- Fostering entrepreneurship in agriculture needs intensive capacity building efforts at all levels. Hence, SAUs ought to design customized programmes and build platforms for training, knowledge sharing and policy dialogues.

- SAUs and their KVK system need to build the capacity of entrepreneurs to effectively use the internet and social media for accessing information and knowledge, marketing, advertising, partnering, exporting and networking. The proposed 'Innovation counsellors' may be utilized for this purpose.

Policy Interventions

- SAUs may plan to do need-based research, capacity building, education and outreach activities to cater to the diverse needs of entrepreneurs depend upon agriculture.
- Entrepreneurial promotional initiatives of SAUs system need to have continuous liaison with the developmental agencies of Ministry of Agriculture (MoA), Ministry of Food Processing Industries (MoFPI), Ministry of Micro Small and Medium Enterprises (MoMSME), National Bank for Agriculture and Rural Development (NABARD), Agriculture and Processed Food Products Export Development Authority (APEDA) and other financial organizations, business associations and institutes like National Institute for Entrepreneurship and Small Business Development (NIESBD), National Institute of Entrepreneurship (NIE), Mahatma Gandhi Institute for Rural Industrialization (MGIRI), National Institute for Food Technology and Entrepreneurship Management (NIFTEM), Indian Institute of Management (IIM), Institute of Rural Management (IRMA), National Institute of Rural Development (NIRD), National Institute for Agricultural Extension Management (MANAGE) etc. To augment this, a coordination committee may be established at ICAR.
- SAUs that have strong and traditional work on entrepreneurship promotion may be identified as Regional Institutes of Entrepreneurship in Agriculture (RIEA) to cater to the specific regional level research, capacity building, education and outreach needs of entrepreneurs and coordinate such fostering efforts by various agencies at the regional levels.
- SAU sought to develop need-based and customized courses through online (*webinar, teleconferencing systems and online resources*) and distant education modules effectively coupled with contact programmes to cater to the needs of entrepreneurs.

Conclusions

India has its own tradition of fostering entrepreneurship in agriculture. However, relevant interventions are to be incorporated in order to make the system more effective. US system, especially land grant universities has rich tradition and experience in fostering entrepreneurship through institutional innovations. These innovations throw meaningful implications for piloting by SAUs in India.

Community support is very crucial for fostering entrepreneurship. Innovation counsellors, concept and venture development, client services and HI-VAT models, business library model, PPP in agricultural research and incubation model of MSU may be piloted in Indian scenario.

Initiating network research programmes through selected SAUs and identifying Regional Institutes for Entrepreneurship in Agriculture (RIEA) are suggested to coordinate the research, education and outreach activities. The SAUs system ought to support the government initiatives through sharing expertise, experiences and active participation.

References

Bonny, P. Binoo and Mathew, Joy (2010). Farm Transition Network –A Participatory Approach to Reinvent University Agricultural Extension in India. *Journal of Social Sciences*. 23(3): 171-178.

Chand, Ramesh (2014). *From Slowdown to Fast Track: Indian Agriculture since 1995. Working Paper 01/2014*. National Centre for Agricultural Economics and Policy Research, Indian Council of Agricultural Research, New Delhi, pp.1-26.

Fails, Barbara (2011). *Building entrepreneurial communities: Examples in Michigan*. Rural Research Report, Illinois Institute for Rural Affairs, 22 (5): 6.

http://asia.ypard.net/news/icar-focus-retaining-rural-youth-agriculture/ accessed on 20.10.2014

http://foodsystems.msu.edu/ accessedon 8/11/2013

http://global.broad.msu.edu/ibc/ accessedon 7/11/2013

http://ie.broad.msu.edu/ accessedon 7/11/2013

http://innovation.msu.edu/about.html/ accessedon 8/11/2013

http://productcenter.msu.edu/about/ accessed on 7/11/2013

http://www.businessconnect.msu.edu/ accessedon 9/11/2013

http://www.landpolicy.msu.edu/ accessedon 7/11/2013

http://www.law.msu.edu/clinics/sbnp/ accessedon 9/11/2013

http://www.spartaninnovations.org/ accessedon 9/11/2013

Jon-Gillespie-Brown (2007). Understanding the stages of Entrepreneurship development. Entrepreneurship.

Rehman, Anis Ur. and Elahi, Yasir Arafat (2012). 'Entrepreneurship Education in India – Scope, challenges and Role of B-schools in Promoting Entrepreneurship Education'. *International Journal of Engineering and Management Research*, 2(5):5-14.

Saxena, Sandeep (2012). 'Problems Faced by Rural Entrepreneurs and Remedies to Solve it' *IOSR Journal of Business and Management* (IOSRJBM), 3(1):23-29.

Sulaiman, V.R., van den Ban, A.W. (2000). Agricultural extension in India; The next step, *Policy Brief 9*,New Delhi: National Centre for Agricultural Economics and Policy Research.

Extension Management Strategies for Sustainable Agriculture, pp 367-375

New India Publishing Agency, New Delhi, India

41

Factors Affecting Managerial Role of Farm Women

Srividya Rani, N., T. Lakshmi and S.V. Prasad

The prosperity and growth of a nation depends on the status and development of women as they constitute about 50.00 per cent of the human resource of the country. Women play a significant role in agriculture and allied fields. She also performs several management and decision making roles in farming and home making practices with her male counterpart and sometimes alone.(Sreenivasulu and Punna Rao, 2005). Khelkar (1995) found that 64.55 per cent had medium level of farm management followed by high and low 18.18 per cent and 17.27 per cent respectively. Shilpa (2001) reported that majority (62.00%) of farm women were grouped into medium level of managerial role followed by high and low (20.67%) and (17.33%) respectively. Premavathi and Netaji Seetharamam (2002) reported that 8 to 13 per cent of the women had taken their own decisions. While 37 to 53 per cent had consulted their spouse/elder to take operational decisions on farm activities. Justification for improving farm women's access to agricultural extension system must begin with analysis of women participation in managerial aspects of agricultural production process.

With this back ground, the present study was conducted to unearth the relationship between profile characteristics of farm women and their managerial role.

Material and Methods

A study was conducted with and expost-facto research design to access profile characteristics of farm women in Chittoor district of Andhra Pradesh which was purposively selected for the study because maximum number of farm women were involved in agricultural operations. Chittoor district was selected as the researcher hails from the same area and had the familiarity with social conditions, local language and culture of the people. Chittoor district comprises of 66 mandals out of which four mandals namely Ramanchandrapuram,

Vadamalapeta, Puttur and GD Nellore were purposively selected for the study. From each of the selected mandals, two villages were selected based on random sampling procedure. Thus, totally eight villages were selected for the study. A total sample of 120 farm women were selected by selecting 15 farmers from each village through simple random sampling procedure. Keeping the objectives of the study in view, a semi structured interview schedule was developed and pretested. This was administered to sample respondents through personal investigation. The data obtained were coded, classified and tabulated. Finally statistical tools such as arithmetic mean, standard deviation, frequency and percentage were used for the analysis of the data, so that the finding could be meaningfully interpreted and conclusions drawn.

The dependent variable managerial role was operationally defined as the role of women in the management of farm activities i.e. in planning, organizing, supervision, communication, coordination and controlling the farm activities. To find out the managerial role of farm women, a scale developed by Shilpa (2001) was used for the study. Twenty nine items were incorporated under six managerial functions. Each item of the schedule was rated on 5 point continuum namely always, frequently, sometimes, rarely and never with weightages of 5, 4,3,2,1, respectively. Suitable measures were determined to quantify the selected independent variables of the study. The data were collected with the help of a pretested interview schedule. The data thus obtained was processed through frequency, percentage, correlation and MLR analysis for interpretation.

Results and Discussion

An attempt was made to understand the relation between profile characteristics of farm women and their managerial role. It could be noticed from Table 1 that 55.00 per cent of the respondents were grouped under medium level of managerial role while 23.33 per cent and 21.67 per cent were in high and low groups respectively. Thus, there is a scope to increase the farm women participation and level of farm management. The findings of Khelkar (1995) and Shilpa (2001) were also in conformity with the present finding.

Table 1: Distribution of farm women based on their managerial role **(n=120)**

S.No.	Category	Frequency	Percentage
1.	Low	26	21.67
2.	Medium	66	55.00
3.	High	28	23.33
	Total	**120**	**100.00**
		Mean=127.57	SD=11.36

Women were consulted more for selected agricultural decisions like quantity of grains to be sold, used and stored, getting credit and its repayment, employment of labour for operations like sowing, weeding, harvesting and buying equipment etc. Their participation in matters like adoption of practices and farm credit was found to be supportive in nature. Farm women participation was found to be better in managerial roles like planning, organizing, supervision, communication, co-ordination and controlling and all were under the medium category. With a view to understand the nature of relationship between independent and dependent variable, the data were subjected to correlation coefficient and presented in Table 2. The relational analysis revealed that the computed 'r' value of education (0.5707), extension contact (0.4650), value orientation (0.5277), mass media exposure (0.4981), achievement motivation (0.3236), innovativeness (0.5265), scientific-orientation (0.4510) and risk orientation (0.3475) were found to be significant at 0.01 level of probability. This indicates that education, extension contact, value orientation, mass media exposure, achievement motivation, innovativeness, scientific orientation and risk orientation exhibited a positive and significant relationship with managerial role of farm women.

Education is the means for development. Education facilitates for the understanding and interpretation of facts. The possible reason for the existence of positive and significant relationship between education and managerial role might be due to the fact that educated women had more knowledge about cultivation. They possess more grasping power to identify pest and disease incidence. Educated women can easily get the idea of cost of cultivation in agricultural operations. So education affects the managerial role of farm women. Sreedevi (1996) and Sreenivasulu and Punna Rao (2005) also reported similar results.

Table 2: Relationship between selected profile characteristics of farm women and their managerial role

S. No.	Variable	Correlation co-efficient'r' value
X_1	Age	-0.2651**
X_2	Education	0.5707**
X_3	Farm size	0.0155NS
X_4	Family income	0.1340NS
X_5	Extension contact	0.4650**
X_6	Value orientation	0.5277**
X_7	Participation	0.1229NS
X_8	Mass media exposure	0.4981**
X_9	Decision making	0.1580NS
X_{10}	Achievement motivation	0.3236**
X_{11}	Innovativeness	0.5265**
X_{12}	Scientific orientation	0.4510**
X_{13}	Risk orientation	0.3475**

* : Significant at 0.05 level of probability
** : Significant at 0.01 level of probability
NS : Non-significant

Extension agents are the best and reliable sources of information for farming community. Farm women with good extension contact have more knowledge which develops favourable attitude towards agricultural activities which in turn leads to more managerial role of farm women. Of late many government programmes were introduced keeping in view of small and marginal farmers. The extension functionaries were making some efforts to meet the farmers through agricultural and non agricultural programmes which were mainly focused on the poverty alleviation and improving the socio economic standards of farm women. Hence positive and significant relationship was observed. The findings were in accordance with the findings of Sreedevi (1996).

Farm women who are cosmopolite, liberal and hold scientific values are likely to acquire more information on recent technologies by exposing themselves to cosmopolite and mass media sources of information and adopt new innovations. They will also be enthusiastic and self driven to seek information on advanced technologies which facilitates adoption of advanced crop production technologies. Thus contributes to better managerial role of farm women. Hence positive and significant relationship was established between value orientation and managerial role of farm women. This finding was in conformity with the findings of Kiran (2000) and Begum (2008).

The positive relationship between mass media exposure and the managerial role of farm women might be due to increase in the telecast of agricultural programmes through different channels and number of news papers publishing a full column on agriculture the women farmers were becoming aware of specific practices with regard to crop cultivation and this led to the managerial role of women farmers on these specific aspects of cultivation and post harvest technology of the crop production.Here age of the information explosion and several new technologies were being diffused into agricultural communities through mass media. Farm women with good mass media exposure are in a position to pick up right technologies at right time and implement them, so better management of farm women can be seen. The finding was in conformity with the finding of Kiran (2000) and Begum (2008).

Majority of the respondents had achievement motivation to some extent as they were living on subsistence economy they had to get some income which led to increased managerial role of women farmers in recent times. Individual with high achievement motivation would be determined to reach his goal with concentrated efforts. In this process, she knows the importance of recommended practices and this leads to efficient managerial role of farm women.Hence it is concluded that there was a positive and significant relationship between achievement motivation and managerial role of farm women.The results were in conformity with the findings of Begum (2008).

Innovativeness was associated with the individual's earliness in use of new practices. An innovative farmer always reaped windfall profits from new technologies. Farm women with this trait had better managerial role. Hence there was a positive and significant relationship between innovativeness and managerial role of farm women.The results were in conformity with the findings of Kiran (2000).

The probable reason might be that farmers with high scientific orientation will prefer to cultivate the crops as per the production recommendations given by the scientists and extension personnel. This will give ample scope for the farm women to think logically and scientifically. Farm women having good scientific orientation will naturally prefer to know advanced technologies in agriculture. Therefore, they are very much interested in knowing about latest agricultural practices. In this process, they might have acquired more knowledge. Hence the above trend was noticed and there was a positive and significant relationship between scientific orientation and managerial role of farm women. The findings were in conformity with the findings of Sreenivasulu and Punna Rao (2005).

Risk orientation is expressed as the degree to which a farmer is oriented to take risk. Farm women with this particular trait have a tendency to try new innovations and would naturally prefer to know about advanced technologies and practice them. It is one of the characteristics of innovators/ early adopters. Farm women with high risk orientation acquire more knowledge and would have better managerial role. Hence, positively significant relationship was observed.The results were in line with the findings of Sreenivasulu and Punna Rao (2005).

Multiple regression analysis was carried out in order to determine the combined effect of all the independent variables towards variation in dependent variable *viz.*, managerial role of farm women and result was presented in Table 3. It could be understood that out of 13 independent variables fitted in regression equation, four independent variables *viz.*, education, value orientation, participation and scientific orientation were found to be positively significant. Coefficient of multiple determination (R^2) was found to be 0.5823 indicating that the above said independent variables could explain variation towards dependent variable viz. managerial role of farm women to the extent of 58.23 per cent. An increase of one unit in education, value orientation, participation and scientific orientation would result in an increase of 3.9947, 3.3969, 3.5159 and 2.1784 units respectively in the managerial role of farm women.

The findings revealed that 23.33 per cent of farm women were in high managerial role category. As management provides maximum profit from farm business by effective use of all resources, the present investigation will be useful as guideline for the farm women, extension workers and policy makers.Efforts should be made to adopt suitable communication technology and use of modern ICTs for spreading of farm information at a faster pace to the farm women. Considering the substantial roles played by women farmers in overall agricultural process, the extension service must be directed with special efforts towards them. In order that women accept the change and contribute fully to development it is essential to develop first their potentiality by improving their knowledge, attitude and skills with much concern for long range consequences. Adequate extension facilities preferably by female workers should also be made available to acquaint them with latest agricultural developments along with necessary inputs.

Table 3: Multiple Linear Regression analysis of the selected independent variables with the managerial role of farm women

S.No.	Variable	Std. error	'b' values	't' values	'P values'
X_1	Age	0.0846	0.0413	0.4883^{NS}	0.6262
X_2	Education	0.6772	2.7054	3.9947**	0.0001
X_3	Farm size	1.3893	-1.5450	-1.1121^{NS}	0.2686
X_4	Family income	0.0206	0.0050	0.2445^{NS}	0.8072
X_5	Extension contact	0.3928	0.6598	1.6796^{NS}	0.0959
X_6	Value orientation	0.4363	1.4822	3.3969**	0.0009
X_7	Participation	0.2673	0.9401	3.5159**	0.0006
X_8	Massmedia exposure	0.3845	-0.1187	-0.3088^{NS}	0.7580
X_9	Decision making	0.8467	1.3361	1.5779^{NS}	0.1175
X_{10}	Achievement motivation	0.6021	-0.7798	-1.2951^{NS}	0.1980
X_{11}	Innovativeness	0.3072	0.2871	0.9347^{NS}	0.3520
X_{12}	Scientific orientation	0.4615	1.0054	2.1784*	0.0315
X_{13}	Risk orientation	0.6167	0.3850	0.6243^{NS}	0.5337

$R^2 = 0.5843$

* Significant at 5% level of probability.

** Significant at 1% level of probability.

NS: Non Significant

The variables education, value orientation, participation and scientific orientation were positively and significantly contributing to most of the variation in managerial role of farm women. So these variables should be given proper attention while framing any training or developmental programmes intended for improvement of living standards of farm women.

Implications of the Study

1. The present study revealed that the independent variables *viz*. education, extension contact, value orientation, mass media exposure, achievement motivation, innovativeness, scientific orientation and risk orientation were positively and significantly related with managerial role of farm women. Hence, the above variables should be taken into consideration to improve the managerial role of farm women.

2. Multiple regression analysis revealed that out of the 13 selected independent variables education, value orientation, participation and scientific orientation were positively and significantly contributing to most of the variation in managerial role of farm women. So these variables should be given proper attention while framing any training or developmental programmes intended for improvement of living standards of farm women.

Suggestions for Future Research

1. Gender analysis should be done for future research and for development of farm women.
2. Women research studies should be done to gain more knowledge on the managerial abilities of farm women.
3. Systematic analysis of training needs of different categories of women farmers should be done in different agro climatic zones to improve their technical skills may be taken up.
4. Similar studies can be repeated after a conclusive period of 5 years to assess the impact of knowledge and managerial role of the respondents in improving the profile characteristics of the farm women.
5. There is need to support women managed rural production ventures in horticulture, floriculture and post harvest processing of various commodities. Rural women need to be shifted from low paid jobs to better paid jobs.

Conclusion

Majority of the farm women were in the medium category. The farm women should be able to improve their conditions and they have to be educated well and have good mass media exposure and extension contact and more training programmes should be conducted by the extension agents. Adequate extension facilities preferably by female workers should also be made available to acquaint them with latest agricultural developments along with necessary inputs.The transfer of technology to farm women would have to synchronize with the effort in technology generation to reduce drudgery in women life. There is an equal need to conduct researches in socio economic implications of the new technologies.

References

Begum, M.K. (2008). A study on participation and decision making of woman farmers in rainfed groundnut cultivation. *M.Sc. (Ag.) Thesis,* Acharya N.G. Ranga Agricultural University, Hyderabad. India.

Khelkar, H.B. (1995). Managerial behaviour of farm women. *M.Sc. (Ag.) Thesis.* Dr. Punjabrao Deshmukh Krishi Vidyapeeth, Akola.

Kiran, S.M. (2000). A study on the participation of farm women in agriculture and allied activities in Chittoor district of Andhra Pradesh. *M.Sc. (Ag.) Thesis*, Acharya N.G. Ranga Agricultural University, Hyderabad, India.

Premavathi, R. and Seetharamam, R.N. (2002). Decision making pattern of rural women in farm related activities. *Journal of Extension Education*, 13(3):3382-3386.

Shilpa, D.K. (2001). Managerial role of women in farm activities. *M.Sc. (Ag.) Thesis*. Dr. Punjabrao Deshmukh Krishi Vidyapeeth, Akola.

Sreedevi, R. (1996). Gender analysis of managerial abilities and farming performance in Krishna Godavari Zone of Andhra Pradesh. *Ph.D. Thesis,* Andhra Pradesh Agricultural University, Hyderabad.

Sreenivasulu, M. and Punna Rao, P. (2005). Managerial role of farm women. *Karnataka Journal of Agricultural Sciences.* 18(4):1026-1030.

Extension Management Strategies for Sustainable Agriculture, pp 377-381

New India Publishing Agency, New Delhi, India

42

Women Empowerment Index (WEI) to Measure Level of Women Empowerment

Sakunthalai. A and R.Ganesan

Women hold up half the sky' according to Chinese saying. This is true with the Indian scenario as well, since women make up half the country's population. The rural women in India constitutes nearly 77.00 per cent of the total female population. About 79.00 per cent of rural women workers are involved in agriculture related pursuits as cultivators and labourers. Farm & home are integral parts in the rural society in India and the 'house wifes' in the farmstead often plays multiple roles as producer, entrepreneur, workers, consumers, and homemaker. Women are more sensitive than men to the extremes of poverty and its consequences. For poor households the women's capacity to work, her health, her knowledge and her skill endowments are often the only resources to fall back for survival. The rural women in India are, therefore, critical to the process of moving their families out of poverty. The multilevel contribution of women has largely been ignored and inadequately understood in the Indian context. The developmental professionals and the policy makers also divert their attention towards the empowerment of women. But the feminists lacking yardstick to measure their empowerment critically. It is felt necessary to conduct an empirical research on the involvement of women in the empowerment activities so as to tone up the existing systems of developmental scheme on women upliftment.

Research Methodology

Women Empowerment can be operationalised as for the purpose of the present study that, it is an active multidimensional process which should enable the women to realise their power/ identity in the process of decision making, social participation, economic independence, market intelligence, political participation, extension participation, information seeking behaviour practicing sciences – based technologies, leadership quality, awards won and maintenance of health status. It should encompass all the provisions of Laws of equal remuneration, property rights, one – third reservation of seats for women in Panchayat Raj system and equal opportunity for nutritional care and education for the female child. It would create greater access to knowledge and resources and liberate them from all oppressions either in home or in the society.

Developing Women Empowerment Index (WEI)

Definition for Index

The term index has been operationalised as a numerical scale used to compare variables with one another. It is a statistical indicator providing a representation of the value of the securities which constitute it. Indices often serve as barometers for a given market (or) industry and benchmarks against which financial (or) economic performance is measured. For example, broad base index, composite index, high and low index, market index, market share weighted index, market value weighted index, r-square, Russell employment cost index *etc.* are some of the commonly used indices.

Procedure for Index development

Women Empowerment Index was arrived by adding the total scores of each indicator *viz*, decision – making, social participation, economic independence, market intelligence, political participation, information seeking behaviour, science-based technologies, leadership quality, awards won and maintenance of health status. It has been calculated by adopting the following mathematical formula. The procedure had been followed with slight modifications as adopted by Kaliappan (1985) Ganesaan (1989) Mansingh (1993) and Palmurugan (2002). Cumulative frequency was used to classify the level of empowerment of the women. They were classified into five categories *viz*., very low, medium, high and very high.

$$WEI = \frac{(I_{1x} + I_{2x} + I_{3x} + \dots\dots\dots I_{11x})}{(I_{1y} + I_{2y} + I_{3y} + \dots\dots\dots\dots I_{11y})} X\ 100$$

Where,

WEI = Women Empowerment Index

I_{1x} = Total score obtained by 1st indicator

I_{2x} = Total score obtained by 2nd indicator

I_{11x} = Total score obtained by 11th indicator.

I_{1y} = Total maximum possible score for 1st indicator by an individual.

I_{2y} = Total maximum possible score for 2nd indicator by an individual.

I_{11Y} = Total maximum possible score for 11th indicator by an individual.

Thus arrived scores of each indicator was used to calculate WEI. (Women Empowerment Index).

Results & Discussion

It could be discerned from the Table 1 that 40. 60 per cent of total farm women had medium level of empowerment which was followed by low (21.36%), high (18.40%), very high (11.28%) and very low (8.90%) categories. It could be inferred from the Table 1 that the variation in the results that is in the empowerment level of the farm women was mainly due to the differential level of participation in various development programmes. It is quite interesting to note that nearly one- fifth of farm women belonged to high level category of empowerment and another 11.28 per cent belonged to very high category. The women belonging to medium, high and very high level of empowerment worked out to nearly 70-00 per cent. This is an encouraging tread. This would have resulted due to concerted efforts taken by the extension officials, particularly Through TANWA and TNWDP programmes. Another possible reason may be majority of the respondents belonging to illiterate groups which was possible through the intensive adult education programme called Arivoli Iyakkam (Total Literacy campaign).

Table 1: Distribution of respondents according to women empowerment **(no = 337)**

Sl. No	Category	No	%
1.	Very low	30	8.90
2.	Low	70	21.36
3.	Medium	135	40.06
4.	High	62	18.40
5.	Very high	38	11.28
Total		**337**	**100.00**

The Long experience in the social organisations of political participation paves way for aware of their rights and privileges available for them and built up their managerial capabilities which would enhance the leadership qualities (or) to be a good leader. In addition to this the family members would have supported and encouraged them to undergo training programmes they might have given freedom in mobility and financial aspects. Further the farm women might have been motivated and acknowledged by way of issuing awards for their distinguished activities. Above all maintenance of health status was equally important as that of education and other factors. The improvement in health status may be through proper medical facilities. If the farm women is physically healthier which reflects in their efficiency and progress of their every work. They will be able to give disease free healthier babies. This result is corroborated with NSSO (2003) which reported that the health of an individual farm women is an important factor in determining the over all health of society. If pregnant women are not well nourished they are more likely to give birth to weaker babies. Leading to a higher infant mortality rate (Fig. 1).

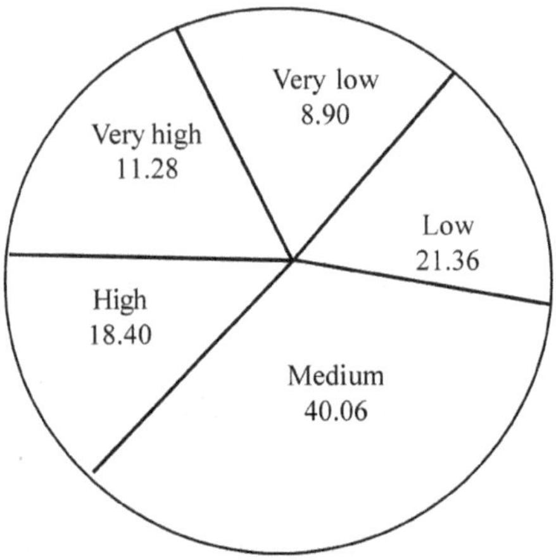

Fig. 1: Distribution of respondents according to women empowerment (%)

This research findings who derives support from Palmurugan (2002) revealed that more than half (66.307) of the form women had medium empowerment level followed by low level (18.90 %) and the remaining 14.80 per cent with high level of empowerment.

Sailaja (2002) study findings also give support that the 50.00 per cent of the total women respondents were empowered through cooperatives by availing the benefits being as member (or) office bearer in the same.

The empowerment level was studied by Shantha Sheela (2002). The study identified the performance of farm women who involved in the developmental activities. That is cent per cent of the respondent presidents performed the roles by conducting Grama Sabha meeting as per schedule and passed resolutions in the same. More than ninety per cent of respondents performed the role in executing decisions on Grama Sabha, in public distribution systems, the activities related to primary and middle school, rural electrification works, in providing transport facilities, in providing drinking water facilities and also in rural housing activities. Considering the overall performance, 41.10 per cent fell in the medium category followed by 32.88 and 26.02 per cent in low and high categories respectively. These were the reasons behind the farm women who were in the high to very high category of empowerment. The farm women distributed in the other categories might be due to the variation in these developmental activities.

Conclusion

The critically examined empirical data on overall empowerment of women would be tangible to the policy makers of Govt, Private firms, NGOs and Live departments. The data obtained for a particular period would reflect the efficiency made on women empowerment through all stakeholders of all empowerment activities. Thus it would assist the programme / project / scheme evaluations. Further it is observed in the category of low to very low category. If the development agents consider the aforesaid indicators in the developmental schemes, these kinds of variations could be managed positively.

References

Poornima Advani (2004). Index lies in empowerment: Make women's commission autonomous statutory body.The Hindu, dt:Feb.20,2004.

National commission for women (1994). *In:* Literacy: Gateway to women empowerment. Padmanabhan, K. Serials Publications, New Delhi, pp.126-131.

www.ministry of population and environment.com.2003.

Extension Management Strategies for Sustainable Agriculture, pp 383-396

New India Publishing Agency, New Delhi, India

43

Family Farming for Success Through Integrated Farm

P. Marimuthu

World-wide, both in developing and in developed countries;family farming is one of the most predominant forms of agriculture. It includes a wide spectrum of farm sizes and types. In high-income economies, it ranges from very large land holdings that are easily cultivated by one or two family members with the use of labour-saving machinery and hired labour. In low-income economies,with small holdings of a few hectares or less and that are often oriented towards subsistence with low marketable surplus.

There are approximately 500 million small family farms, of which 280 million are in China and India alone (IFPRI, 2007).

Family farmers produce more than half of the world's food.Yet, 70% of the world's most impoverished people live in rural areas and belong to family farming or pastoralist communities.

Some countries are experiencing strong economic growth, even as food and nutrition insecurity increase. Conflicts over resources (land, water, and biodiversity) and degraded resource bases often threaten the poor and the marginalized who are directly dependent on these resources.

Poverty is not just about a lack of money. It has as much to do with people's asset base in a much broader sense, and thus with power. Poverty is also about people's capability to deal with situations of shock or stress – whether social, economic, political or physical. Poor people lack the *resilience* to effectively cope with these shocks. Over the last decade climate change has added to these problems, resulting in ever growing *resilience deficits* in rural communities.

There are powerful forces pulling agriculture into a very different direction, as can be seen in the focus on agribusiness in rural policy and practice. In more and more places, young family farmers are discovering a future in agriculture by working with, rather than against, nature.

Agro-ecological approaches, agro ecological farming practices and the social dimensions of family farming strengthen the resilience of family farmers, sustainable resource management and help them break out of poverty. Family farmers also connect with urban consumers, building new local and regional food systems that are transparent, healthy, fair, efficient and sustainable.

Meaning of Family Farming

Family farming is understood to include crop, livestock, forestry, fishery and aquaculture production by producers who, despite their great heterogeneity among countries and within countries, have the following key characteristics *viz.,* limited access to land and capital resources, predominantly family labour is used in the production process and agriculture/forestry/aquaculture/fishery activity is the main source of income for the family (FAO).

Family farmers' inherent motivation

Even though family farming may not appear as neat and scientific as commercial farming, there are significant gains to be made by supporting family farming more effectively. One of the main reasons is that as the family and the farm are linked and co-evolve, they combine not only economic functions but also a range of other 'hidden' functions, including environmental, reproductive, social and cultural ones, often in lieu of state institutions or the private sector.

Environmental functions include soil enrichment, carbon sequestration, water purification, pest control, pollination and biodiversity enhancement. Reproductive and social functions include childcare, nutrition, water and energy provisioning, education, health, social security, insurance and risk management. Cultural functions include transmission of identity, symbolic and religious values of resources and territories, knowledge and technologies.

In performing all these functions, family farming is also often a means of maintaining family patrimony, cultural heritage, territories, landscapes, and communities. As a result, the motivations of family farmers often go far beyond profit maximization, to encompass social, cultural and ecological aspects.

Importance of Family farming

Family farming ensures guarantee of food supply as 70% of the world's food production is provided by family farmers which is key to fight hunger and malnutrition. Family farms are often more productive and sustainable per unit of land and energy consumed.

Characteristics	FAO	USA	Colombia	Brazil	Paraguay	Uruguay	Argentina	Chile	Europe	Central America	Russia	Asia	East Africa
Own and control by family		■	■	■					■	■	■		■
Predominant use of family labour	■		■	■	■		■	■	■	■		■	
Access to land	■		■	■	■	■	■	■		■		■	■
Farming, main source of income	■			■		■		■	■				■
Multiple income source										■			
Total value of the farm			■										
Family lives on the farm					■	■			■				
No permanent non-family workers							■						

Characteristics of Family farming worldwide

It generates welfare for 40 % of world households who depend on family farming and out of the 3,000 million rural people in developing countries; 2,500 million belong to families engaged in Family Farming. It is at least twice as effective as other production sectors in the prevention of poverty.

Family Farming, besides being a source of genetic agro-diversity, can ensure their preservation through the use of seed varieties and native livestock breeds well adapted to various environments. Women contribute a significant proportion of the agricultural labor force in family farming in developing countries (FAO estimates this figure at 43%).

Family farming is linked with food security. It preservcs traditional food products and knowledge, contributes to a balanced diet, safeguards agro biodiversity, contributes to sustainable natural resources use and boosts local economies. Family farming is associated with family values such as solidarity, continuity and commitment, specific entrepreneurial skill, business ownership and management, choice and risk behavior, resilience and individual achievement.

Family farming preserves traditional food products, while contributing to a balanced diet and safeguarding the world's agro-biodiversity and the sustainable use of natural resources. Family farming represents an opportunity to boost local economies, especially when combined with specific policies aimed at social protection and well-being of communities.

Family farming carries the promise of creating agricultural practices that are highly productive, sustainable, receptive, responsive, innovative and dynamic. Given all these features, family farming can make a strong contribution to food security and food sovereignty. It can strengthen economic development in a variety of ways, creating employment and generating income. It strengthens the economic, ecological and social resilience of rural communities. It offers attractive jobs to large parts of society and can contribute considerably to the emancipation of downtrodden groups in society. Family farming can also consistently contribute to the maintenance of beautiful landscapes and biodiversity.

Qualities of Family farming

Family farming is not just about the size of the farm, it is more about the way people farm and live. This is why family farming is a way of life. The reality of family farms is far richer than the two single aspects that are most commonly used to describe them: that the farm is owned by the family and that the work is done by the family members.

Qualities	Characteristics
Balance of farm and family	• Controls its main resources • Provides main part of labour force • Family-farm nexus
Linking past, present and future	• Provides income, food and nutrition • Home to the family, place of belonging • Links past, present, future • Place for learning, knowledge building
Tied to its environment	• Active part of rural economy • Keeps culture alive • Tied to rural landscape, its environment

A Balance of Farm and Family

The farming family has control over the main resources that are used in the farm. This includes the land, but also the animals, the crops, the genetic material, the house, buildings, machinery and, in a more general sense, the know-how that specifies how these resources need to be utilised and combined. Access to networks and markets, as well as co-ownership of co-operatives, are also important resources. Family farmers use these resources not to make a profit, but to make a living; to acquire an income that provides them with a decent life and, if possible, allows them to make investments that will further develop the farm.

In addition the family farm is a place where the family provides the main part of the labour force. This makes the farm into a place of self-employment and of progress for the family. It is through their dedication, passion and hard work that the farm is further developed and that the livelihood of the family is improved. The farm is there to meet the many needs of the family, whilst the family provides the possibilities, the means and also the limitations for the farm.

This nexus between the family and the farm is at the core of many decisions about the development of the farm. Each particular farm has its own specific balances, for instance between the mouths to be fed and the hands to do the work. These balances tie family and farm together and make each family farm into a unique constellation.

Linking Past, Present and Future

But there is more than ownership and labour. Family farms provide the farming family with a part (or all) of its income and food. Having control over the quality of self-produced food is becoming increasingly important for farmers around the world.

However, the family farm is not only a place of production. It is also home to the farming family. It is the place they belong to, as much as the place that gives them shelter. It is the place where the family lives and where children grow up.

The farming family is part of a flow that links past, present and future. This means that every farm has a history and is full of memories. It also means that the parents are working for their children. They want to give the next generation a solid starting point whether within, or outside, agriculture. And since the farm is the outcome of the work and dedication of this and previous generations, there is often pride. And there can also be anger if others try to damage or even destroy the jointly constructed farm.

The family farm is the place where experience accumulates, learning takes place and knowledge is passed on, in a subtle but strong way, to the next generation. The family farm often is a node in wider networks in which new insights, practices, seeds, *etc*., circulate.

Tied to its environment

The family farm is not just an economic enterprise that focuses mainly, or only, on profits, but a place where continuity and culture are important. The farming family is part of a wider rural community, and sometimes part of networks that extend into cities. As such, the family farm is a place where culture is applied and preserved, just as the farm can be a place of cultural heritage.

The family and the farm are also part of the wider rural economy; they are tied to the locality, carrying the cultural codes of the local community. Thus, family farms can strengthen the local rural economy: it is where people buy, spend and engage in other activities.

Similarly, the family farm is part of a wider rural landscape. It may work with, rather than against nature, using ecological processes and balances instead of disrupting them, thus preserving the beauty and integrity of landscapes. When family farmers work with nature, they also contribute to conserving biodiversity and to fighting global warming. The work implies an ongoing interaction with living nature – a feature that is highly valued by the actors themselves.

Freedom and Autonomy

The family farm is an institution that is attractive as it allows the farm family a relative degree of autonomy. It embodies a "double freedom": there is freedom from direct external exploitation and there is freedom *to* do things in your own way. In short, family farming represents a direct unity of manual and mental labour, of work and life, and of production and development. It is an institution that can continue to produce in an adverse capitalist environment.

External threats

However, it may turn out to be impossible to effectively realise all these promises. This is particularly the case these days, when family farming is being squeezed to the bones and becoming impoverished. When prices are low, costs are high and volatility makes medium or long term planning impossible, and when access to markets is increasingly blocked and agricultural policies neglect family farmers, and when land and water are increasingly grabbed by large capital groups – in such circumstances it becomes impossible for family farmers to provide positive contributions to society at large. This is why we have now ended up in the dramatic situation that the land of many family farmers is laying idle. Or, to use a macro indicator, that 70% of the poor in this world today, are rural people.

Internal threats

There are internal threats as well. Nowadays it is *en vogue* to talk about the "need to make family farming more business-like", in other words that it should be oriented towards making profits. Some even argue that this is the only way to keep young people in agriculture. According to this view, family farming should become less "peasant like" and more "entrepreneurial". This has effectively turned the family farm into a mere supplier of labour, at the expense of all the other features.

Formally, these entrepreneurial farms are still family farms, but in substance they are quite different. One major difference is that "real" family farms grow and develop through clever management of natural,economic and human resources, and through (intergenerational) learning. Entrepreneurial farms mostly grow through taking over other family farms. This tendency to enter into entrepreneurial trajectories is a major internal threat to the continuity and dominance of family farms.

Experiences of Family farming in Theni District

Family farming through Integrated farming system (IFS)

The farmer Mr. P. Kottaichamy @ Sekar is hailing from Kutchanur village in Uthamapalayam Taluk of Theni District in Tamil Nadu. He developed 15 acres of land holding with family farming. Interventions of KVK made him to adopt technologies of Integrated Farming System for reducing the dependence on external inputs and increasing net profit from his farm.

He started practicing the IFS with available resource potential. This approach not only saved lot of investment but also ensured several direct and indirect benefits. He has started dairy farm as a base in his farm and then expanded various enterprises for increasing the opportunity to effectively recycle the waste and utilize it in the farm for soil fertility maintenance. The farmer has decided to use only organic inputs in his farm.

Components of IFS

He has integrated cattle rearing, vermicompost unit, organic banana and rice cultivation, minor millets cultivation, cultivation of tree fodder and cattle feed mill in his farm. The production of Dried FYM, Enriched Farm yard manure, Jeavamirtham and Pancha Kavya are done in the farm itself for regular utilisation. His wife maintains all the dairy and allied activities. All the crop wastes are effectively utilized as mulching materials. He solely depends for 90% of the inputs in his farm and only 10% of the inputs are purchased from outside.

Cattle farming

The farmer have decided to use organic inputs in his farm and established a cattle farm with 10 cows initially. Now it has grown up to 50 numbers of cows. Cattle manure is the primary source of soil organic input and hence the farmer decided and established the cattle unit. The growth and development of the farm, income from the farm, reduced input cost, increased crop yield, local indigenous inputs and resource utilization have induced the intension of other farmers to practice IFS. Cattle shed with 50 cows are located in the farm itself. Sufficient cattle fodder crops and cattle feed mill is located in the farm site. The raw materials for production of cattle feed *viz*. Maize, Sorghum, Cumbu and pulses dust are produced in the farm land.

Vermicompost production

There are 24 number of vermi pit established with 30 x 5 x 2 feet L x B x D size to produce vermicompost. The construction is made with cement and bricks on all sides including the platform. The bottom and sides, inside and outside lined with cement and sand plastering. The farm waste *viz*. dried leaves, coconut refuse, banana leaves, waste cattle fodder and Paddy straw are effectively recycled for vermicompost production. About 20 ton of vermi compost is produced per month. Out of this, 5 ton is used for the farm use to banana and coconut crops and 15 ton is sent to external market at Kerala for cardamom and tea Gardens. The farmers of Iddukki district are directly purchasing the Vermicompost since the village is nearer to Idukki district of Kerala State.

Production of Enriched Farm Yard Manure

Cowdung is separately collected and Enriched Farm Yard Manure is prepared through aerobic and anaerobic process. Such produced manure is rich in nutrient content which reduces 90% at the fertilizer application or 60% of the FYM application in the field. Enriched FYM have lot of fungi, bacteria and other beneficial pathogens which helps to improve the Soil health status and the quantity of application is enormously reduced by one by tenth of direct application of FYM. Before application of EFYM, it should be mixed with Jeevamirtham 1 lit, Azospirillum 100 g + Phosphobacteria 100 g + Pseudomonas 25 g + Trichoderma viridi 25 g per every 100 kg of EFYM.

Production of Jeevamirtham

Jeevamirtham is used as insect pest repellant as well as liquid nutrient. About 10 lit of cow urine + 10 kg cowdung + 2 kg dhal powder + 2 kg jaggery + local soil 100g + oil cake 100 g are mixed together in 100 lit, of water and allowed for fermentation. After 30 days, the mixure is used as crop spray or root application through drip to banana and coconut crop. Each 1 lit of Jeevamirtham is mixed with 100 lit of water and used in drip irrigation.

Family Farmers as change agent

As the farmer is associated with Banana growers association, fellow farmers visited this family farm and observed the day to day activity of this family farm, source of income, quantity produced, quality of production, marketing strategies and market place. He used to share his experience in a practical way to other farmers as a change agent. His motivational experiential sharing has benefitted 120 farmers of Theni and other neighbouring districts by introducing integrated family farm.

Local department officials also arranging tour programme to this farm as exposure visit. There are about 200 farmers, farm women and rural youth and students from various parts of India have come and visited this family farm and undergone training on various organic inputs preparation. This farmer also goes for consultancy programme to various places inside and outside the state on Integrated Farming system.

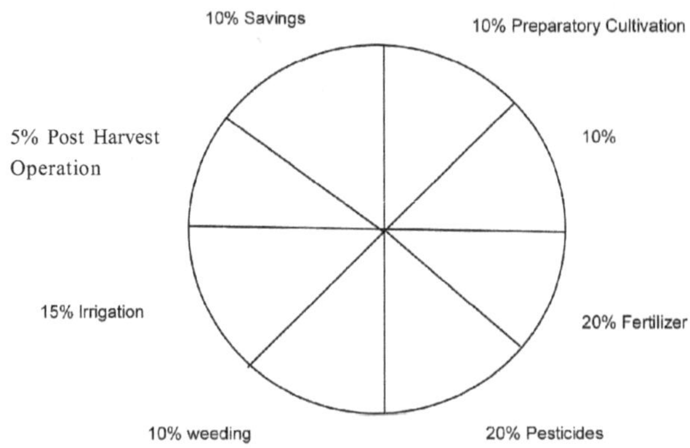

Input Cost before Integrated Farming System

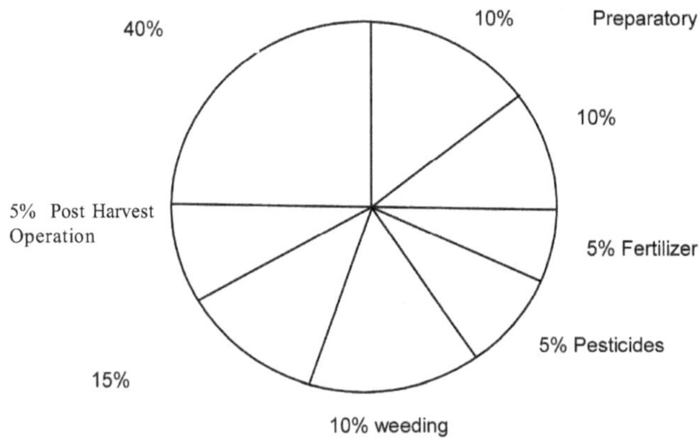

Input Cost after Integrated Farming System

Family Farming by Paliyar Tribal families

Paliyar Tribals practice indigenous way of family farming. About 40 Paliyar tribal families of Paliyankudi village practice family farm. Each and every person in the household divides and shares the responsibility for successful running of the family farm. The family farming helped to revive minor millets cultivation, reduced dependence on forest for fodder through fodder cultivation, revived herbal and kitchen garden, revitalized local health traditions and healing practices and improved social value.

The family farming helped to identify two indigenous minor millets varieties *viz*., Sadai Samai and Kirukku Varagu and tribals have started cultivation in family farm. The family farm and the cohesive bonding among family members and the Paliyar tribal community helped in sharing indigenous knowledge on medicinal plants, medicinal tubers and traditional healing.This helped them to conserve diversity of medicinal plants, medicinal tubers and minor millets.

Strategies for Strengthening Family farming

Re-peasantisation

Many family farms are strengthening their position and their income by following agro-ecological principles, by engaging in new activities, or by producing new products and new services – often distributed through new, nested markets. Analytically these new strategies are defined as forms of re-peasantisation. Re-peasantisation is way of defending and strengthening family farming.

Revitalising family farming

Traditional practices in family farms can strengthen cultural identity, build solidarity among farms, and assign greater value to traditional ethics. Revitalising and enhancing traditional practices is a learning opportunity that has practical implications for climate change mitigation and building resilience to disasters and crises. Finally, family farming needs to be made more attractive for future generations, using social marketing and the media to counter the ongoing devaluation of family farmers. Rather than framing farming as a last resort for the uneducated and unemployable, we need to promote family farming as a noble vocation.

The environment policy is extremely important for the fate of family farming. Although family farming can survive in highly adverse conditions, positive conditions can help family farming reach its full potential. This is precisely the reason that the policies of state apparatuses, multinational forums (such as the FAO, IFAD and other UN organisations), but also of political parties, social movements and civil society as a whole, are so important.

Policy can ensure that family farmers' rights are secured and that sufficient investments are made in infrastructure, research and extension, education, market channels, social security, health and other areas. This provides farmers with the security to invest in their own futures, as recently reconfirmed by the prestigious High Level Panel of Experts on Food Security and Nutrition.

Strengthening rural organisations and movements is also of great importance. We have to keep in mind that, all around the world, family farmers are trying to find and unfold new responses to difficult situations.Thus, identifying successful responses, building on novel practices, communicating them to other places and other family farmers and linking them into dynamic processes of change must be central items on our agenda. In short, much needs to be done. The good news though, is that every step, no matter how small, is helpful.

In contrast to the dominant calls for commercialization and intensification, farmer organisations have to implement initiatives to support family farmers by sensitizing on "Preserve the best and adopt the new". The organisations can supply services and improve access to existing services, including markets and value-adding processes. The organisations can help showcase effective farming practices and try to build a strong positive image of family farmers and their products.

The strategies should aim to strengthen production systems, to rescue and revive ancient knowledge, and to allow the participation of grassroots organisations and peasant families in decision making.

Agro-ecological centres can be established which will enable young people to contextualise and practice agro-ecology in the communities.

International Year of Family Farming (IYFF)

On September 2011, the 66th Session of the General Assembly of the UN formally declared 2014 to be the International Year of Family Farming (IYFF). It stresses the need for appropriate investment in sustainable family farming. Family farming can feed the world and strengthen resilience. The International Year of Family Farming brings hope to many small-scale farmers, who are counting on firm steps at international, national and local levels.

The major aim of the IYFF is to raise the profile of family farming and smallholder farming by focusing the world's attention on their key role in alleviating hunger and poverty, providing food and nutrition security, improving livelihoods, managing natural resources, protecting the environment, and leading towards more sustainable development, in particular in rural areas.

The goal of the 2014 IYFF is to reposition family farming at the centre of agricultural, environmental and social policies in the national agendas by identifying gaps and opportunities to promote a shift towards a more equal and balanced development. The 2014 IYFF will promote broad discussion and cooperation at the national, regional and global levels to increase awareness and understanding of the challenges faced by smallholders and help identify efficient ways to support family farmers.

Hopefully the International Year will support family farmers by working with all stakeholders, women, men, young and old, to identify new and better ways to enable them to enhance their prosperity, sustainability and freedom to achieve their own aspirations for a better future.

Family farmers and the many ways in which they contribute to food security, healthy landscapes and thriving rural communities can be supported in a number of ways. Supporting family farmers will both protect them and help them reach their full potential. Appropriate policies and measures like easy access to resources, such as credit and land, support for farmers who practice agro-ecological techniques, access to fair markets, including adequate roads and transportation, and protection against dumping, better inclusion of women farmers in financial and technical support schemes, and appropriate facilities for food storage and processing, to add value and avoid post-harvest losses.

As the family and the farm are linked and co-evolve, they combine not only economic functions, but a range of other 'hidden' functions, including environmental, reproductive, social and cultural functions. As a result, the motivations of family farmers often go far beyond profit maximization.

Education for Family farming

Family farmers can make a large contribution to increasing food production and enhancing sustainability. And yet, conventional agricultural education pays very little attention to sustainable family farming, the way it works, or its potential. There is a huge gap between the daily realities of hundreds of millions of family farmers and what is being taught to agricultural students around the world.

Men and women play different, but complementary, roles and strengthen the farm as a multifunctional system. But sometimes their roles can clash. This can happen when farmers move to more entrepreneurial value chains. This may have negative implications for women's autonomy with respect to food production, and for food security at the household level. Therefore, women need to be centre stage in the IYFF, and in decision-making on the future of our global food system.

Furthermore, family farming relies upon family members with different labour power, skills, capacities, opportunities and constraints, which vary in part depending upon gender and age. These characteristics influence intra-household relations, which in turn influence the distribution of resources, roles and responsibilities. Put simply, the intra-household distribution of resources and responsibilities in family farms is often not equitable, especially with regard to women and children.

The multiple motivations that confront family workers, in contrast to hired workers, and especially the strong incentive to work for the sake of their own families' well-being, greatly reduce the supervision costs associated with agricultural labour. Moreover, because family farmers often have intergenerational bonds with the holdings they work, their production also frequently provides continued ecosystem services and care for the natural resource base. Because of this, family farming is particularly well suited to holdings characterised by a highly diverse set of economic activities and mosaic-type landscapes, in which the supervision and knowledge required for numerous small and on-the-spot production management decisions is most efficiently and cost-effectively devolved to family workers.

Supporting family farming would therefore also be giving preference to a specific type of landscape and set of traditions that are more conducive to biodiversity preservation, ecosystem balance, and good environmental stewardship.

Family farming and the support it receives need to adjust in ways that can respond to these changing conditions. Innovation is key to make this happen: family farmers need to innovate in the systems they use; governments need to innovate in the specific policies they implement to support family farming; producers' organisations need to innovate to respond better to the needs of family farmers; and research and extension institutions need to innovate by shifting from a research-driven process predominantly based on technology transfer, to an approach that enables and rewards innovation by family farmers themselves.

References

Jan Douwe van der Ploeg (2013).Ten qualities of family farming. Farming Matters 12/2013-29.4

Biplab Sarkar (2014). Concepts of Family Farming. FAS, Bangalore.

Dorian Kalamvrezos (2013). Family farming – the backbone of sustainable rural development. www.DorianKalamvrezos.Navarro(at)fao.org

http://www.fao.org/family-farming-2014

http://www.agriculturesnetwork.org/news/india-launch-iyff

http://www.agriculturesnetwork.org/what-we-do/advocacy/2014-iyff

http://www.agriculturesnetwork.org/magazines/global/family-farming

http://www.fao.org/family-farming-2014

http://en.wikipedia.org/wiki/Family_farm

http://opinion.inquirer.net/79690/yes-to-family-farming-farmers-land-rights

http://jamaica-gleaner.com/gleaner/20141027/int/int1.html

http://www.trinidadexpress.com/commentaries/Family-farming-nourishing-the-world-281395651.html

http://www.thepigsite.com/swinenews/38138/family-farmers-play-significant-role-in-eradicating-world-hunger

http://realtruth.org/articles/100607-006-family.html

http://www.timeanddate.com/year/2014/family-farming.html

http://www.familyfarmingcampaign.net/en/family-farming/concept

http://www.astc.org/blog/2014/04/17/family-farming-feeding-the-world-caring-for-the-earth/

http://rustikmagazine.com/importance-family-farming/#.VFoTume1-t8

Extension Management Strategies for Sustainable Agriculture, pp 397-401

New India Publishing Agency, New Delhi, India

44

Livelihood Security of Tribal Women in Eastern Ghats

K. Dhanasree, B. Vijayabhinandana, S.V. Prasad
P.B. Pradeep Kumar and D. Sekhar

There are 437 tribes in India and Andhra Pradesh is inhabited by 33 tribes. Of these, 13 tribal groups inhabit in high altitude zone of Visakhapatnam. Livelihood security is the ability of a household to meet its basic needs like food, health and shelter, basic education and community participation. Livelihoods can be made up of a range of on-farm and off-farm activities which together provide a variety of procurement strategies for food and cash. Thus, each household can have several possible sources of entitlement which constitute its livelihood comprises the capabilities of people, their assets and activities required for a means of living. A livelihood is a system, which can cope with and recover from stress and shocks, enhance its capabilities, assets and provide sustainable opportunities. Each household combines its livelihood resources within the limits of their social and political content to persue a number of livelihood strategies such as various types of production and income generating activities (Frankellis, 2007). A holistic analysis of livelihood security begins with the understanding of the dimensions, the factors that influence the range of possibilities for each livelihood system and should include the household livelihood analysis to determine the information on the resources held by each household and how these are used to earn adequate income and critical outcomes achieved in terms food security, nutrition and health status and access to resources like water, shelter and safe environment. Households combine their livelihood resources within the limits of their content and use their institutional connections to persue a number of different livelihood strategies. Such strategies can include various types of production and income generating activities like farming, non-farm activities, formal sector employment or a combination of all these.

In this context, an understanding of livelihood security of tribal women with a multidimensional approach was undertaken with the following objectives.

1. To examine the livelihood systems and livelihood security of tribal women.
2. To design a strategic model for enhancing livelihood security.

Methodology

The study was carried out in 6 tribal dominated Mandals of high altitude and tribal zone of Andhra Pradesh during 2010-11. Six Mandals namely Chintapalle, G.K.Veedhi, Hukumpeta, G.Madugula, Araku and Pedabayalu were selected for investigation. Three villages from each of the six mandals were selected and from each village 10 respondents were selected thus making a total of 180 respondents. Data was collected through structured and pre-tested interview schedule developed covering all the facts of livelihood security. The collected date was coded, tabulated and analyzed statistically and the results were interpreted accordingly.

Results and Discussions

Livelihood systems and livelihood security of tribal women

Livelihood systems in the area of study are primarily dependent on combinations of agriculture, forest and wage earners. Due to very small holdings and very low productivity of the land most households earn a living by maintaining a diversified pattern of occupations, no single activity provides sufficient resources to entirely ensure their livelihood. Women among tribal population have very important role in agriculture cum forest based economies.

Table 1: Distribution of the respondents according to their livelihood systems

S. No.	Livelihood systems	Frequency	Percentage
1.	Forest based activities + wage earners	76	42.22
2.	Agriculture + Forest based activities + wage earners	49	27.22
3.	Agriculture + wage earners+ Animal husbandry	27	15.00
3.	Agriculture alone	16	8.88
4.	Agriculture + petty business	12	6.66
	Total	**180**	**100**

It was found from Table 1 that majority (42.22%) of the respondents were involved in Forest based activities + wage earners followed by Agriculture + Forest based activities + wage earners (27.22%), Agriculture + Wage earners + Animal husbandry (15.00%), Agriculture alone (8.88%) and Agriculture +

Petty business (6.66%). Most of the households supplemented their household income through Forest based activities + wage earning occupation for their livelihood. This might be because of two reasons (i) most of the tribes were landless and (ii) those who possessed land, hold only limited land holding. They were not in a position to meet the expenditure required to run the family with only one source of income. Hence most of the households earn a living by maintaining a diversified pattern of occupations viz.,on-farm activities, wage employment, forest activities, small enterprises. Tribal women in the study area are involved in seasonal collection of forest products like leaves, hill-brooms, tamarind, honey, medicinal plants and barks, thatching grass. Majority of the respondents agreed with the fact that they were getting employment under the scheme of National Rural Employment Guarantee Scheme (NREGS). The scheme provides alternative employment opportunities to earn wage income and reducing the drudgery (Shrinidhi, 2006). Respondents indicated that income from this scheme is sometimes utilized as a coping strategy to meet household expenses.

The overwhelming conclusion is that women not only bear a greater proportion of the burden and their efforts are critical in contributing up to 50 per cent of the household income. Women in tribal households often have a greater say because of their enormous contribution to the welfare of the household. Once the sale of the forest produce is made in the weekly market or at the forest cooperative, women generally purchase food and other basic necessities for the household. The results indicate that appropriate steps need to be taken to improve upon livelihood of tribal women.

Livelihood Security of Tribal Women

Almost in all the sample villages, besides agriculture the tribal women including children go to forest for collection of leaves, hill-brooms, tamarind, honey, medicinal plants and barks, fuel wood, roots, tubers, leafy vegetables and medicinal plant species, whereas wage labour provides engagement for 100 days, which has been possible after implementation of NREGS and other poverty alleviation schemes. Majority of the households have poor access to safe drinking water. Health measures and village sanitation are unsatisfactory in most of the villages.

On the basis of livelihood security index score, the respondents were categorized into three groups *viz.*, Rich, Better and Poor.

Table 2: Distribution of respondents according to their livelihood security **(n = 180)**

S.No	Category	Frequency	Percentage
1.	Rich	22	12.23
2.	Better	95	52.77
3.	Poor	73	35.00
	Total	**190**	**100.00**

From the above Table 2 it is evident that 52.77 per cent of the households in the study area had better livelihood security while (35.00%) had poor livelihood security and (12.23%) of the households had rich livelihood security. Women among the tribal population have very important role and women's work is regarded as crucial for the survival of tribal households in terms of provisioning for food, income, earning and management of financial resources (Singh, 1993). Therefore it can be said that in order to contribute to their household needs women must have exposure to areas of new developments which would relevant in opening the windows of livelihood opportunities which would be able to ensure their livelihood security.

Based on the findings it is suggested that, it is endeavor of all the line departments like agriculture, horticulture, animal husbandry, extension agencies and other non-governmental organizations have to give more emphasis on women participation. This will make them to get aware of day to day technological developments and the impact of adoption of those scientific practices on their enterprises and in turn enhances their livelihood.

Strategy for Enhancing the Livelihood Security of Tribal Women

The empowerment of tribal women and development of tribal women and sensitization of research and extension systems about their roles and contributions are important for sustainable growth and management of livelihood security. It is this reason that emphasizes the need to develop understanding about tribal women, their potentials, awareness of household food security management and extent of participation in planning and decisions of different farm and home management works towards food security and the constraints inhibiting them to efficient food security management at household level. A strategy in this study was conceived as a plan of action in order to develop the tribal women and sensitize the development agencies for greater contribution for ensuring livelihood security. Suggestions made by the sample were listed and filtered them based on their feasibility and retained only the workable suggestions for developing a strategy as an action plan of activities to be undertaken by the administrators.

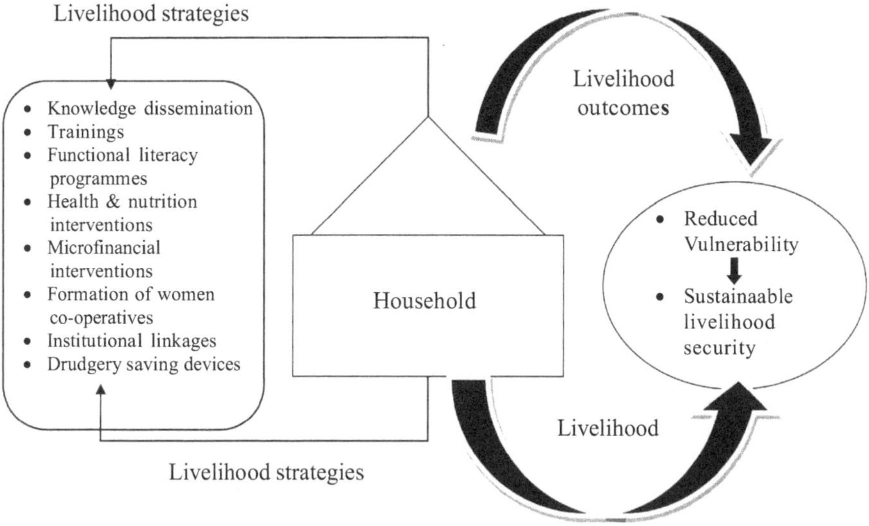

Fig 1: Strategy for enhancing the livelihood security of tribal women

Conclusion

The study revealed that holistic approach is desirable for success of tribal development and for a sustainable livelihood pattern a strong resource base is a pre-requisite. It is most desirable to extend sustainable socio-economic status, empowering women, better health care, programmes aiming to improve nutritional status, providing transport and communication facilities to tribal households. Efforts should be strengthened by conducting group trainings in a systematic manner, providing continues facilitation support by exchange of information and ideas, mutual trust between banks and SHG's motivation campaigns and skill oriented trainings for empowerment of tribal women which also facilitates in terms of better livelihood support, poverty reduction, food and nutritional security of the family.

References

Frankellis (2007). Household strategies and rural livelihood diversification. *Journal of Development Studies*. 35: 1-38.

Shrinidhi (2006). Differential impacts of commodification of agriculture on people's livelihoods and the environment in the Western Ghats of India: An extended environmental entitlement analysis, *Ph.D thesis*, the Florida State University, College of Social Sciences.

Singh (1993). Status of tribal women in India. *Social Change*. 23(4): 3-18.

Extension Management Strategies for Sustainable Agriculture, pp 403-412

New India Publishing Agency, New Delhi, India

45

Personal, Socio-economic and Self Employment Aspiration of Food Technology Students

P. R. Deshmukh and R. P. Kadam

The College of Food Technology was established in the year 1976 at Parbhani as one of the constituent College of VNMKV, Parbhani. Food Technology College involves the knowledge of application of different fields of food Science and also includes transfer of Food Science technology to the students to make them well equipped with knowledge and skills required for employment in public and private sector.

Food technology education offers many opportunities to plan and choose a carrier to its recipients. However, the faculty students do have some plans to persue and specific carrier after getting their degree. They do aspire for remunerative job as a fruit of their hard labours and have their own liking and disliking for various carrier avenues open to them after graduation. Every student has some aspirations *i.e.* the desired conditions they think proper and they are always directing their energy and efforts to attain the desired aim in relation with major function of Agricultural Universities or to impart under graduate and post graduate education to the students of Agriculture universities.

Prime importance of institutionalized education is to produce qualified man power to uphold social development aspiration of individuals is the goal statement concerning future level of achievements. People wish to thinks for which experience towards a goal.

The study of aspiration of students in Food Technology would help to understand potential of man power development which will be responsible for influence the social development through agriculture. In this modern age, it is imperative to know the choice of carrier of food technology students and to prepare them for satisfied living to avoid frustration in their life. In view of the above it was necessary to study aspiration and employment opportunities of food technology students with the following specific objective.

1. Objectives

1. To study the profile of Food Technology Students.
2. To study the motivation of the students for securing admissions in food technology.
3. To study the aspiration of food technology students.

2. Methodology

The present study was conducted in the two food technology colleges under Vasantrao Naik Marathwada Krishi Vidyapeeth, Parbhani. *i.e.* Food Technology College, Parbhani and Rajiv Gandhi Food Technology College, Parbhani. In this study the data were collected from randomly selected 45 under graduating students of third year of each college admitted in the academic year 2013-14. Thus a sample of total 90 students was selected for this study.

The data gathering tool was questionnaire paper based survey was used in order to allow survey to achieve high coverage as questionnaires could be given directly to the students in the class and collected at the same time, which was structured around two parts. The first part was related to profile of the students and the second part which is most important section of questionnaire. In particular the research question for study includes educational, social, economic, professional, job and self employment aspiration.

The data were subjected to simple statistical analysis like frequency, per centage, mean and standard deviation etc.

3. Findings

The finding of the study is reported below.

3.1 Profile of the Respondents

3.1.1 Family background

It is seen from table that significant percentage of the respondents (66.67 per cent) were from rural areas where as 33.33 per cent were from urban areas.

3.1.2 Parents occupation

It is observed that majority of the respondents (47.78 per cent) were having farming while 28.89 per cent were having service followed by 16.67 per cent having their own business; 4.44 per cent and 2.22 per cent were engaged in labouring and other occupation respectively.

3.1.3 Family type

It is noticed that majority (66.67 per cent) of the respondents were from nuclear families whereas 33.33 per cent were from joint families.

3.1.4 Family size

The present data indicated that family size of majority (53.34 per cent) of the respondents was medium. There were 33.33 per cent respondents having small size family and 13.33 per cent had big size family.

3.1.5 Land holding

It is evident from table that 25.56 per cent of respondents parents were having medium land holding while, 24.44 per cent of there were having no land followed by 18.89 per cent respondents parent were having small land holding and others possessed semi medium land holding (17.78 per cent) Big land holding (10.00 per cent) and marginal land holding (3.33 per cent).

3.1.6 Annual income of family

It is seen from table that most (46.67 per cent) of the respondents family had low level of income upto Rs. 75984. Where as 28.89 per cent of respondents family had medium income level *i.e.* Rs. 75985/- to Rs. 245855 and 24.44 per cent respondents' annual family income was above Rs. 245856/- higher income level.

3.1.7 Family educational status

It is noticed from table that 26.67 per cent respondents' mother education was secondary school level, 25.55 per cent respondents' mother was found illiterate, while 16.67 per cent and 14.45 per cent respondents mother education was higher secondary school and middle school level respectively. However, 8.88 per cent and 7.77 per cent respondents' mother education was graduate and primary school level.

However, 44.44 per cent respondents' father education was graduate level, 21.11 per cent respondents' father education was higher secondary school, while 14.45 per cent and 13.33 per cent respondents father education was secondary school level andilliterate respectively. Whereas, 4.45 per cent and 2.22 per cent respondents' father education wasmiddle school level and primary school level respectively.

3.1.8 Scholarship

The data revealed that (41.11 per cent) of respondents availed the GOI scholarship, followed by 32.22 per cent of respondents were not holding any type of scholarship. whereas 14.45 per cent respondents belongs to merit cum scholarship while 8.89 per cent and 3.33 per cent respondents availed other scholarship and ICAR scholarship respectively,

3.1.9 Academic performance

It is clear from table that majority (44.45 per cent) of the respondents' secured first class with distinction in 12th standard while 41.11 per cent were secured first class and 13.33 per cent respondents were secured second class. Only 01.1 per cent respondents were passed without any class.

3.1.10 Other curricular activities

It is evident from table that 75.55 per cent of respondents were participated in NSS activities, 60.00 per cent of respondents were participated in cultural activities, while 53.33 per cent respondents were interested in tree plantation and 30.00 per cent respondents were reported that the have donated blood.However, 18.88 per cent, 13.33 per cent and 11.11 per cent respondents were interested in other curricular Activities, student forum and NCC activities respectively.

3.2 Students motivation for securing admission

The data in table 2 indicated that (38.89 per cent) of the respondents were motivated for securing admission in food technology due to availability of job, while 35.56 per cent respondents were motivated because of their own interests, followed by 18.89 per cent respondents get motivated to start own business, where as same per cent i. e. 3.33 per cent of respondents were get motivated due to willingness of parent and near to native place.

3.3 Aspiration of the respondents

3.3.1 Educational aspiration

The Table 3 revealed that majority (70.00 per cent respondents were aspired to complete post graduation, followed by 28.88 per cent of the respondents were aspired to complete MBA, whereas 15.55 per cent respondents were aspired to complete post graduate diploma, while 12.22 per cent, 10.00 per cent, 8.88 per cent and 3.33 per cent respondents were aspired to complete short term course, doctoral studies in foreign country, complete under graduation and other aspiration respectively.

3.3.2 Social aspiration

The data regarding social aspiration indicated that 40.40 per cent respondents were aspiring to develop own family and 38.88 per cent of them aspired to work for development of village, while 30.00 per cent of the respondents have aspiration to work for development of farmers; 23.33 per cent respondents were aspired to become recognized social worker; 20.00 per cent respondents were aspired to get social recognition; 17.77 per cent respondents were aspired to become ideal husband / wife, where as remaining 14.44 per cent, 12.22 per cent and 11.11 per cent respondents were aspired to develop their caste / community, other aspiration and to make effort to remove traditionalism respectively.

3.3.3 Economic aspiration

In respect of economic aspiration it is noticed that 31.11 per cent respondents were aspired to earn Rs.3, 00,001 to Rs. 5, 00,000 per year followed by 30.00 per cent respondents were aspired to earn Rs. 7, 00,001 /- and above per year and 16.66 per cent respondents were aspired to earn Rs. 5, 00,001 to Rs. 7, 00,000/- per year where as 12.12 per cent and 11.11 per cent respondents were have economic aspiration of below Rs. 1,000,000/- and Rs. 1,00,001/- to Rs, 3,00,000/- per year respectively.

3.3.4 Professional aspiration

It is revealed from the table that 42.22 per cent respondents were aspired to become a successful businessman, while 21.11 per cent of the respondents aspired to become successful entrepreneur, followed by 20.00 per cent respondents were aspired to become a well known administrator and remaining i.e. 14.44 per cent, 08.88 per cent, 07.77 per cent and 05.55 per cent respondents were aspired to become well known academician, well known scientist, as innovative farmer and other aspirations respectively.

3.3.5 Job aspiration

The table 7 indicates that 42.22 per cent respondents were aspired to do job in multinational company, whereas 34.44 per cent of respondents were aspired to do job in administrative position in public sector, while 20.00 per cent respondents were aspired to do job in agricultural university. The other job aspiration of the respondents were to secure job in National Bank (11.11 per cent), to secure job in private organization (10.00 per cent), to secure job in corporate sector (05.55 per cent), to secure scientist post through ASRB (03.33 per cent), to secure job in Panchayat Raj Institutions (02.22 per cent) and to secure job in voluntary organizations (01.11 per cent).

3.3.6 Self employment aspiration

Majority of the respondents (63.33 per cent) were aspired to start their own business, while 24.44 per cent and 12.22 per cent respondents were aspired with self employment to start small teaching institute / Research laboratory and to start or development farming as a business, whereas 07.77 per cent and 05.55 per cent respondents were aspired to with other self employment aspiration and to start consultancy services.

Conclusions

1 Most of the respondents wear from rural background, farming a major occupation of family, size having medium land holding (*i.e.* 4.1 to 10.00 ha.), sought GOI scholarship, having low level of income father education was graduate while mother education secondary school level, possessed first class with distinction in academic performance, participated in NSS activities.

2. Findings revealed that respondents have considerable motivation for securing admission in food technology due to availability of job.

3. The Majority of the respondents have aspired to complete their post graduation, develop own family, earn Rs. 3 to 5 lacks per year, to become a successful businessman, to do job in multinational company and start their own business.

Recommendations

Considerable per cent of respondents get motivated for securing admission in food technology is due to availability of job and their own interest. It is therefore, recommended that university should strengthening the placement cell.

Majority of the respondents were aspired to start their own business and about half of them were as personal to become a successful business man, it is therefore recommended that government departments and other agencies like DIC's should emphasis on establishment of cooperative organizations of the food technology graduates / post graduations which would guide & held them to establish their own business.

Most of the students aspired for post graduates education leading to M. Tech and Ph.D. degrees. Looking to this finding it is recommended that PG programme of the university may be strengthened to accommodate more number of students.

In respect of job aspiration it was noticed that the respondents were aspired to do job in multinational company and in administrative position in public sector, it is therefore, recommended that the university may start consultancy orientation loading classes for them.

Table 1: Distribution of respondents according to their profiles

Sr. No.	Category		Frequency		Percentage
1	**Family Background**				
	Rural		60		66.67
	Urban		30		33.33
2	**Parents Occupation**				
	Farming		43		47.78
	Service		26		28.89
	Dairy		00		00.00
	Labour		04		04.44
	Business		15		16.67
	Other		02		02.2
3	**Family type**				
	Joint		30		33.33
	Nuclear		60		66.66
4	**Family size**				
	Small (Up to 04)		30		33.33
	Medium (05 to 07)		48		53.34
	Big (08 & above)		12		13.33
5	**Land holding**				
	Landless (No land holding)		22		24.44
	Marginal (Up to 1.0 ha.)		03		03.33
	Small (1.1 to 2.0 ha.)		17		18.89
	Semi-medium (2.1 to 4.0 ha.)		16		17.78
	Medium (4.1 to 10.0 ha.)		23		25.56
	Big (Above 10.1 ha.)		09		10.00
6	**Family Annual Income**				
	Low (Up to Rs.75984)		42		46.67
	Medium (Rs.75985 to Rs. 245855)		26		28.89
	High (Rs.245856 & above)		22		24.44
7	**Family Educational Status**		**Father**		**Mother**
	Illiterate	12	13.33	23	25.55
	Primary School	02	02.22	07	07.77
	Middle School (5th -7th)	04	04.45	13	14.45
	Secondary School (8th-10th)	13	14.45	24	26.67
	Higher Secondary (11th -12th)	19	21.11	15	16.67
	Graduate	40	44.11	08	08.88

Sr. No.	Category	Frequency	Percentage
8	**Scholarship**		
	Govt. of India	37	41.11
	ICAR	03	03.33
	Merit-cum	13	14.45
	OtherNo.	08	08.89
	Scholarship	29	32.22
9	**Academic performance** (12th Std)		
	Pass (35- 45%)	01	01.11
	Second class (45.01 to 60.00%)	12	13.33
	First class (60.01 to 75.00%)	37	41.11
	First class with distinction (75.01%)	40	44.45
10	**Other curricular activities**		
	NSS	68	75.55
	NCC	10	11.11
	Blood donation	27	30.00
	Tree plantation	48	53.33
	Student forum	12	13.33
	Cultural activities	54	60.00
	Other	17	18.88

Table 2: Distribution of respondents according to their motivation for securing admissions in food technology

Sr. No.	Particulars	Frequency	Percentage
1.	Jobs are available	35	38.89
2.	To start won business	17	18.89
3.	Own interest	32	35.56
4.	Willingness of parent	03	03.33
5.	Next to native place	03	03.33

Table 3: Distribution of respondents according to their Educational aspiration

Sr. No.	Levels of Educational aspiration	Frequency	Percentage
1.	To complete Post graduation	63	70.00
2.	To complete Doctoral studies	11	12.12
3.	To complete Short term course	11	12.22
4.	To complete Degree course other than Food Technology	05	05.55
5.	To complete Doctoral studies in foreign country	09	10.00
6.	To complete Post Graduate Diploma	14	15.55
7.	To complete MBA	26	28.88
8.	To complete Under graduate	08	08.88
9.	Other	08	03.33

Table 4: Distribution of respondents according to their Social aspiration

Sr. No.	Nature of Social Aspiration	Frequency	Percentage
1.	To develop own family	36	40.00
2.	To work for development of village	35	38.88
3.	To work for development of farmers	27	30.00
4.	To develop our caste/community	13	14.44
5.	To get social recognition	18	20.00
6.	To make efforts to remove traditionalism	10	11.11
7.	To become an ideal Husband/wife	16	17.77
8.	To become recognized social worker	21	23.33
9.	Any other	11	12.12

Table 5: Distribution of respondents according to their Economic aspiration

Sr. No.	Levels of EconomicAspiration	Frequency	Percentage
1.	Below Rs. 1,00,000/-	11	12.12
2.	Rs. 1,00,001 to Rs. 3,00,000/-	10	11.11
3.	Rs. 3,00,000 to Rs. 5,00,000/-	28	31.11
4.	Rs. 5,00,000 to Rs. 7,00,000/-	15	16.66
5.	Above Rs. 7,00,000/-	27	30.00

Table 6: Distribution of respondents according to their Professional aspiration

Sr. No.	Nature of Profession	Frequency	Percentage
1.	To become a well know scientist	08	08.88
2.	To become an innovative farmer	07	07.77
3.	To become a well known administrator	18	20.00
4..	To become a successful business man	38	42.22
5.	To become a well known academician	13	14.44
6.	To become a successful entrepreneur	19	21.11
7.	Any other	05	05.55

Table 7: Distribution of respondents according to their Job aspiration

Sr. No.	Nature of Job	Frequency	Percentage
1.	Administrative position in public sector	31	34.44
2.	Job in Agricultural University	18	20.00
3.	Job in Private Organization	09	10.00
4.	Job in National Bank	10	11.11
5.	Job in Corporate Sector	05	05.55
6.	Job in Panchayat Raj Institution	02	02.22
7.	Job in Voluntary Organization	01	01.11
8.	Job in Multi National Company	38	42.22
9.	To secure scientist post through ASRB	03	03.33
10.	Any other	05	05.55

Table 8: Distribution of respondents according to their Self Employment aspirations

Sr. No.	Nature of Self Employment	Frequency	Percentage
1.	To start own business	57	63.33
2.	To start / develop farming as a business	11	12.22
3.	To start small Teaching institute / Research Laboratory	22	24.44
4.	To start consultancy services	05	05.55
5.	Any other	07	07.77

Extension Management Strategies for Sustainable Agriculture, pp 413-427

New India Publishing Agency, New Delhi, India

46

Coping Behaviors with Extremely Heavy Rainfalls in Southern Zambia–Comparison between 2007/08 and 2009/10

Miyazaki Hidetoshi, Ishimoto Yudai, Yamashita Megumi, Tanaka Ueru and Umetsu Chieko

In the semi-arid tropics, where small farmers depend heavily on rain-fed agricultural systems, variability in precipitation has a dramatic impact on crop production and livelihoods. Because small-scale agriculture depends on rain-fed cultivation, it is vulnerable to factors such as flooding and drought. Unfortunately, the frequency of unexpected heavy rainfall or flood events has increased in many areas of Africa, including Southern Zambia. About 75% of farmers in Zambia engage in small-scale farming, with an average farm size of 2 ha (VAM Steering Committee Zambia, 1998). Most of these farmers cultivate maize, which is a staple food in Zambia, and stable maize production and increased productivity are important both for food security and to provide a livelihood to farmers.

Resilience represents a system's ability to experience shocks while retaining much of its original function, structure, and feedbacks, thereby preserving its identity (Walker, 2004). Adger (2000) defined *social* resilience as the ability of a group to resist external stresses and disturbances caused by social, political, and environmental change. Shimada (2009) notes that among the strategies for coping with risk, access to diverse resources is particularly important for resilience.

Ellis (1998) defined "Livelihood diversification" as a strategy for developing a more diverse range of activities, including social support resources, thereby improving both the ability to survive sudden shocks (whether environmental or social) and one's livelihood. In this context, a *livelihood* comprises an individual's capabilities, resources (both material and social), and the activities performed to earn a living. A livelihood is sustainable when it helps the individual to cope with and recover from stresses and shocks, and to maintain or improve their capabilities and assets, without damaging the resources on which the livelihood depends (Scoones, 1998). In rural Africa, sustainability depends on diversification of the individual's assets, activities, and income sources (Barrett et al., 2001). Where these strategies are insufficient to withstand a given shock, poor households may be forced to switch from "asset smoothing" (balancing the flow of benefits from the assets) to "consumption smoothing" (reducing the consumption of threatened assets) to ensure their survival (Dercon, 2000). Non-farm activity, such as seeking alternative forms of employment in nearby cities, is playing an increasingly important role in rural household income (Zhu and Luo, 2006). Because environmental and social shocks cannot always be predicted. Thus, it is necessary to understand how small farmers develop resilience and sustainable livelihoods.

Unexpected heavy rainfall has become increasingly frequent in many parts of the tropics, and has tested the resilience of small farmers. In Southern Zambia, severe rainfall events occurred during the 2007/08 and 2009/10 cultivation seasons, forcing local small-scale farmers to adopt various coping behaviors. In the present study, we investigated the coping strategies they adopted in response to these shocks. Our goal was to understand the coping with the environmental shocks that are expected to become more frequent and severe in the context of global climate change.

2. Materials and Method

2.1. Study sites

We chose three study sites in Zambia's Southern Province (Fig. 1), which is in the semi-arid tropics. The elevation increases gradually from the shore of Lake Kariba to the highlands. The three sites were located on the lower terrace near Lake Kariba (hereafter, the lower-terrace site), the middle of the escarpment (hereafter, the mid-escarpment site), and the upper terrace of the escarpment (hereafter, the upper-terrace site). The lower-terrace and mid-escarpment sites each consisted of two villages, versus one at the upper-terrace site. The sites were separated by more than 10 km, and the elevations of the lower-terrace, mid-escarpment, and upper-terrace sites were about 500, 750, and 1000m, respectively. Therefore, the agro-ecosystems were different because of differences in rainfall, temperature, and vegetation. The lower-terrace site is almost flat. There were many steep slopes at the mid-escarpment site. At the upper-terrace site, there were gentle slopes between the mountain ridge and at the valley bottom, were *dambo* which is land near river/stream bank. All residents were of the Tonga ethnicity.

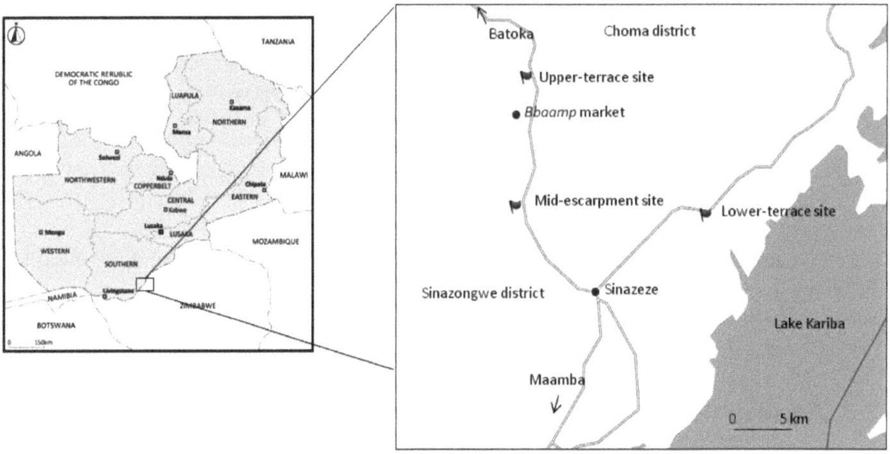

Fig. 1: Locations of the three study sites in Zambia's Southern Province

2.2. Rainfall variability in the study area

The rainy season occurs from November to April. Figure 2 shows the variations from mean annual rainfall and mean maize yields from 1986/1987 to 2006/2007. Rainfall data at Choma, the meteorological station nearest to our study area (about 60 km northwest of lower-terrace site), was obtained from the Zambia Meteorological Department. Crop production data were obtained from the Central Statistical Office (CSO, 2010). The annual precipitation from 1986 to 2006 averaged 757.6 mm (Fig. 2). The maize yields in Southern Province averaged 228.7 Mt/year. The long term mean annual rainfall was clearly variable (757.6±190.2 mm). From 1990 to 2005, Zambia experienced six droughts: in 1991/1992, 1994/1995, 1997/1998, 2000/2001, 2001/2002, and 2004/2005 (Lekprichakul, 2008).The maize yields exhibited similarly variation (228.7±129.0 Mt), and followed the same trend as rainfall. Maize yields of 2006/2007 was equal with long term maize yields. Therefore, in this paper, 2006/2007 was defined as normal year.

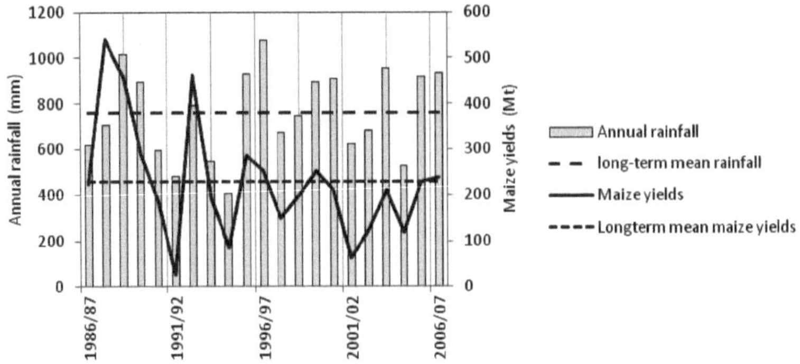

Fig. 2: Annual rainfall from 1960 to 2006 at the Choma station in Zambia's Southern Province.

Sources: Rainfall data were provided by the Zambia Meteorological Department. Maize yields were provided by the Central Statistics Office, Lusaka, Zambia.

2.3. Agriculture

Small-scale farmers in the study area commonly cultivate maize as their staple food crop under rain-fed conditions. Other crops (e.g., pumpkins, sorghum, pearl millet, beans) are typically planted in the same field as the maize, at the same time. After planting the maize, farmers plant other crops (e.g., cotton, groundnuts, sweet potato, sweet sorghum) in other fields. Cotton is planted as a cash crop. Some "green" maize is harvested in March, and the remaining maize is harvested after it dries in the field. The dried maize is used to make flour, usually beginning at the end of April. Farmers also plant maize during the dry season. The typical sowing time differs among the sites. At the lower-terrace site, planting takes place from late February to early March, but at the mid-escarpment and upper-terrace sites, maize is planted from late August to October. Other crops are also grown during the dry season (Miyazaki *et al.*, 2013). The dominant dry-season crop is okra at the lower-terrace site and sweet potatoes at the upper-terrace site. No specific dry-season crop is grown at the mid-escarpment site. Vegetables such as cabbage, tomatoes, and rape are planted in gardens at all of the sites. These crops are planted in March or April by farmers with suitable fields, such those near riverbanks or *dambo* (seasonal wetlands). Some of the farmers grow and harvest vegetables all year round to sell as a cash crop, particularly at the upper-terrace site.

2.4. Methods

To determine the area cultivated for each crop, we measured all of the cultivated fields at the three sites from 2007 to 2010 during the rainy season using hand-held GPS units. We also measured the dry-season fields from 2007 to 2010.

When other crops were planted with maize in the same field, we considered the entire planted area to represent the primary crop, which was defined by the field's owner. In addition, we measured fields damaged by heavy rains in both years to quantify the damage from heavy rains at the study sites. We focused on two heavy rainfall events (see section 3.3 for details). To clarify the activities farmers used to earn income, we interviewed the household head or their spouse for all households at the three sites from 2006 to 2010.We interviewed 77 farmers at the lower-terrace site, 37 at the mid-escarpment site, and 85 at the upper-terrace site. Our field assistants conducted semi-structured interviews using questionnaires that had been tested and revised before the field study. Rain gauges were installed at the three sites and connected to a data logger (Hioki Inc., Nagano, Japan) to record precipitation data during the rainy season from 2007 to 2010.

3. Results

3.1. Agricultural distribution

Maize was the dominant crop grown during the rainy season (Table 1). The percentage of the area under maize cultivation exceeded 50% at the lower-terrace site, and exceeded 75% at the mid-escarpment and upper-terrace sites. Cotton production accounted for a large amount of the area (42.1%)at the lower-terrace site, versus small percentages at the mid-escarpment site (11.7%) and the upper-terrace site (8.7%). The upper-terrace site had the largest percentage of land devoted to sweet potato cultivation (9.3%). Other crops included sorghum, pearl millet, cowpea, sugarcane, sunflower, and groundnut, but none of these crops was grown in more than 6% of the total area at any site. Clearly, maize dominated the planted area, but the farmers planted a variety of crops.

Table 2 presents data for the dry season. At the lower-terrace site, the dominant crop was garden vegetables, followed by maize and sugarcane. At the mid-escarpment site, maize remained the dominant crop, but garden vegetables accounted for 19.7% of the production area. At the upper-terrace site, maize was also dominant, accounting for 90.7% of the total area, and garden vegetables accounted for the remaining 9.3%.

Table 1: Area of cultivated crops during the 2007/08 rainy season at the three sites

	Crop		Lower-terrace site		Mid-escarpment site		Upper-terrace site	
	Scientific name	Local name	-	(ha)	(% of total)	(ha)	(% of total)	(ha)	(% of total)
Maize	*Zea mays*	*Mapopwe*	-	66.1	53.3	80.5	76.2	115.5	77.9
Cotton	*Gossypium* spp.	*Buluba*	-	52.2	42.1	12.3	11.6	12.9	8.7
Sorghum	*Sorghum bicolor*	*Maila*	-	3.3	2.7	5.7	5.4	0.5	0.4
Pearl millet	*Pennisetum glaucum*	*Nzembwe*	-	0.0	0.0	4.2	4.0	0.0	0.0
Sunflower	*Helianthus annuus*	*Malangazuba*	-	0.0	0.0	1.3	1.2	0.0	0.0
Groundnut	*Arachis hypogaea*	*Indongwe*	-	0.0	0.0	1.2	1.1	3.8	2.5
Cowpea	*Vigna unguiculata*	*Nyangu*	-	0.8	0.7	0.0	0.0	1.6	1.1
Sweet potato	*Ipomoea batatas*	*Chimbwali*	-	0.0	0.0	0.1	0.1	13.8	9.3
Sweet sorghum	*Sorghum saccharatum*	*Munsale*	-	0.7	0.6	0.0	0.0	0.0	0.0
Garden vegetables	-	-	-	0.1	0.1	0.4	0.4	0.1	0.1
Other	-	-	-	0.7	0.5	0.0	0.0	0.0	0.0
Total	-	-	-	**123.9**	**100.0**	**105.7**	**100.0**	**148.2**	**100.0**

Table 2: Area of cultivated crops during the 2007/2008 dry season at the three sites.

Crop	Lower-terrace site		Mid-escarpment site		Upper-terrace site	
	(ha)	(% of total area)	(ha)	(% of total area)	(ha)	(% of total area)
Maize	0.6	16.7	2.4	80.0	17.2	90.5
Sweet sorghum	0.5	13.9	0	0	0	0
Garden vegetables	2.5	69.4	0.6	20.0	1.8	9.5
Total	**3.6**	**100**	**3.0**	**100**	**19.0**	**100**

3.2 Income-generating activities

There were many income-generating activities, ranging from 11 activities at the lower-terrace and upper-terrace sites to 14 activities at the mid-escarpment site (Table 3). Maize sales were the dominant source of income at the lower-terrace site, whereas vegetables were dominant at the mid-escarpment and upper-terrace sites. More than 50% of households sold maize (the staple food) at all sites. Cotton and vegetables were also sold at all sites. Sales of animals, such as cattle and goats, and piecework were common cash income activities at all sites. Poultry sales were higher at the upper-terrace site than at the other sites. Maize, vegetables, and other commodities (e.g., poultry) are sold by farmers within the villages and at roadside markets or markets in the town. After Zambia's economic liberalization program began in the early 1990s, some types of crops and vegetables were increasingly sold to earn income along the main roads in rural areas of Zambia (Hanzawa, 2010). The sale of lumber and stones for house construction was undertaken at the mid-escarpment site, but not at the other sites (except for one family at the upper-terrace site). At the lower-terrace site, fish sales includes catching fish in Kariba Lake and selling them at local markets, or trading fish for other products; at the upper-terrace site, fish sales only represents trading of fish or dried fish. Piecework mostly refers to work at the kapenta companies or a fishing camp near the Kariba Lake at the lower-terrace site during the dry season. Kapenta is a small freshwater fish that was introduced from Tanganyika Lake at the end of the 1960s (Nakamura, 2009). Kapenta-processing companies employ many workers in the rural areas around Kariba Lake. At the upper-terrace site, piecework refers to work in another farmer's fields during any season. These characteristics of the income-generating activities depend on the site location because the location determines what resources are available to residents. There were some bricklayers and carpenters in each village, which represent special skills.

Table 3: Income-generating activities and participation rates in 2007 at the three sites. Totals do not add to 100% because most households participated in multiple activities.

	Activities	Lower-terrace site (***n*** **= 62**)		Mid-escarpment site (***n*** **= 27**)		Upper-terrace site (***n*** **= 66**)	
		Number of households	Participation (%)	Number of households	Participation (%)	Number of households	Participation (%)
Cultivation	Maize	44	71.0	15	55.6	61	92.4
	Cotton	42	67.7	5	18.5	31	47.0
	Vegetables	17	27.4	21	77.8	62	93.9
Other activities	Animals	19	30.6	13	48.1	33	50.0
	Poultry	0.0	0	1	3.7	60	90.9
	Gathering wild plants	10	16.1	12	44.4	2	3.0
	Fish sales	6	9.7	0.0	0	4	6.1
	Bricklaying	5	8.1	2	7.4	2	3.0
	Carpentry	3	4.8	5	18.5	4	6.1
	Lumber production	0.0	0	6	22.2	1	1.5
	Stone production	0.0	0	7	25.9	0	0.0
	Shop operation	0.0	0	2	7.4	0	0.0
	Piecework	32	51.6	8	29.6	46	69.7
	Alcohol sales	2	3.2	1	3.7	0	0.0
	Other	15	24.2	1	3.7	0	0.0

3.3. Heavy rainfall

Daily and cumulative precipitation at the three sites during three rainy seasons is shown in Figure 3. Annual rainfall at all sites in each of the three rainy seasons totaled more than 1300mm/year. This was 1.71 times the long-term average rainfall from 1986 to 2006.Two heavy rainfalls occurred during the study period: one each in the 2007/08 and 2009/10 rainy seasons. The first heavy rainfall occurred at the end of December 2007, whereas the second occurred at the end of February 2010. In the 2008/09 rainy season, daily precipitation was gentle and continuous, even though the annual total was high.

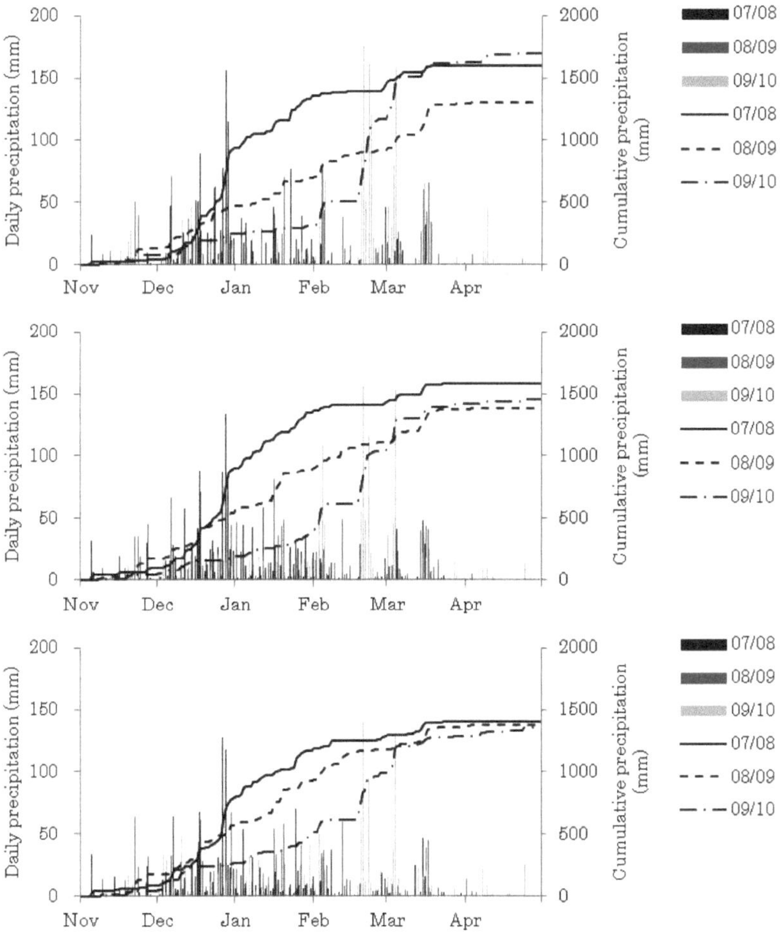

Fig. 3: Mean daily and cumulative precipitation during the 2007/08, 2008/09, and 2009/10 rainy seasons at (a) the lower-terrace site, (b) the mid-escarpment site, and (c) the upper-terrace site.

On 28 December 2007, unexpectedly heavy rain fell at the lower-terrace and mid-escarpment sites (Fig. 3). Daily rainfall at the lower-terrace and mid-escarpment sites was 156.4and 133.5mm, respectively, versus 127.5mm at the upper-terrace site on 27 December. The rainfall for the period from 2 days before the heavy rainfall to 2 days after the heavy rainfall event totaled 399.8, 349.6, and 327.4 mm at the lower-terrace, mid-escarpment, and upper-terrace sites, respectively. These totals amounted to more than 40% of the long-term average rainfall of this area.

Heavy rainfall events occurred at all three sites in 2010 from late February to the beginning of March (Fig. 3). During this period, daily precipitation exceeded 100 mm four times at the lower-terrace site, three times at the mid-escarpment and two times at upper-terrace sites. Total precipitation during this period at the lower-terrace, mid-escarpment, and upper-terrace sites was 991.6, 685.7, and 576.4 mm, respectively. This amounted to more than 76.1% of the annual average.

3.4. Damage caused by the heavy rainfall

The two heavy rainfalls described in the previous section damaged many crop fields. Because maize seeds are planted at the beginning of the rainy season, the two heavy rainfalls damaged the crops at different growth stages. In 2007/08, lodging of young maize occurred everywhere at the lower-terrace site where drainage was bad. Young maize was washed away by running water and a landslide occurred at the mid-escarpment site. At the upper-terrace site, young maize was washed away by running water on the gentle slopes and lodging caused by flooding water occurred at the valley bottom and in *dambo* areas. In 2010, similar damage occurred at all sites, but in addition, many of the ears of pre-harvest maize rotted. Table 4 summarizes the area of crop damage and percentage of the total area affected in both years. The proportion of the field area that was damaged was again much larger in 2007, affecting roughly 3 times the area affected in 2010 (overall averages of 18.1% and 6.5% of the total area, respectively). The proportions of the area damaged at the lower-terrace, mid-escarpment, and upper-terrace sites were 27.1%, 28.0%, and 3.5% in 2007, respectively, versus 7.9%, 1.0%, and 8.3%, respectively, in 2010. These results indicate that the degree of the damage was affected by the timing of the heavy rainfall because of the different crop growth stages and site conditions. The decreased maize yields in 2007/08 resulted in increased food prices before the next harvest season, and directly affected poor households that had already depleted their food stocks and that were forced to purchase food with their available cash (Sakurai *et al.*, 2011).

Table 4: Area of crop damage caused by the heavy rainfalls at the three sites, and the corresponding proportion of the total area.

Site	December 2007			February 2010		
	Total field area	Damaged area		Total field area	Damaged area	
	(ha)	(ha)	(%)	(ha)	(ha)	(%)
Lower-terrace site	123.9	33.6	27.1	109.6	8.6	7.8
Mid-escarpment site	105.7	29.6	28.0	62.6	0.6	1.0
Upper-terrace site	148.2	5.2	3.5	108.1	8.9	8.2
Total	**377.8**	**68.4**	**18.1**	**280.3**	**18.1**	**6.5**

3.5. Agricultural coping activities and income-generating activities

In this section, we analyze and discuss how the small-scale farmers coped with the heavy rainfall damage through agricultural coping activities and income-generating activities.

3.5.1. Agricultural coping activities in 2007/08

A lot of damaged fields were abandoned after the heavy rainfall damage, but at least some farmers at all three sites were able to switch to different crops or replant maize and other crops in some of the fields that had been damaged. At the lower-terrace site, the damaged field area was replanted with maize and cotton. At the mid-escarpment site, damaged fields switched to planting groundnuts or sorghum, and then planted maize again during the dry season. At the upper-terrace site, damaged fields switched to sweet potato during the rainy season and planted maize again in the dry season.

3.5.2. Agricultural coping activities in 2009/10

A lot of fields were abandoned after the heavy rainfall at the lower-terrace, mid-escarpment, and upper-terrace sites. At the lower-terrace site, quarter damaged areas were replanted maize or switched to planting okra, sugarcane, and green beans after the heavy rainfall event. At the mid-escarpment site, a few damaged areas were replanted with maize. At the upper-terrace site, some damaged fields were switched to sweet potato.

3.6. Income-generating coping activities

Many households could not sell maize because of the heavy rainfall damage. As a result, they were forced to find new income-generating activities. Therefore households were selected to analysis of new income-generating activities. They sold maize in normal year, however they could not sold maize after heavy rainfall event. And also the reason of new activities selected to analysis, because it is clear that coping activities were engaged by them to mitigate shortage of maize yields. Of source, to cope heavy rainfall damage, farmers also changed load of regular income-generating activities. New income-generating activities households used to cope with the damage from the heavy rain. There were several new income-generating activities at each site, ranging from three activities at the upper-terrace site in 2007/08 to six activities at the lower-terrace site in 2007/08 and six at the lower-terrace and upper-terrace sites in 2009/10.

3.6.1. Income-generating coping activities after the 2007/08 heavy rainfall

In 2008, animal sales were a common strategy at all sites. Sales of vegetables and poultry were performed at the lower-terrace and mid-escarpment sites. Fish selling was performed at the lower-terrace and upper-terrace sites. At the lower-terrace site, fish selling includes both catching the fish and trading them for other products; at the upper-terrace site, it means trading of fresh or dry fish. Piecework mostly means work for companies, mostly kapenta-processing companies and fishing campsor road-construction companies because some bridges and roads were washed out by the rain. One widowed household head started operating a bar. Sales of lumber and stones for house construction increased at the mid-escarpment site in both years. These activities were available because they depended on site-specific resources. At the lower-terrace site, four households of elderly individuals and widows asked for financial assistance from their relatives.

3.6.2. Income-generating coping activities after the 2009/10 heavy rainfall

The income-generating strategies in 2010 were similar to those in 2008 at the lower-terrace and mid-escarpment sites, although there were fewer activities in 2010 at the latter site. However, vegetable sales, gathering wild plants, and "other" activities increased, and animal sales, fish sales, and bar operation decreased at the upper-terrace site. Other activities consisted mainly of money transferred from relatives' at all three sites.

The coping behavior varied among the sites based on differences in the available assets and resources among sites.

4. Discussion and Conclusions

4.1. Differences in coping activity between the two heavy rainfall events

Farmers used agricultural strategies such as switching to a new crop or replanting a crop to change their land-use patterns in 2007/08. However, most of the damaged land was abandoned in 2009/10. This difference can be explained by the timing of the heavy rainfall. In 2007/08, the heavy rainfall occurred at the end of December; near the beginning of the rainy season. As a result, farmers expected that there was enough time for a new crop to grow and that enough rain would come to support crop growth, because 3 months remained in the rainy season. As a result, they responded to the shock by trying to grow crops again. In contrast, the heavy rainfall occurred late in the rainy season in 2009/10. Because little time remained to grow a new crop, and the farmers expected there would be insufficient rain, they chose alternatives to agricultural activities.

The nature of the responses to the rainfall shock differed among the sites based on the locally available resources. However, other factors also affected the choice of an activity. For example, the site conditions affected the severity of the damage; landslides and removal of the corn plants by running water were more common on steeper slopes, whereas lodging of corn was more common on flat sites with poor drainage. As well, older residents had less ability to perform heavy labor, and instead chose activities such as operating a bar or requesting money from relatives. Some farmers took advantage of the damage to homes and roads by providing lumber and stones for rebuilding or working on road construction.

5. Conclusions

Our key findings were that the nature and extent of the damage varied according to a site's natural and social conditions, the timing of the extremely heavy rainfall (thus, the growth stage of the crops), and the coping behavior available to the farmers (due to the available resources at a given site).

Acknowledgements

This study was funded under the "Vulnerability and Resilience of Social-Ecological Systems" Project and the "Desertification and Livelihoods in Semi-Arid Afro-Eurasia" Project, both administered by the Research Institute for Humanity and Nature, and by a Japanese Society for the Promotion of Science KAKENHI (Grant-in-Aid for Scientific Research) (B) Program (No. 23310027).

References

Adger, W.N. (2000). Social and ecological resilience: are they related? Progress in Human Geography 24:347-364.

Barrett, Christopher. B., Thomas Reardon and Patrick Webb (2001). Nonfarm Income Diversification and Household Livelihood Strategies in Rural Africa: Concepts, Dynamics, and Policy Implications. Food Policy 26, no.4: 315-31.

CSO (Central Statistical Office) (2010). Crop Forecast Survey data 1986-2010 Survey, Lusaka: Zambia Central Statistical Office, Unpublished Survey.

Dercon, S. (2000). Income Risk, Coping Strategies and Safety Nets. Background paper World Development Report 2000/01, Centre for the Study of African Economies, Oxford University.

Ellis, Frank (1998). Household Strategies and Rural Livelihood Diversification. Journal of Development Studies 35, no.1 : 1-38.

FAO/WFP (Food and Agricultural Organization and World Food Program) (2003). FAO/WFP Crop and Food Supply Assessent Mission to Zambia, Special Report: Rome, United Nations (10 June), ftp://ftp.fao.org/docrep/fao/005/y9664e/y9664e00.pdf Accessed 12 May 2014.

FAO/WFP (Food and Agricultural Organization and World Food Program) (2005). FAO/WFP Crop and Food Supply Assessment Mission to Zambia, Special Report: Rome, United Nations (24 June), http://www.fao.org/docrep/008/j5511e/j5511e00.htm Accessed 12 May 2014.

Hanzawa, K. (2010). Use of dambo resource and agricultural change: 18 years in a village of Central Province, Zambia. Journal of Arid Land Studies, 19(4): 579–583. (in Japanese).

Lekprichakul, T. (2008). Impact of 2004/2005 Drought on Zambia's Agricultural Production: Preliminary Results. Working Paper on Social-Ecological Resilience Series No. 2008-002 :1-26.

Miyazaki, H., Y. Ishimoto, U. Tanaka and C. Umetsu (2013). The Role of the Sweet potato in the Crop Diversification of Small-scale Farmers in Southern Province, Zambia. African Study Monographs, 34(2): 119-137.

Nakamura, T. (2009). The Livelihood of 'Escarpment Tonga': A Case Study of One Village, Southern Zambia. Working Paper on Social-Ecological Resilience Series No. 2008-005: 1-33.

Sakurai, Takeshi, Akiko Nasuda, Akinori Kitsuki, Ken Miura, Taro Yamauchi and Hiromitsu Kanno (2011). "Vulnerability and Resilience of Households: The Case of Zambia" The Economic Review, Vol. 62, No. 2, Institute of Economic Research, Hitotsubashi University, Kunitachi, Tokyo, Japan, pp. 166-187, in Japanese.

Scoones, Ian (1998). Sustainable Rural Livelihoods: A framework for analysis. IDS, Working Paper 72, IDS, Brighton, UK, June 1998.

Shimada, S. (2009). Introductory Analysis of Social Vulnerability in Rural Africa. E-Journal GEO 3(2):1-16. in Japanese, Refereed.

Siamwiza, B.S. (2006). Famine and hunger in the history of the gwembe valley, Zambia, c. 1850-1958, in The Tonga-Speaking Peoples of Zambia and Zimbabwe: Essays in Honor of Elizabeth Colson, edt Lancaster C and Vickery K P, Univ Pr of Amer, pp. 392.

Umetsu, Chieko (2013). Resilience of social-ecological systems for food security. ARDEC (48) :8-12. (in Japanese).

VAM Steering Committee Zambia (1998). A comparative analysis of the vulnerability conditions of rural Zambia. Prepared by the Food, Health and Nutrition Information System, Ministry of Agriculture, Food and Fisheries Programme Against Malnutrition, Meteorological Department World Food Programme, FEWS/USAID Vulnerability Assessment and Mapping.

Walker, B., Holling, C.S., Carpenter, S.R. and Kinzig, A. (2004). Resilience, adaptability and transformability in social-ecological systems. Ecology and Society, 9(2):5.

Zhu, N. and Luo, X.B. (2006). Nonfarm Activity and Rural Income Inequality: A Case Study of Two Provinces in China, World Bank Policy Research Working Paper 3811, http://elibrary.worldbank.org/doi/pdf/10.1596/1813-9450-3811 Accessed 12 May 2014.

Extension Management Strategies for Sustainable Agriculture, pp 429-442

New India Publishing Agency, New Delhi, India

47

Role of Institution in Tank Water Management in Tamil Nadu–India

M. Jegadeesan and Koichi Fujita

Tank irrigation is passing through a critical juncture in India today. Tank irrigation contributes significantly to agricultural production in India in general, and in Andhra Pradesh, Tamil Nadu and Karnataka, in particular. The tank irrigation system is one of the vast networks of thousands of water bodies that constituted a distinctive landscape which was medieval in origin but still the basis of livelihood in the dry Southern plains(Mosse, 2003, P-4). Tank is a small reservoir constructed across the slope of the valley to catch and store water during rainy season. Water is controlled by sluices attached to the tank bank and is delivered to paddy field by distributing channels. Tank is considered as a common property resource. The National Sample Survey Organization defines common property resources as the resources that are accessible to and collectively owned, managed by identifiable community and on which no individual has exclusive property rights (NSSO, 1999). The tank does not only provide irrigation water but it also provides biomass, fuel wood, fodder for livestock and other forms of economic livelihood sustenance of villagers (Chopra and Dasgupta, 2008). Tank irrigation is of special significance as it provides livelihood support to large number of marginal, small farmers and landless agricultural labours (Palanisami, 2000, P-1). This system once well maintained by villagers, slowly disintegrated over the period of time due to various reasons like changes in land holding pattern, development of large scale irrigation project and ground water development and change in preference of livelihood strategies among villagers and so on (Sakthivadivel *et al.*, 2004).

Despite the tank irrigation had significant impact on rural livelihood; it is now losing its importance day-by-day. The farmers cannot rely on the poorly-managed, fragile tank irrigation system. Most of the problems associated with tank irrigation today are management problems by the government or ignorance of the local people. The government attempted to solve the tank irrigation problems through various programs like European Economic Community assisted tank modernization during 1989-1999. Most of the programs initiated by the government did not consider the importance of participation of the local people with the management of the programs. The government and the agencies involved in tank modernization had thought that merely physical improvement of tank system with machines even without participation of the local villagers can do wonder. But local population has a greater interest in the sustainable use of resources than does the state or distant corporate manager; that local communities are more cognizant of the intricacies of local socio-ecological process and practices and they are more able to effectively manage those resources through local or traditional forms of access (Brosius, Tsing and Zerner, 1998 and Li., 2002). In recognition of this fact, the Government and the Non Governmental Organizations (NGOs) tried to motivate farmers to rebuild the institution beset with innumerable problems. But things did not happen as expected.

The main Focus of this Paper is to

i. Compare and contrast the effectiveness of traditional irrigational institution with the Government-sponsored and NGO-sponsored ones.

ii. Analyze the different modes of operation.

iii. Explore the reasons for the disintegration of the tank irrigation system.

Methodology

The study was conducted in three tank villages in Madurai district of Tamil Nadu, India. These study villages were selected purposively as they represent different kinds of irrigation institution. Given the fact that all the villages were receiving more or less same rainfall, consideration was given to availability of water in selecting the three sample villages.

Of the three villages selected, tank village 1 represent traditional institution, tank village 2 represent institution promoted by Government, and tank village 3, represent the institution promoted by NGO.

The data were collected through pre-tested, semi-structured interview method paying personal visit to the villages. Simple random sampling was employed to identify sample respondents (farmers). The data were collected through personal interview; focus group interaction and discussion with opinion leaders. The study was undertaken in 2007.

General Characteristics of the Sample Villages

The Table 1 below presents general characteristics of the selected study villages. The study villages Kadaneri, Kovalapuram and Menachipuram are located in Peraiyur *Taluk* of Madurai district. All the selected villages almost depend on agriculture and allied activities for their livelihood. The fate of agriculture is determined or influenced through rain fed tank irrigation system in the villages. The major crops cultivated are paddy, cotton and pulses. Mostly single season crop and rarely, farmers are going for second crop. In the last 10 years there was no intervention on these tanks to improve its performance. As a result, employment generated through tank irrigated agriculture is in terminal decline. In recognition of this, the government of Tamil Nadu, brought this villages under the National Rural Employment Guarantee Scheme (NREGS) to provide supplementary non-farm employment to assist them (BDO, 2007). Out of three tanks, two tanks are managed by Public Works Department, and one is coming under *Panchayat* Union management regime.

Table 1: General characteristics of study tank villages

Characteristics	Tank village 1	TankVillage 2	Tank village 3
Total population	2234	520	440
Command area (Ha)	41.60	62.26	7.94
Management authority	Public Works Dept.	Public Works Dept.	*Panchayat* Union
Type of Institution	Traditional	Govt. Sponsored	NGO Sponsored
Basin Location	Vaipar	Vaipar	Vaipar
Tank capacity (mcft)	14.0	17.66	9.20
Source of water supply	Rain fed	Rain fed	Rain fed
No. of sluices	1	2	1
No. of supply channels	2	2	2
Extent of encroachment (Ha)	0.21	Not available	Not available
No. of wells in command area	12	18	1
No. of castes in village	9	5	2
Total No. of households	387	133	110
Farming households	214	67	53
Landless Agricultural labors	148	43	42
Non farming households	25	23	15
Major cropping pattern	Paddy, pulses	Paddy, cotton, pulses	Paddy
Tank intervention in last 10 years	No	No	Yes. By NGO (2006)
Tank performance (farmer's perception)	Moderate	Poor	Moderate

Source: Water Resource Atlas of Madurai district, and author's field survey in 2007

Tank water Institutions

The villagers generally have traditional, informal association other than village *Panchayat*. These associations have a leader who is respected by villagers, some of them by virtue of their age and service rendered in the past and their social status, wield considerable influence in the village.

Traditional Irrigation Institution

Traditional irrigation institutions may be referred to the evolution of principles for collective action of users, for broad spectrum of social responsibilities such as system maintenance, watcr sharing and conflict resolution (Coward, 1980; Vaidyanathan, 1985 and Janakarajan, 1993). Even today villagers have traditional institutions in many villages to manage the tanks effectively as common property resources. Traditional system of water distribution was based on their beliefs, customs and the concept of equality. The water allocation ensured smooth sharing to all its members without any default. The performance of these tank irrigation systems depends on collective decisions made by them. These institutions characterized by socio-cultural and contextual arrangement in order to provide services to village community. These institutions have rules and regulations in the form of ethics and norms as it is resultant of complex pattern of behavior of large number of people over protracted period of time (Basu, 2000).

Government Sponsored Institutions

Effective functioning of tank system is simply based on how its different components like physical, technical and institutional parameters are managed. In the earlier days, villagers considered the tank irrigation as a system. Over the period of time, when government took over these structures, it failed to consider the institutions as a system. The tank irrigation system was managed by five different departments and acting as separate entity in different directions. After some period, government concentrated only on physical improvements of the tanks. But still they did not yield fruitful result as there are no institutional structures for maintenance. Thus institutional problems crop up and it was aggravated by changing social structures, land holding pattern and population pressure on the lands. During the 1980s, the international donor agencies wanted to modernize the tank irrigation system through water user association at tank levels. The government has also shown interest to form institution at tank levels as desired by the donor agencies.

NGO Sponsored Institution

Following the principle of *putting the last first* and to promote participatory management, many NGOs in India are working with the rural people in tank command area. To manage the irrigation system effectively, the NGOs follow different methods to organize farmers and develop institutions in the community level. They employ local people as negotiators to motivate farmers to participate in the institutional initiatives.

Field Observations

The research demonstrates some specific observations about the differences in strategy, notion, structure and functioning style among all three institutions in the study villages. The overall aim of all the stakeholders involved in this campaign was to create a local, independent and self-organized institution at community or village level. But notably, these groups varied tremendously in their values, attitudes and beliefs towards the cooperation and the best means to achieve their desired ends. All initiatives look for the active participation of rural people in working out a better livelihood access for themselves. New policies and schemes have been set in the place both by the Government and NGOs to facilitate this process of involvement.

Table 2 shows the functioning structure of tank institutions. In all the three types of institutions, irrespective of their mode of operations, the performance efficiency was at the lowest level. The most important ingredient for the institutional building is a sense of belonging, mutual trust and empathic cooperation. But unfortunately these ingredients are missing or not given due importance to create or enhance those ingredients of social capital.

Table 2: Functioning structure of tank institutions

Criteria	Traditional	Govt. Sponsored	NGOs sponsored
Responsibility of organizing villagers	Villagers themselves	Govt. official in charge of village	Facilitator appointed by NGO
Selection of leaders	Villagers	By election	By group opinion & rotational
Functioning style	Informal	Formal	Semi formal
Financial support	Collective contribution	Villagers & Govt.	Villager, NGO and Govt.
Work execution	Regular	Demand based	Regular
Activeness	Relatively Active	Inactive	Relatively Active

Source: Field survey by authors in 2007

Trust building, sense of belonging and social affiliation towards institutions will come when the villagers are convinced that their participation yield good livelihood base for them. It is important to distinguish between different kind of faith or involvement that people have within their socio-economic and cultural context such as bonding, bridging and linking with these institutions.

Generally, bonding relationship is viewed as strong or thick, while bridging relation is weak or thin (Narayanan, 1999; Onyx and Bullen, 2000; Putnam, 2001; Woolcock, 2001). Thus, bonding relationship existed in traditional institutions, which means that villagers have close relationship with this institution. These people tend to make close relationship as they have similar interests and common affiliation. Ann Dale and Jennie Sparkes (2007,P-143) argued that adhesiveness within "bonding" network is a sense of deep trust held among members, which is often highly relational, personalized and thus, has potential for conflict when their trust and commonalities break down. Once, the tank irrigation system had been considered as a sole livelihood provider, almost all villagers depended on it. During 1980-81, the number of population depended on agriculture in the study villages was 92 percent, but in 2007, it was 67 percent. (Block Statistics, 2007). Over the period of time, due to changes in government policy and spread of education opened various avenues for the villagers. Different industries like match box, fire work and cotton industries were localizing in nearby towns and these opened opportunities for the youngsters. These industries also offered relatively higher wages than agriculture. Attractive salary packages acted as *push factor* for the migration to the towns. Gradually, the youngsters migrated to the towns for better opportunities. Consequently, farmers faced with labor shortage as they were not able to attract laborers through competitive wages. Most farmers leased out their lands or left fallow. They were also looking for non-agricultural employment in the vicinity of the villages and relied on the remittances from their sons or daughters who had migrated to the towns. The government also introduced some programs like Sampoorna Grama Rozgar Yojana (SGRY), National Rural Employment Guaranty Scheme (NREGS), Swarna Jayanthi Grama Swarozgar Yojana (SGSY) and Ananithu Grama Anna Marumalarchi Thittam (AGAMT). The basic objectives of all these programs were to provide supplementary wage employment to rural labourers. The upper caste farmers having control over lower caste people, lost their control due to changes in social structure and land holding pattern. The unity of the people of different castes was no longer there. The role of people's participation in institution is much diminished now than before.

In case of the government sponsored institution, the cohesive force could be termed as "Bridging". "Bridging" occurs when someone from the government try to connect with local people through some agenda (Granovetter, 1973). Here, the trust among members are often thin and tend disappear when the bridger from the government side left the village or once his agenda or program is completed. This type of institutions tends to provide comprehensive solutions that have tried to exercise the factors which hinder the progress and simply do not work as expected. It is often conceived as designed to provide temporary or comic relief but not relief on a sustained basis. This system failed to understand the fact that villagers are divided into many groups, based on their caste, income status and land holding etc. Connecting thread is diluted by communal force and widespread social disparitywhile connecting or bringing them into one group as tank command areas farmers, Government sponsored institution did not take note of this problem of heterogeneity of the people. They try to identify all the farmers as tank farmers. They have time limit to implement program and within these time limit, they couldn't or were not interested to address this problem.

The core principle followed in NGOs sponsored institutions is "Linking". The NGOs try to mobilize the farmers and link them with government agencies and other financial institutions. The prime objective of this "Linking" is to get accustomed to use government program for the benefit of common people. It is also considered as opportunistic ties and viewed as the capacity provider for institution to lever resources, ideas and information from the formal institution (Woolcock,2001). NGOs showed interest to operate in villages only when favorable conditions prevailed. In case of any difficulty, the NGOs withdraw from these villages automatically from institution building process. In our experience, in the study villages from 1992-2002, one NGO called ASSEFA (Association of Sarva Seva Farm) came to create sound institutional and regulatory framework as well as enabling environment for people's participation by providing loans. But after the initial involvement they exhibit, they failed to imbibe a sense of self-help and a sense of sustainable progress. With the disappearance of the enabling environment, the ASSEFA withdrew from the villages. It is easy for the NGOs to operate in new villages rather than operate in one village for longer time. Another NGO called DHAN Foundation came to operate in this village after the ASSEFA had withdrawn. Considering that relatively small village with single community, the basic platform to launch its program was already initiated by earlier one. This NGO also did its level best to organize the farmers to form tank institution called "Vayalagam". They showed substantial and positive improvement in tank performance surpassing initial hurdles. But there were differences of preferences between the NGOs and the villagers. Bolding (1994) argued that any external involvement, no matter how well intentioned, can be perceived as meddling and even be feared. Hence, what they need to do is not bringing expert from outside, but an awakening of the expertise within the villagers.

Modes of Operations of the Institutions

Traditional irrigational institution is functioning as a two tier system. In the top level, there will be commanding position called "Nattamai"(informal village leader) usually occupied by upper caste people. In the lower level, there will be an executing position as irrigation worker called "Neerkatti"(water manager) "Neerpachi"(water distributor) and "Thotti" (field assistant) are employed. They are usually hired from scheduled caste household on rotation basis. In government sponsored institution, they organize water user association with membership of all the *ayacut* (tank command) farmers. They are expected to elect three positions like president, secretary and treasurer. Based on the number of villages included in association, they will select members also. Apart from this elected body, this system also employs irrigation workers from scheduled caste households. In case of NGO sponsored institution, the NGO appoints one person as negotiator to motivate farmer to join in irrigational institution. The member farmers elect or select their president, secretary and treasurer. The NGO provide staff to help the farmers to maintain their accounts.

Role Execution

Traditional tank water institution is existing here from the time immemorial. These institutions have complete control over the common resources. The way they approach to the problems are perhaps most incisive and provide constructive contribution to its better performance. These institutions took the responsibility of supply channel maintenance, de-silting tank bed (farmers are allowed to remove top fertile layer of silt for their manure need), strengthening of tank bund, maintaining of tank physical structure (sluice and surplus weir), water distribution, resolving dispute and conflict resolution. However, the present situation is that most of the functions are not performed as external environment changed significantly. Farmers are not allowed to take silt from the tank as social forestry program implemented by the government. Due to this misplaced priority, regular de-siltation did by farmers are stopped. As a result, every year about 2 percent of tank capacity is lost due to silt accumulation. Supply channels and catchment area are also encroached and but these institutions have no power to deal with them. Most of the tank water institutions have only limited responsibilities that too are also not regularly (Janakarajan, 1993 and Palanisami, 2006).

Table 3: Role execution of Institutions

Roles assumed	Traditional	Govt. Sponsored	NGO Sponsored
Supply channel cleaning	Occasionally	Occasionally	Yes
De-silting tank bed	No	No	No
Strengthening tank bund	No	Yes	No
Sluice and weir maintenance	Yes	Occasionally	Yes
Outlet channel maintenance	Yes	No	Yes
Water distribution	Yes	No	No
Conflict resolution	Yes	No	No

Source: Field survey by authors in 2007

Table 3 shows that the gap between perceived roles and performed roles is large and illuminating. In government sponsored institutions, water user association was active only during tank rehabilitation program in 1996-1998. After completion of this European Economic Community assisted program, officials responsible for water users association, failed to maintain the tempo of their members (Palanisami *et al.*, 2007, P-11). Farmers also complained that they spent much money on tank structures. The main problem is that its catchment and supply channel has been encroached upon, and nothing has been done about it. Farmers participating in the ongoing process were also not given any role to play. These kinds of participation are often criticized as tokenist, giving participant no power (Smith, 1998, P-197). This kind of reluctant partnership does not ensure active participation of the people. Pearce and Stiefel(1980) concluded that the promotion of participatory institutional building may be regarded as no more than rhetoric unless communities have some degree of power over the services. Smith (1998) also argued that passive participation in the name of consultation is the weakest form of participation in decision making, is often said to be a means of indoctrinating the public in the values and priorities of the planner to ensure that they obtain public endorsement of their decision, rather than understanding of local needs and priorities.

As said earlier, the government policy change in terms of transferring land from the upper caste to the lower caste does not only mean transfer of ownership only. It does also mean power transfer. This power sharing is not supported by the upper caste, apparently though they agreed to transfer the land ownership. Villagers in the lower caste being deprived from jobs in the enterprises of the upper caste people became self-employed. In the mean time, the entry of more and more caste based political party into the village system damaged the village cohesiveness and consequently wiped off cooperative attitude within and between farmers and villages. This could be a possible reason for dismantling traditional institutions. Disintegration of joint family, promotion of education, development of cottage industry hastened the process. As Agarwal (2001) rightly put if farmers have earning activities that are not reliant on common resources, their incentives to the collective management will be reduced. The degree of dependency on small scale irrigation will depend both on farmer's capacity to exploit it and on what alternative livelihood options are available to them. Our observation confirmed that farmers are slowly losing their ability to exploit potential benefit from tank irrigation system because of their weak institutional power. Compared to government sponsored institutions, traditional and NGO sponsored institutions showed incremental increase in the delivery system. In these two organizations farmers strive continuously to subjugate impossibility and then try to succeed.

Role Execution of Irrigation Functionaries

An institution, irrespective of its nature or governance, is assisted by a group of irrigation workers called "Neerkatties" (water man) who are generally hired from scheduled caste house- hold in rotation in the tank village. If a particular tank village does not have that particular schedule caste community, they employ "Neerkatties" from nearby villages. The discussion about "Neerkattis" becomes important, considering the services they render to tank institution. They are the specialists in water management, having rules to allocate water in the time of scarcity, on the basis of detailed knowledge of the needs of individual wetland fields, thus mitigating usual tension between head and tail-enders(Mosse,2006) The "Neerkatties" are omnipresent who work almost in all the tank villages making their livelihood based on their services like sluice operation, irrigation to the field, protecting tank resources and so on. In the mean time, tank, like any other institution, has also changed a lot and profiles of these functionaries also changed. In many cases, our field experience showed that, such changes have made their lives more precarious despite the fact that many are trying their best to survive (Vasimalai, 2003, P-2). Among the study villages, two villages had "Neerkatti" community and one village did not have any "Neerkatti" community. By custom, the "Neerkatties" are expected to perform some responsibilities (Table 4).It is clear from the Table 3 that mere existence of "Neerkatti" family in the village is no guaranty for execution of the expected work. During our interview with "Neerkatties" in the village, they informed that they are not doing jobs what their father or grandfather as a "Neerkatti" did.

Table 4: Role execution of Neerkatti

Assumed Roles	Traditional	Govt. Sponsored	NGO Sponsored
Mobilize village farmer	Yes	No	Yes
Watch and ward of tank asset	No	No	No
Water management	Yes	No	No
Farm management	No	No	No
Arranging religious ceremony	Yes	Yes	Yes
Sluice operation	Yes	No	Yes
Moderator of dispute between farmers	Yes	No	Yes
Common fund collector	Yes	No	Yes
Announcer	Yes	Yes	Yes
Directing Neerpatchi and Thotti	No	No	No

Source: Field survey by authors in 2007

Dependency : In the past 10 years, because of the uncertainty and insufficient rainfall the tank did not receive enough water to cater farmers' need. Studies showed that only 2 years in the last 10 years the tank received water at its full capacity. As a result, most of the farmers ended up with crop failure or left fallow. One "Neerkatti" needs to work for atleast 30 acres of farmers' field as water man to get reasonable income. Below this income, the "Neerkatti" finds it difficult to maintain his standard of living. This low income also forces the "Neerkatti" to leave the job and seek some other employment.

Payment : Usually after the crop harvest, the "Neerkatties" are entitled to have 12 kg of grain per acre. This type of payment is applicable only during normal tank season. When tank fails or partially performed they are not sure about their payment. Again some farmers, even if they reap good harvest, they are reluctant to come forward to pay the wages of the "Neerkatties". This type of problem is increasing day by day. Sometimes there were disputes over the nature of work of the "Neerkatties". So there are differences between the amount of work perceived and the actual work performed. Majority of them could not earn enough income from agriculture and as such they were looking for some jobs in the non-farm sector, leaving agricultural sector behind.

Conclusion

The thinking of community was of lowest level of aggregation at which people organize for common efforts; *i.e.* a small, homogenous, harmonious and territorially bound unit (Kumar, 2005). Many researches showed that the rural or traditional communities are in harmony with local customs and demonstrate long established patterns of sustainable and equitable resource use (Li,1996). Traditional or institutional approach to common property received wide spread acceptance and resulted successful for quite a long period. It is proved that community can own, manage, sustain and enhance common resources such as tank irrigation system (Berkes,1989 and Ostrom,1990). But present situation, tank irrigation system as an institution failed to deliver what it is capable of. Reasons are multifold and deep rooted.

The main flaw in today's approach to tank institution is its fragmented approach and the urgent need is a holistic approach. Tank irrigation system involves physical structures, technical aspects and institutional factors. All the attempts made so far to modernize or rehabilitate the tank system fully concentrated only on physical improvements. That too was not as good as farmers expected. The institutional aspects completely ignored until international donor agencies was asked to do so. Even then reports showed that government spent 71 percent of money in physical improvement and 27 per cent spent towards administrative purpose. Meagre 2 per cent was spent on institutional aspects and after maintenance (ADB, 2006). Importantly, the institutional factors and physical factors do not act in isolation; they are so complex and often interact with each other. Hence, it is recommended that due importance need to be given to address institutional aspects. About 10 per cent of the costs could be spent towards institutional and system maintenance.

The farmers asserted that government induced participation is often purely exploitative. Such participation offers some temporary palliatives to address much deep rooted problems. As a result things would not happen the way one would have expected. The minor irrigation system is to be treated as one integral holistic unit comprising catchment, water spread, tank structures and tank command. As experience showed that most of the encroachment occurred in catchment and supply channel by some other village *Panchayat*. So the institution could not exhibit its power on this chronic problem. These institutions are provided with power to evict encroachment and safeguard its resources. In our view, true attempt could be made for revival of traditional irrigation institution with its original vibrant. The policy should be underpinned by the principles of sustainability and equity. Women are widely encouraged to participate in the institution. Like in the case of Pudhucherry, women and men from every agricultural household could become member in the institution. A meaningful partnership, not just reluctant partnership, between the villagers and the tank institution will go a long way to meet the ever increasing demand of the farmers. A sustainable tank irrigated agriculture with all its uncertainties and complexities cannot be envisaged without all the actors being involved with real enthusiasm in all aspects of planning, execution and management process.

Acknowledgement

We would like to express our sincere thanks to our respondents for their cooperation and also we sincerely thank the Japan Society for the Promotion of Science (JSPS), Japan and the Suntory Foundation for their financial assistance. M. Jegadeesan is JSPS Post-Doctoral Fellow, Center for Southeast Asian Studies, Kyoto University, Japan. K. Fujita is Professor, Center for Southeast Asian Studies, Kyoto University, Japan Address for correspondence: jega@ cseas. kyoto-u.ac.jp and smjegadeesh@gmail.com

References

Agarwal, A. (1999). Community in conservation: tracing the outlines of an enchanting concept, in R. Jeffery and N. Sundar,eds *A New moral economy for India's forests? Discourse of community and participation*, Sage Pulication, New Delhi.

Ann Dale and Jennie Sparkes (2008). Protecting ecosystems: network structure and social capital mobilization, *Community Development Journal*, 43:143-156.

Asian Development Bank (2006). The rehabilitation and management of tanks in India: A study of selected states.

Basu, K. (2000). Prelude to political economy: A study of the social and political foundations of economics, OUP, Oxford.

BDO (2007). Block development office, policy note.

Berkes, F. (1989). Common Property Resources: Ecology and community based sustainable development, Belhaven Press, London.

Block statistics (2007). Block Statistical Office, Govt. of Tamil Nadu.

Bolding, A. (1994). We thought we knew it all, *Zinwesi News letter*, University of Zimbabwe and Wageningen Agricultural University, Mutare, (3).

Brosius, J.P., Tsing, A.L. and Zerner, C. (1998). Representing communities: histories and politics of community based natural resource management, *Society and Natural Resources*, 11(2): 157-168.

Chopra, K and P. Dasgupta (2008). Nature of household dependence on common pool resources: An Empirical study, *Economic and Political Weekly*, Feb 23, pp 58-66.

Coward Jr, E Walter (1980). Irrigation development: institutional and organizational issues in Coward Jr E Walter (ed), Irrigation and Agricultural Development in Asia: Perspectives from social sciences, Cornell University Press, London.

Granovetter, M. (1973). The strength of weak ties, *The American Journal of Sociology*, 78(6): 1360-1380.

Janakarajan, S. (1993). In Search of tanks: Some hidden facts, *Economic and Political Weekly*, June 26, pp A53-A60.

Kumar, C. (2005). Revisiting community in community based natural resource management, *Community Development Journal*, 40(3):275-285.

Li, T.M. (1996). Images of community: Discourse and strategy in property relations, *Development and Change*, 27(3):501-528.

Li, T.M. (2002). Engaging simplification: Community base natural resource management, market processes and state agendas in upland Southeast Asia, *World Development*, 30(2): 265-283.

Mosse, David (2003). The rule of water: Statecraft, ecology and collective action in South India, Oxford University Press, Delhi.

Mosse, David (2006). Collective action, common property and social capital in South India: An Anthropological commentary, *Economic Development and Cultural Change*, 54(3): 695-724.

Narayanan, D. (1999). Bonds and Bridges: social capital and poverty, World Bank, Washington, DC.

NSSO (1999). Common property resources in India, NSSO 54th round survey (Jan 98- June 98), Govt. of India.

Onyx, J. and Bullen, P. (2000). Measuring social capital in five communities, *The Journal of Applied Behavioral Science*, 36 (1):23-42.

Ostrom, E. (1990). *Governing the commons: The evolution of institutions for collective action*, Cambridge University Press, Cambridge, UK.

Palanisami, K. (2000). *Tank irrigation: Revival for prosperity*, Asian Publication Services, New Delhi.

Palanisami, K. (2006). Sustainable management of tank irrigation system in India, *Journal of Developments in Sustainable Agriculture*, 1:34-40.

Palanisami, K., M. Jegadeesan; K.Fujita and Y. Kono (2008). Impacts of tank modernization programs in Tamil Nadu state, India. Working Paper Series, CSEAS, Kyoto University.

Pearse, A. and Stiefel, M. (1980). Enquiry into participation. A Research Approach (eds). United Nations Research Institute for Social Development, Geneva.

Putnam, R. (2001). Social capita measurement and consequences, *ISUMA*, 2(1):41-52.

Sakthivadivel, R., P. Gomathinayagam and Tushaar Shah (2004). Rejuvenating irrigation tanks through local institutions, *Economic and Political Weekly*, July 31:3521-3526.

Smith, B.C. (1998). Participation without power: Subterfuge or development, *Community Development Journal*, 33(3):197-204.

Vaidyanathan, A. (1985). Water control Institution and agriculture: A comparative perspective, *Indian Economic Review*, Vol. XX. No 1.

Vasimalai, M.P. (2003). *Neerkatties: The Rural water manager* (eds. by Seenivasan. R), DHAN Foundation, Madurai, India.

Woolcock, M. (2001). The place of social capital in understanding social and economic outcomes, *ISUMA*, 2 (1):11-17.

Extension Management Strategies for Sustainable Agriculture, pp 443-447

New India Publishing Agency, New Delhi, India

48

Effectiveness of TNAU and CTCRI Interventions in Cassava Value Addition Techniques in Tamil Nadu

A. Janaki Rani and Preethi Priya Dharshini

Cassava is one of the leading staple food crop which is consumed as food and rest is used in industries and as feed to animal. Cassava has the potential to increase farm income, reduce rural and urban poverty and help to close the food gap (Nweke *et al.,* 2001). It is used in many sectors like food processing, textile industries, glucose production etc. Salem stands first in area and has about 650 sago and starch industries acting as a prime reason for the intensive area under the cassava cultivation in Salem also 60 per cent of the starch produced in India comes from Salem District alone. Nearly 200 products can be made from cassava which leads to implementation of several interventions in the district which helped in imparting knowledge on cassava value addition techniques and improve adoption. Present study was undertaken to analyse the perceived impact by focusing on the effect of interventions of CTCRI and TNAU among the women growers in enhancing group approaches and empowerment through entrepreneurship activities.

Methodology

The study was materialized in the Atturtaluk of Salem district, Tamil Nadu in the villages namely Pedanaikenpalayam, Umayalpuram and Ramanaike-npalayam. The selection of the study area was done on focusing the population of socially backward cassava women growers, area under cassava cultivation and also considering cassava project functioning in the district. A sample size of 25 women beneficiaries was selected. The measurement of the impact was made through the Participatory Rural Appraisal (PRA) technique of simple scoring method.

Simple scoring method adopted by Andrew *et al.* (2010) was used in this study, in which visual aids (photos) were arranged and the women were asked to place stones or twigs based on the self perceived changes due to value addition intervention. Each stone was given score of five per cent by which maximum of 20 stones can be used for highest preference (100.00%). The range of the score will be 1 to 20. Scores were given for each stone with minimum of 5 to maximum of 100%. Later the stones or twigs in each item were counted and ranked according to highest score and analysed. The items were then ranked according to their highest scores. The results obtained through this procedure were discussed below.

Results and Discussion

Perceived impact in this study refers to the self-perceived changes among the respondents on the value addition techniques. The results of the study were detailed below.

Under this simple scoring method the women farmers kept the stones based on their importance. Table 1 shows that self confidence (18 score) was given first preference by the women growers. To start a venture on value addition first the members should gain confidence and maintain positive attitude to get success.This might be the reason for highest perceived impact on self confidence.

The data predicts that knowledge and skill enhancement about value addition was ranked as second. Because the women beneficiaries felt that to start any activity the knowledge of that aspect needed along with skill specific to that activity should be enriched for better and sustainable outcome in this way knowledge on value addition was imparted through trainings and demonstrations lead to enhancement of skill. Further, free distribution of chipping machineries added to the perceived impact. Group co-operation ranked third in position with 16 score since it is essential to make a venture to grow or fall. In order to taste good outcome co-operation among the members is indispensable.

Deciding product for value addition, financial position, and infrastructure facilities to start up the venture plays crucial part this makes the decision making power score 15 and rank in fourth position. Income level ranked fifth with 14 scores. This might be the result of the factors like supply of free chipping machines, knowledge on marketing network, training and demonstration provided on preparation of value added products leading to gain additional income. This lead to improve family maintenance by scoring 13 score.

Table 1: PRA-Simple scoring method to assess perceived impact **(n=25)***

S.No.	Content	No. of stones	Scores	Percentage	Rank
1.	Self confidence	******************	18	90	1
2.	Decision making power	***************	15	75	4
3.	Knowledge and skill enhancement about value addition	*****************	17	85	2
4.	Self employment and Income generating activity	************	12	60	7
5.	Family maintenance	*************	13	65	6
6.	Gained respect from family and village	*********	9	45	10
7.	Improvement in social participation and networking	********	8	40	11
8.	Awareness about value addition and new development schemes	*****	5	25	14
9.	Income level	**************	14	70	5
10.	Saving habit	***********	11	55	8
11.	Group co-operation	****************	16	80	3
12.	Information sharing	******	6	30	13
13.	Market linkage with local institution	****	4	20	15
14.	Debt clearance	**********	10	50	9
15.	Nutritional status	*******	7	35	12

The entrepreneurship skills and regular income had created much impact on women growers. The women SHGs were formed based on the best performance during the training period. At an initial stage the SHG leaders were given special training to enhance the entrepreneurship skills which gives self employment and income generating activities. Hence, growers ranked it as seventh important impact.

Savings habit (score-11) and debt clearance (score-10) were ranked as 8th and 9th perceived impacts by majority of the women growers. Due to the income generation the saving habit and the debt clearance to some extent were possible by the women growers. They have gained respect from the family and village people (score-9) due to the interventions and entrepreneurship skill development. Further, they could involve more in social activities (score-8) and development programmesorganised by Department of Agriculture, University and other Institutions. Hence they ranked it as 10th and 11th perceived effect.

Due to frequent intake of cassava value added products the women felt that the nutritional status especially the carbohydrate level little bit increased might be the possible reason for the women to rank it as 12th perceived effect.

Transfer of technology is the major focus of the SHGs, hence the women shared their information got through various trainings and demonstrations to the fellow women growers in their own village. So they have given the score of 6 and ranked it as 13th perceived impact.

The women got awareness about the value added products and new development programmes through this intervention might be the reason and given 14th rank. The value added products like wafers, vadagam prepared by the socially backward women growers were linked with WAFER association present in Namagiripettai for marketing.

Conclusion

The study highlights that perceived impact can be utilized to understand and visualize the grass root level needs and preference of the women growers. This helps in improving the diffusion process and focus indepth on the most preferred and least preferred aspects separately to attain sustainable and fruitful result. The present study act as a stepping stone for future promotion and for enhancement of the interventions according to their need.

References

Andrew *et al.* (2010). Participatory Impact Assessment a Guide for Practitioners. Tufts University, USA.

Aniedu, C., O.C. Aniedu and N. Nwakor (2012). Impact and Adoption of Value Added Innovations in Root and Tuber Crops Among Farmers in Imo State, Nigeria. Global Journal of Science Frontier Research Agriculture and Veterinary Sciences,12(11).

Arunkumar, A. (2002). Retrospects and Prospects of Commercial Cassava Cultivation.Unpub. M.Sc (Ag) Thesis, AC & RI, TNAU, Coimbatore.

Ekwe *et al.* (2008). Building Farmers' Knowledge in Cassava Value Addition for Improved Rural Livelihoods in Abia State: Issues for Policy Consideration. Journal of Agriculture and Social Research, 8(2).

http://faostat.fao.org accessed on March 2014.

Janakirani, A. (2012). Yield Gap Analysis of Cassava Growers in India. Lambert Publications, Germany.

Nweke (2001). The Cassava Transformation: Africa's Best Kept Secret. Michigan State University Press, East Lansing.

Sakthivel, K. (2000). Knowledge, Adoption of Tapioca Growers and their Attitude towards Value Addition.Unpub. M.Sc (Ag) Thesis, AC & RI, TNAU, Madurai.

Extension Management Strategies for Sustainable Agriculture, pp 449-455

New India Publishing Agency, New Delhi, India

49

Sustainability Through Local Storage Structures

N. Jayakumar and M. Sundaramari

Sustainability issues have infused sanity in utilizing interventions without much damage to the environment. The wisdom of the practitioners, who had over generations developed their own eco-friendly practices, is being appreciated. Recent researches have aimed at documenting, rationalizing, and blending the Indigenous Knowledge with the so called Scientific or Western knowledge so as to take care of the sustainability issues. Many researchers have contributed to this pool of knowledge by documenting many of the technologies, crop wise, area wise and operation wise and so on.

Sustainable agricultural practices that have evolved over the years could provide answers to the current crisis in the agriculture sector. Sustainability does not limit itself to the practices of production and protection but extends beyond even to storage and availability for future uses.

Food storage is an essential development for food production and farming, and represents a major evolutionary threshold for human civilization (Testart,1982). People were able to regularly store food beyond their annual consumption needs, including banking grain to overcome spoilage, and to provide seed for planting and potential years of crop failure. In some cases storage necessitates, or is necessary for, changes in social systems, invoked both in increasing corporate activities and for the development of hierarchical structures. Storage also represents a critical form of risk management and economic intensification (Ian Kuijt and Bill Finlayson, 2009).

It has been estimated that storage losses alone contribute to a loss of more than 20% of the total food grains produced in India (Khan, 2005). Even after the strenuous effort of producing crops, the farmers are struggling hard to protect the food grains from various conditions causing damage to them. Until the entry of modern techniques of food production and storage, farmers were possessing excellent storage structures and practicing indigenous storage techniques (Sundaramari *et al.*, 2010).

Paddy, being the major food grain has to pass through too many middlemen such as commission agents, wholesalers, millers cum wholesalers, retailers in the chain of distribution (Prakash, 2012).

Storage is particularly important in agriculture because agricultural production is seasonal while demands for agricultural commodities are more evenly spread throughout the year. In this circumstance, there is a need to meet average demand by storing excess supply during the harvesting season for gradual release to the market during off-season periods.

"The farmer, in general, sells his produce at an unfavorable place and at an unfavorable time, and usually gets very unfavorable terms." So, in such circumstances, it is common to find that the agricultural producers as a class are being exploited by the purchasers. There are possibilities that they could get better rates, if they could store their produce conveniently, so that they could sell it at an advantage. Here comes the importance of local storage structures, which are convenient to a farmer.

Methodology

A study on the awareness and adoption of Indigenous technologies in rice cultivation was carried out in Vellore district of Tamil Nadu state. Rice growing farmers of Arakkonam, Kaveripakkam, Nemili, Sholingur and Thimiri blocks were selected for the study. These blocks were purposively selected because of the large area under Rice cultivation.

During this study, grain storage structures in Vegamangalam village of Kaveripakkam block were found to be interesting. The Kaveripakkam block is a revenue block in the Vellore district of Tamil Nadu, India. It has a total of 55 Panchayat villages.Vegamanagalam village is situated on the banks of Palar River and a very fertile agricultural belt. The village has a total population of 1258 people with approximately 121 farmers predominantly growing paddy, traditionally. The rice storage structures in this village provide evidence about the legacy of their rich, flourishing agriculture and bountiful harvests.

Traditional storage structures for storing grains have been documented from different parts of the state by many researchers. Karthikeyan *et al.* (2009) in their study on Indigenous storage structures have documented structures like Kuthir, Thombarai, Kalanjiyam, Macchu, Paanai, Urai, Sakkupai, Arisipetti, Saal, Marathombai. Similarly, Sundaramari *et al.* (2010) documented structures like Kulumai, Kudhir, Modappanai, Kaambara and Underground pits. All these structures however were found to store small quantities of grains possibly for seed purpose or for domestic use, whereas the storage structures found in Vegamangalam could store huge quantities of paddy, possibly the whole harvest itself. Case Study approach was followed for studying these Indigenous Rice Storage Structures.

Map showing the study area – Vegamangalam village, Kaveripakkam, vellore district, Tamil Nadu

Rice storage structures in Vegamangalam village

These storage structures have been built 100 years ago using lime mortar. Lime mortar is a type of mortar composed of lime and an aggregate such as sand, mixed with water. There are two types of storage structures available in this village to store paddy.

1. Nell Veedu
2. Macchu

1. Nell Veedu

The Nell veedu located in the Mr. R. Shanmugharaman's home is a traditional storage structure that has been built as part of their home, within the same building adjacent to their living room. The nellveedu has a dimension of 11 x 15 feet with a height of 18 feet. The whole structure has been plastered using lime mortar both inside and outside. There are no openings on any side except the front portion. The front door is made up of wood which is thick and has a bulky wooden frame, which prevents the entry of rodents in to the room. The door is designed in such a way that it can be split in to two horizontally. Once a considerable amount of grain has been stored in the room to fill it to the fullest capacity, the bottom half of the door is closed and the grains are filled in to the room. Further the top portion of the door is closed and now filling is done from the opening in the top portion of the front wall. The roof is made up of wooden tress paved with country tiles and plastered from the bottom. The room has been designed in such a way that its fullest capacity could be utilized for storage purpose.

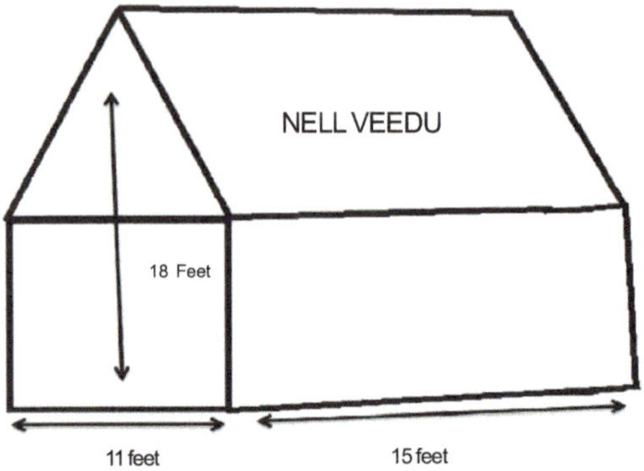

Salient Features of Nell Veedu

Opening at the top or storing the grains

Stone marking to place the vessel for grain collection

Door that could be split horizontally for storing the grains

Illustratration of the dimensions of the Indigenous storage structure (Nell Veedu)

The ingenuity of the design is evident from the small pigeon hole with a sliding door, provided on the top portion of the door. This is a special arrangement to access the grains in case of emergency. Small quantities of grains could be removed from the room without opening the door. There is a marking in front of the door using a granite stone, to keep a vessel to collect the grains.

2. Macchu

Another type of storage structure found in this village was locally called as "Macchu". Though macchu had been documented before (Karthikeyan *et al.*, 2009), this structure has got some other special characteristics worth mentioning.

Similar to the Nell Veedu, these structures are also made of lime mortar and have crossed 100 years of existence. The building materials and roof are similar. The house of Mr. R. Selvamani is of a dimension of 210 feet by 42 feet. The maximum height of the Macchu (*ie.*) at the centre is 4.5 feet. The Macchu is not confined to one room but to the whole house. The Macchu has been made using wooden planks that are firmly and permanently fixed on to wooden beams.

In one of the common rooms there is a provision for filling the grains into the Macchu. An elevated opening is given at the top with a provision where a person could stand and transfer the grains inside.

Inlet for storing the grains on to the Macchu **Storage space on the Macchu**

Salient Features of Macchu

The ingenuity of the design is by providing access to the grains. Though filling can be done only through one opening in a common room, grains can be removed from each room without much labour. For this purpose, one wooden plank forming the Macchu in each room is kept loose and this plank has been marked from the bottom using a nail. To remove the grains from the Macchu, this plank could be removed from the bottom and the grain collects into a vessel which is kept below.

Present Scenario

The present situation is that though these structures have been maintained in good condition till today, presently they do not serve the purpose of storing grains. Recent problems of rice cultivation like shrinking area under cultivation and reduced production due to multifarious reasons and the changing trends whereby farmers dispose of whatever they produce, immediately following harvest, have resulted in keeping these storage structures empty.

Though there were many houses with this type of storage system, at present only four houses still have them in perfect condition, though unused. Others have renovated their houses with the changing trends and many of these storage structures have vanished.

Conclusion

These traditional storage structures give us a very clear picture of the agricultural practices that were prevalent in earlier times. They throw light on the quantum of paddy that was harvested, much before the introduction of the so called modern technologies.

Secondly, these structures build around their living area provides security for their produce. No travelling, no paper works, always easily available as compared to the recent trend of storing the grains in warehouses.

The cost involved in building such structures is not high and once built could be used for generations. They are made up of locally available materials and amenable to alteration in their size and shape according to the requirements of the users (Sundaramari *et al*, 2003).

These structures indirectly provide the food security and self-sufficiency apart from reviving the pride and prestige of the farming community. They provide opportunities for the producers to tide over the disadvantages of glut during the time of harvest and take advantage of the spiraling prices in the market during off seasons. Hence, these indigenous storage structures need to be conserved and propagated for use among the farming community, helping in securing food security and self-sufficiency.

These traditional structures which are making a landscape of Vegamangalam village with pride, boasting of the agricultural legacy of this village will be lost in due course, unless their advantages understood and propagated. It is the reform imperative that this knowledge needs to be disseminated among the farming community, researchers, academicians, administrators and policy makers so that the potential of this structure for promoting sustainability is adequately recognized.

References

Ian Kuijt and Bill Finlayson (2009). Evidence for food storage and predomestication granaries 11,000 years ago in the Jordan Valley. *Proceedings of the National Academy of Sciences of the Unites States of America*, 106:(27).

Karthikeyan, C., D. Veeraraghavadattam, D. Karpagam and S. Ayisha Firdouse (2009). Indigenous storage structures. *Indian Journal of Traditional Knowledge*, 8(2): 225-229.

Khan Amir Ullah (2005). Working paper on the domestic food market: Is India ready for food processing?India development Foundation, Gurgoan, website.

Prakash, C. (2012). Problems and expectations of the farmers in marketing paddy in Tiruvarur district, Tamil Nadu, *Asian Journal of Management Research*, 3(1): Online Open Access publishing platform for Management Research Article ISSN 2229 – 3795.

Sundaramari, M., G.S. Kannan, S. Ganesh, T.T. Ranganathan and R. Udhayakumar (2003). A study on Indigenous practices and structures for storage of rice. Paper presented in the international seminar on "Downsizing technology for rural development" Conducted by Regional Research Laboratory, Bhubaneswar, 7-9.Oct. 2003.

Sundaramari, M., S.Ganesh, G.S. Kannan, M. Seethalakshmi and K. Gopalasamy (2011). Indigenous grain storage structures of South Tamil Nadu. *Indian Journal of Traditional Knowledge*, 10 (2):380-383.

Testart, A. (1982). The significance of food storage among hunter-gatherers: Residence patterns, population densities and social inequalities. *Curr. Anthropol.* 23:523–538. Cross RefWeb of Science Google Scholar.

Extension Management Strategies for Sustainable Agriculture, pp 457-461

New India Publishing Agency, New Delhi, India

50

How Equipped are Farmers for Climate-Smart Agriculture? A Case Study of Paddy Cultivators in the Erode and Trichy Districts of Tamil Nadu

P. Sivaraj, H. Philip and B. Swaminathan

Climate change referred as, a change of climate which is attributed directly or indirectly to human activity that alters the composition of the global atmosphere and which is in addition to natural climate variability observed over comparable time periods (IPCC, 2001). In other words, it is any long term conspicuous deviation from usual prevailing climate bringing variations in normal temperature, rainfall, atmospheric circulation with abnormal expression in extreme climate such as floods, droughts, extreme temperature etc.

Agricultural sectors must become climate-smart to successfully tackle current food security and climate change challenges. Agriculture, including forestry and fisheries, is crucial for food security and rural incomes as well as other essential products, such as energy, fiber, feed and a range of ecosystem services. Climate-smart agriculture is a pathway towards development and food security built on three pillars: increasing productivity and incomes, enhancing resilience of livelihoods and ecosystems and reducing and removing greenhouse gas emissions from the atmosphere. Climate-smart agriculture contributes to a cross-cutting range of development goals. There are many opportunities for capturing synergies between the pillars of climate-smart agriculture, but also many situations where trade-offs are inevitable.

Agriculture has to address simultaneously three intertwined challenges: ensuring food security through increased productivity and income, adapting to climate change and contributing to climate change mitigation (FAO, 2010a; Foresight, 2011a). Addressing these challenges, exacerbating global pressure on natural resources, especially water, will require radical changes in our food systems.

Climate change has already significantly impacted agriculture (Lobell *et al*, 2011) and is expected to furtherimpact directly and indirectly food production. Increase of mean temperature; changes in rain patterns; increased variability both in temperature and rain patterns; changes in water availability; the frequency and intensity of extreme events; sea level rise and salinization; perturbations in ecosystems, all will have profound impacts on agriculture, forestry and fisheries. The extent of these impacts will depend not only on the intensity and timing (periodicity) of the changes but also on their combination, which are more uncertain, and on local conditions. Anticipating appropriately the impacts of climate change on agriculture requires data, tools and models at the spatial scale of actual production areas.

Methodology

Ex-post facto research design was followed for carrying out the study. Paddy is the staple food crop of Tamil Nadu and is heavily exposed to the extreme and extraneous events of climate change. Erode and Tiruchirapalli districts were purposively selected for the study as the district has high range of variability in both rainfall and temperature. Kalingarayan (Erode) and Ponnaniyar (Tiruchirapalli) basins were then chosen as they have maximum acreage under paddy with majority of the farmers being small (2.5 to 5 acres) and marginal (< 2.5 acres). Canal irrigation was also found to be prominent in these basins resulting in farmers becoming more vulnerable to climate change events. Based on the discussions with the officials and subject matter specialists of the agricultural department one block was selected from each basin. For the selection of villages, an inventory of revenue villages in each block was collected. Then ten villages from each block were randomly chosen. The total sample size was 200 with randomly selecting 100 paddy farmers (comprising 50 male farmers and 50 female farmers) from each of the blocks.

Findings and Discussions

Relationship of independent variables with extent of awareness

The nature and degree of association between the characteristics and Extent of awareness of paddy growers was worked out and the results are presented in the Table 1. The R^2 value of 0.86 which revealed that 86.00 per cent variation in extent of awareness was explained by age, gender, educational status, annual income, family size, occupational status, farm size, farming experience, cropping pattern, irrigation source, training undergone, extension agency contact, mass media exposure, social participation, risk orientation, scientific orientation and exposure to weather advisory services variables selected for the study. The 'F' value was significant at one per cent level of probability. The association between extension agency contact was found to be positively significant. This reveals that greater the degree of contact with extension agency greater would be the extent of awareness. The reason might be that those farmers who contacted the extension personnel more frequently could have become aware of the improved water management practices and also would have been information about climate change.

$Y_{1} = 11.769 + 0.084 (X_1) + 0.006^{**} (X_2) + 0.028^{*} (X_3) + 0.027^{*} (X_4) + 0.901 (X_5) + 0.089 (X_6) + 0.380 (X_7) + 0.283 (X_8) + 0.003^{**} (X_9) + 0.111 (X_{10}) + 0.073 (X_{11}) + 0.901^{**} (X_{12}) + 0.001 (X_{13}) + 0.264 (X_{14}) + 0.403 (X_{15}) + 0.169 (X_{16}) + 0.001^{**} (X_{17})$

Farmers with higher extension agency contact (X_{12}), exposure to weather advisory services (X_{17}), cropping pattern (X_9) and gender (X_2) was positively significant at one per cent level which implies that if extension agency contact, exposure to weather advisory services, and cropping pattern increased by one per cent then the extent of awareness *Ceteris paribus* would increase by 0.901, 0.001, 0.003 and 0.006 respectively. Annual income (X_4) and educational status (X_3) had positive significant at five per centwhich implies that if annual income and educational status increased by five per cent then extent of awareness *Ceteris paribus* would increased by 0.027 and 0.028 respectively.

Table 1: Relationship of independent variables with extent of awareness **(n = 200)**

Variable No.	Variables	Regression coefficient	Standard error	't'
	Constant	0.001	0.739	15.92
X_1	Age	0.084	0.096	1.74
X_2	Gender	0.006**	0.124	2.76
X_3	Educational status	0.028*	0.061	2.21
X_4	Annual income	0.027*	0.090	2.23
X_5	Family size	0.901	0.106	0.12
X_6	Occupational status	0.089	0.066	0.88
X_7	Farm size	0.380	0.240	-1.08
X_8	Farming experience	0.283	0.104	-1.71
X_9	Cropping pattern	0.003**	0.091	3.06
X_{10}	Irrigation source	0.111	0.066	-1.60
X_{11}	Training undergone	0.073	0.236	1.12
X_{12}	Extension agency contact	0.901**	0.020	7.40
X_{13}	Mass media exposure	0.001	0.050	-0.12
X_{14}	Social participation	0.264	0.103	1.80
X_{15}	Risk orientation	0.403	0.018	0.84
X_{16}	Scientific orientation	0.169	0.019	1.35
X_{17}	Exposure to weather advisory services	0.001**	0.035	5.67

* Significance at 0.05 level R^2=0.860 ** Significance at 0.01 level

Contact with extension agencies like agricultural officers and assistant agricultural officers in state departments, dealers and junior research fellows, increased the knowledge of farmers on various technologies so as to follow the coping strategies. The extension functionaries also motivated the farmers to implement the various strategies in farming. Exposure to weather advisory services increased the awareness level of farmers about climatic factors like rain falls, temperature *etc*. So exposure to weather advisory services had significant relation with extent of awareness of farmers on climate change. Cropping pattern increases the farmers' exposure on agriculture and thereby reduces the risk and also by growing more crops the awareness level of the farmers about the climate change aspects would get increased.

Farmers' education level had positive significant relationship with the extent of awareness. The farmers' education status is primary, secondary level to collegiate level. It is an accepted fact that educated farmers always have inclination to the changes in the climate and farming know new technologies very quickly than illiterate farmers they used to take alternate measures to cope with the erratic situation. Annual income had shown positive and significant relationship with extent of awareness the respondents. Higher annual income would have created interest and motivated them to utilize the information sources properly in order to acquire and practice the gained technical knowledge.

Conclusion

The study inferred that farmers had medium to low level of adaptation measures against the climate change in Ponnaniyar and Kalingarayan basin. The role of gender in the adoption of climate-smart agricultural practices by farmers is also considered. Women's rights to property vary within and between countries in India. Land tenure systems and availability of funds to invest in better technologies are gender-sensitive constraints faced by women to the adoption of conservation and climate-smart agriculture. Because of their different social roles, women and men both experience and respond to climate extremes and climate change differently, including through access to development programs. Existing technologies and current institutional structures seem inadequate to achieve the mitigation needed to adequately slow climate change, while also meeting needed food security, livelihood and sustainability goals. We must identify actions that are science-based, utilize knowledge systems in new ways, and provide resilience for food systems and ecosystem services in agricultural landscapes despite the future uncertainty of climate change and extreme events. It is imperative that new modes of science-policy integration transform land management and community action for food security, and for conservation of biodiversity and the resource base upon which agriculture depends.The researcher had certain priorities such as assessing the farmers' perception and increases the adaptation strategies to overcome the effect of climate change by developing new technologies and resistant varieties. Farmers to overcome the constraints were develop less water use technologies like releasing drought resistant varieties, trainings on new adaptation technologies, provision of machineries on subsidiary basis. Development department should make available suitable policy for providing agricultural inputs at right time and measures are to be taken to provide the support price, subsidies to right farmers, insurance to all crops to maintain their livelihood.

References

FAO (2010a). "Climate-smart" Agriculture: Policies, Practices and Financing for Food Security, Adaptation and Mitigation, Rome.

Foresight (2011a). The Future of Food and Farming: Challenges and Choices for Global Sustainability. Final Project Report. London, The Government Office for Science.

IPCC (2001). Climate Change: Impacts, Adaptation and Vulnerability, Cambridge: Third Assessment Report: Cambridge University Press.

Lobell, D., Schlenker, W. and Costa-Roberts, J. (2011). Climate Trends and Global Crop Production since 1980, Science.

Extension Management Strategies for Sustainable Agriculture, pp 463-468

New India Publishing Agency, New Delhi, India

51

Strategies to Strengthen Precision Farming

Padma S.R. and T. Rathakrishnan

In conventional agriculture, spatial variability over the landscape is ignored and fertilizers, herbicides, insecticides, fungicides *etc*. are applied at a uniform rate throughout the crop field. Precision Farming emphasizes on the infield variability which is first recognized, located, quantified and recorded, then managed by applying farm inputs in specific amounts at specific time and at specific locations.

Precision Farming is facilitating the prospects and scope for switching over to modern agriculture leaving the traditional one by utilizing right resources in right time and management, which results an environment friendly sustainable agriculture. Precision farming provides a new solution using a systems approach for today's agricultural issues such as the need to balance productivity with environmental concerns. It is based on the implementation of advanced information technologies. It includes describing and modelling variation in soils and plant species, and integrating agricultural practices to meet site-specific requirements. It aims at increased economic returns, as well as at reducing the energy input and the environmental impact of agriculture.

Precision farming requires some degree of competence in the use of software and hardware on the part of growers and/or crop consultants. Indeed the success of precision farming largely depends on creation of management systems, which will involve some combination of computerized decision support systems and the wisdom of farmers. Growers will adopt information technologies only if they are reliable and easy to use, offer some competitive advantage and can be introduced into farming without too much difficulty or expense.

The advantages of precision farming is that it offers opportunities to improve agricultural productivity and product quality, reduces agro-chemical wastage through efficient application and resulting in minimizing environmental pollution and energy conservation. However adoption of any innovation is hampered or hastened mainly by farmers' characteristics and also the environment in which they live.

An attempt has been made to analyse the extent of support of precision farming in improving the socio-economic status of the precision farming practitioners through the intervention of stakeholders. Based on this, some strategies have been formulated to strengthen the Precision Farming which will definitely uplift the status of the precision farm practitioners to the remarkable level.

Research Methodology

Among the seven agro climatic zones in Tamil Nadu two zones *viz.*, Northern Zone and North Western Zones were purposively selected for conducting the study. In the North western zones the districts Dharmapuri, Krishnagiri and in western zone Coimbatore and Erode were the study areas. Two blocks from each district in which the Precision Farming cultivation is intensive were selected. Thus the study covers eight blocks in four districts. Simple random sampling procedure was followed to arrive 200 precision farming practisioners totally @ 25 from each block and 50 Stakeholders *viz.*, Researchers/Extension personnel, Input dealers, Drip marketers, Produce marketing personnel and Agri clinics personnel @ 25 from each zone.

Dimensions of the study

The study was carried out in the said study areas to assess the technological utilization pattern, perceived effectiveness, market associated activities of the precision farm practitioners, intervention of stakeholders in Precision Farming and the SWOC on Precision Farming as perceived by farmers and stakeholders.

The strategies have been formulated based on the outcome of the study to strengthen the Precision Farming further.

Strategies

- Joint decision making with members of Precision Farm practitioners Association (PFA) and peer group was observed in the study, it has to be fortified by the intervention of stakeholders so that the thorough flow of information could be possible among the PFA members.
- The sample was found with moderate to high level extension participation, Orientation with Research station and its activities. If the stakeholder has more contact with Precision Farmers with frequent intervals, it could be lifted up.

- It was noted that a portion of land was allotted by the farmers for Precision Farming from their total area under cultivation. The size of Precision Farming was semi medium to medium size, despite they could earn Rs. 5 lakh as annual income. If the area under Precision Farming is increased that will have the positive reflection on the livelihood status of the farmers.
- Though there is ample scope for sharing the farm information through PFA, moderate level of sharing pattern was observed. The monitoring team motivates the PFA members to share and discuss every aspect in their areas.
- Among the core technologies recommended for Precision farming, Remote Sensing the prime technology was not at all aware by the sample. The infield variation could be assessed based on remote sensing only. Without considering such in field variation, the other technologies even if followed as recommended will not bring the anticipated output. The Precision Farming Development Centres (PFDC) established at National level have not yet initiated this process. The effectiveness of remote sensing in this regard has been utilized in the developed countries. Such attempt may be taken up by the Regional Remote Sensing Service Centres (RRSSC) located across the country, established by National Natural Resources Management System (NNRMS). The stakeholders especially the service providers of public sectors could obtain the results and adopt the same as a trial in every Precision Farming area. The performance of such approach will enable the farmers to develop awareness and the importance of the use of remote sensing, incidentally it will diffuse into the system.
- The use of quality seed and seedlings is the important component of Precision Farming for producing quality farm produces. The supply of seedlings through Hi-tech community nursery is one of the technologies which was not up to the expected level in the study area. As that of Self Help Groups (SHGs) farm women groups may be organized by the stakeholders in the Precision Farming areas and be involved in the production of seedlings through Hi-tech community nurseries. In order to overcome the initial hurdle, in establishing the nursery, the groups may be provided with a reasonable budget as a seed money from the government as revolving fund. Those groups may be linked with PFA for the supply of seedlings. All these activities will be assisted and monitored by the stakeholders of the public sectors. Through this attempt, in addition to the supply of quality seedlings, entrepreneurial development could also be achieved.

- The Precision Farmers Association (PFA) operating in the Precision Farming areas need to be strengthened. It is ensured as a mandate that every Precision Farm practitioner is to be enrolled as a member. A team may be constituted by the administrators which includes the stakeholders, the progressive farmers of the region and also the office bearers of PFA. PFA activities are to be monitored with fixed time interval by any of the stakeholders alternatively. This will facilitate the following aspects.
- Collective decisions about the choice of crops on demand base.
- Supply of agricultural inputs at lesser price with assured quality by developing direct contact with the manufacturing companies.
- Eliminating middlemen by developing direct contact with exporters.
- Availing common transport.
- In order to lessen the burden of Precision Farmers, the state level administrators may take the initiative to build tie-up with the input production companies to supply Water Soluble Fertilizers and other required agro inputs with high quality on concessional rate to the Precision Farmers Association. Also through this mode the actual demand could be assessed by the units so that the bulk transport could be materialized. Moreover assured return is also possible for the production companies.
- As reported in the study the intervention of stakeholders especially Researchers and Extension functionaries was more frequent than the others. As far as the drip marketers, input agents are concerned they restricted their intervention with the initial establishment. The stakeholders should not work in isolation as observed during data collection. In every activity of the Precision Farming their roles to be made as complementary. The implementing agency of Precision Farming should work out a strategy of involving every one's role by interlinking their activities throughout the process. This will further strengthening and sustaining the Precision Farming activities at all levels.
- The success and experiences of the Precision Farming activities need to be regularly and continuously shared among the stakeholders through publications, websites, newsletter, CDs and other media.
- Stakeholders meet on Precision Farming during crop seasons may be taken up at various levels and their services or technological intervention in time without missing the season make the precision success.
- Stakeholders involved in Precision Farming process are both public and private sectors. Technology delivery process mainly concerned with the

public sectors. The assignment of public sector stakeholders proceeds till the disposal of the produce. In order to perform their roles effectively, they should be competent in their managerial and facilitating skills. Being play the influential role, they have to be trained adequately to equip with the said skills which help them to perform their assigned tasks without delay and deviation as designed. The capacity buildingshould not be a habitual rather the training programmes may be taken up in the reputed management centres periodically.

- Such programmes will help in developing OCTAPACE culture which stands for Openness, Collaboration, Trust, Authentication, Pro action, Autonomy, Confrontation with Problems and Experi mentation among the stakeholders.
- If the said culture is developed among the stakeholders, naturally they will discharge their tasks effectively without lapses.
- The scope of Precision Farming in all its dimensions could be strengthened by involving multidisciplinary team of Researchers in various fields including Agricultural Engineers, Manufacturers, Economists *etc*., Such a team may be constituted by the planners at state and district level.
- The forecast on demand of the market will definitely influence the choice of the crops and production. This could be attended by the PFA and the monitoring team.
- Warehousing centres, Cold Storage facilities and Ripening chambers are need to be established at district level in the major production centres so as to reduce the post harvest losses. These infrastructures may be established jointly in coordination with the drip marketers, NGOs and other agencies working in that zone. This will support the Precision farmers to store the farm produces for a reasonable period to fetch the reasonable price.
- Buffer storage through precision platform ensure the produce available throughout the year in the market so as to get premium price and gaining the confidence of customers with the assurance of its availability.
- The possibility of establishing Fruits and Vegetables Processing centres and value addition centres at Zonal level may be explored to exploit the surplus produces when the market price is low.
- Creating a Logo and having brand name was found with only very few of the PFA in Krishnagiri and Dharmapuri districts. It should be encouraged by the stakeholders so that labelled and packed produce leads to traceability and winning the confidence level of the buyers which will improve the cost of the produces on par with branded goods.

- The administrators may assess the possibility of establishing mobile markets, farmers own mega markets, Farmers retailers markets, so that the farm produces could reach the local customers at their door steps without the middle men intrusions.
- In order to meet the demand for spare parts, Custom service centres may be established at block level. This centres will also attend the repairs and maintenance of farm machineries and tools. Identified potential rural youths, unemployed agricultural diploma holders may be trained on these areas and encouraged to open such centres. Ultimately this will curtail the migration of farm/rural youth. Also it helps for entrepreneurial development among the rural youth.

Conclusion

The strategies have been formulated based on the major outcome of the study. The strategies are focused to further strengthen the activities of PFA, insisting the need for assessing the in field variation through remote sensing, supply of seedlings through hi-tech community nursery, intervention of stakeholders at all levels of Precision Farming, establishment of processing centres, mobile and farmers markets and custom service centres to retain farm youth. The comprehensive approach of attending the strategies by the public sector agencies especially the state government will certainly strengthen the Precision Farming process in the state, ultimately increase the livelihood status of the farmers.

Extension Management Strategies for Sustainable Agriculture, pp 469-477

New India Publishing Agency, New Delhi, India

52

Promoting Climate Resilient Dairy Farming Through Good Feeding and Management Strategies (GOFAMS) : A Farmer Participatory Approach

Sency Mathew, Robert C.P., Sindhu Sadanandan and Ambily Varughese

Livestock sector plays a vital role in the welfare of India's rural population by acting as an important source of income and employment in the rural sector (Dastagiri, 2004). India possesses the largest livestock population in the world and ranks first with (520.6 million), and accounts for the largest number of cattle (world share 16.1%), buffaloes (57.9%), second largest number of goats (16.7%) and third highest number of sheep (5.7%) in the world (FAOSTAT) and sustaining all these in less than 5% of world's and one-third of Asia's agricultural land area (MOA, 2003). The livestock production is an integral part of mixed farming systems practiced in the entire length and breadth of the country. It contributes 9% of GDP and employs 8% of the labour force, thus emerged as an important growth leverage for the Indian economy. Among livestock, dairying has been considered as a major income generating venture among the farming community and which ultimately has been encouraged and endorsed by many agencies. In 2001, India emerged as the world leader in milk production, with a production volume of 84 Mt, closely followed by USA. India has about three times as many 'dairy' animals as the USA, but annual milk yield per dairy animal is about one tenth of that achieved in USA (FAO, 2006).

In the recent years of understanding on global warming leading to climate change, it is viewed that agriculture sector is the second largest contributor of green house gasses after the energy sector. In agriculture sector , worldwide, livestock farming accounts for 37% of anthropogenic CH_4 which is largely produced by enteric fermentation in ruminants as part of their normal digestive processes and 65% of anthropogenic nitrous oxide (FAO, 2006). Methane gas (CH_4) is a natural and unavoidable by product of organic matter fermentation in the rumen of dairy cows and is the second most abundant green house gas after CO_2 accounting for 14-18% of global emissions (Global Methane initiative 2011). Enteric fermentation contributes 91.3% of total CH_4 emissions as compared to only 8.7% by manure management. Globally Enteric Methane Emission (EME) increased by 54% from 61.5 to 94.9 x 10^9 kg annually from the year 1961 to 2010 and is projected to increase to 120 x 10^9 kg by 2050. (Patra AK, 2014). EME from Indian livestock contributed 15.1% of total global enteric methane emission solely because of its enormous livestock population although the emission rate per animal in the country is much lower than in the developed countries (Crutzen *et al.*,1986). A lot of importance is placed on methane because it has a major warming capacity, has a short atmospheric half life and reductions are possible in methane emission. livestock also contributes a small but significant emission of nitrous oxide (N_2O) from animal waste management systems. Nitrous oxide, a potential GHG (global warming potential, GWP: 310) is produced during the microbial processes of nitrification and denitrification of nitrogen contained in livestock waste. The amount of N_2O released depends on the system and duration of waste management (Swamy and Bhattacharya, 2006). Nitrous oxide is also naturally present in the atmosphere as part of the Earth's nitrogen cycle, and has a variety of other natural sources such as fertilizers. Nitrous oxide molecules stay in the atmosphere for an average of 120 years before being removed by a sink or destroyed through chemical reactions. The impact of 1 pound of N_2O on warming the atmosphere is over 300 times that of 1 pound of carbon dioxide. Hence its importance is rightly placed. Nitrous oxide emissions from manure in India is estimated to reach 21.4 x10^6 kg in 2050 from 15.3 x 10^6 kg in 2010 (Patra, A.K., 2014).

Several mitigation options are available for methane emissions from livestock. In India the possibility of capturing or preventing emissions from animal manure storage is limited as it is extensively used as fertilizers and fuel. Hence the scope lies in decreasing the methane production by improving rumen fermentation efficiency. Thus, besides reducing the contribution of livestock to the atmospheric GHG accumulation will also enhance production efficiency of dairy animals. Modest reductions in methane emissions are possible with current technologies, while enhancing productivity. This could be achieved through proper orientation of farming community on technologies adaptable to climate resilient dairy farming practises. According to Abba Chabra *et al.*, 2010 'The CH_4 emission in terms of milk production is less in cows having higher productivity as compared to indigenous cows with low productivity'. Improving diet quality, eliminating nutrient deficiencies and at the same time using growth promoting agents such as probiotics, by pass Protein, UMMB are some of the technologies promoted by CARD-KVK. This paper highlights the GOFAMS (Good Feeding and Management Strategies), a unique participatory approach that comprises of basket of technologies that help in empowering the farmers towards climate resilient dairy farming.

Interventions on climate resilient dairy farming

CARD-Krishi Vigyan Kendra (KVK), the farm science centre of Indian Council of Agricultural Research (ICAR), Govt. of India was established with the objective of transfer of technology in agriculture and allied fields to the farmers, farm women and rural youths of the Pathanamthitta dist. CARD-KVK has been in the forefront in implementation of number of projects of Government of India and Govt. of Kerala. KVK gives thrust to resolve the field problems and advocates viable technological solution to the farmers of the district.

Dairy Sector is an unquestionable part of agriculture forming the backbone of India contributing 9% of the GDP. Considerable quantum of work has been done to transform the dairy sector such as, introduction of Artificial insemination, framing of breeding policies, scientific breeding and feeding strategies *etc* resulting in a substantial increase in milk production thereby helping our country to achieve top position in milk production in the world. As per the latest statistics available, Pathanamthitta district has a crossbred population of 87175, of which 75472 numbers are females'. The average milk production per cow has been found to be less than 9 litres per cow. Studies on assessing the milk production potential of the fourteen district of Kerala indicates that Pathanamthitta district has a high milk production potential which is however not fully tapped. In order to rectify the above mentioned problem CARD KVK has made an attempt to promote Good Feeding and Management Strategies (GOFAMS) in dairy farming in the district (Fig.1). GOFAMS includes the basket of technologies. The technologies were intervened in the district through farmer participatory approach.

Figure 2 illustrates that the main problem of dairy farming in the district was low milk production. Hence KVK intervention focussed on these problems of cattle rearing with suitable technologies to fit the local conditions. Hence Good Feeding and Management Strategies (GOFAMS) which contributes basket of technologies to the farming community for augmenting milk production. Besides augmenting milk production it also contributed substantially in reducing emission intensity of methane. The GOFAMS contribution of technological intervention is as discussed below.

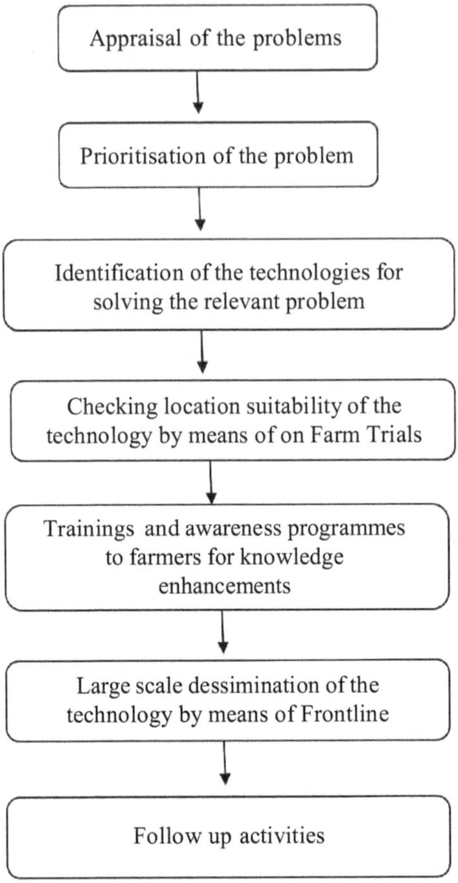

Fig. 1: Technology intervention for problem analysis

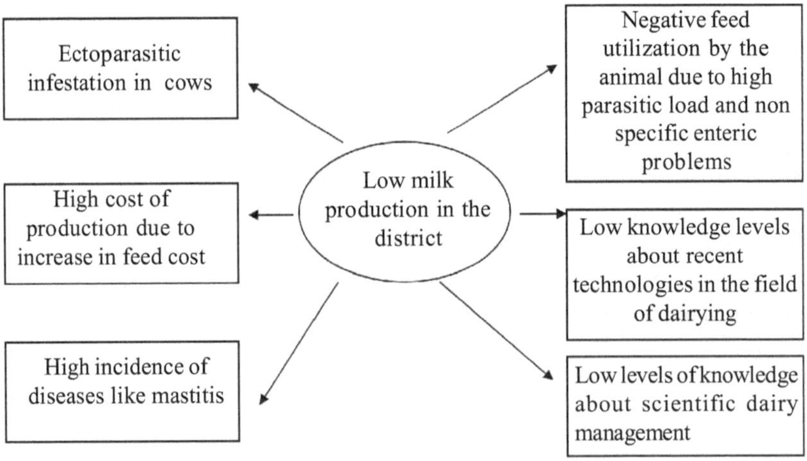

Fig. 2: Problem cause diagram

a) **Use of probiotics** : Probiotics are friendly bacteria which can multiply and colonise in the gut and reduces bacterial count by production of anti-bacterial substances. It causes competitive exclusion of pathogens and thereby regulates gut ecology and thus improves digestion and absorption leading to better utilization of nutrients resulting in better growth rate and higher milk production. Low milk production due to improper feed utilization as a result of non specific enteric diseases in cattle was identified to be one among the major problems in the locality resulting in decreased milk production. Probiotic as part of Good Feeding Strategy was intervened in the district to increase the milk production rate. Table 1 infers that a 19% increase in milk production was found as a result of this intervention.

b) **Use of forages:** Feeding of green forages like fodder and legumes are important since this, besides complementing to concentrate feed also is a source of fibre and protein. High cost of concentrate feed for cows ultimately increases the cost of production. The problem can be allayed to an extent by feeding good quality fodder. Hay which is usually fed to the cattle became scarce due to decline in area under rice cultivation. Hence, the only way out in the area is to incorporate digestible green fodder in the feeding regime. On the basis of identifying the problem, KVK has promoted fodder varities with high palatability such as CO_3 and CO_4 as part of GOFAMS. CO_4 has high palatability and a high crude protein (10.71%), ensures quick generation with early maturity and has a yield potential of 400 tons /ha. Beside this CO_4 is succulent , palatable and farmers observed that feeding one bundle (15 kg) of CO_4 increased milk yield by almost 200ml per cow (Annual Report, 2013-2014, CARD-Krishi Vigyan Kendra, Pathanamthitta). Table 1 infers that an increase in 4.1 % in milk production was noticed as the result of the intervention.

c) **Use of probiotic–yeast mix**: Feeding of probiotic yeast mix (30-50 gm/day/cow) helped to enhance milk production from 6.5 litres to 8.0 litres in cross bred dairy milch cows as part of GOFAMS. 23% hike in milk production was recorded as the result of feeding of probiotic yeast mix to dairy milch cows (Table 1).

d) **By pass protein**: Protein meals normally fed to dairy animals are broken down by 60% to 70% to ammonia, by the rumen microbes. A significant part of ammonia is then converted to urea in the liver and excreted through urine. Thus, a major part of expensive protein meals is wasted, for which animal spends energy. However while using of by pass protein to feed dairy milch cows these protein meals are protected and are saved from degradation in the rumen, getting digested in the lower part of the gastro intestinal tract, for higher milk production. This technology was incorporated in the feeding schedule and resulted in enhanced milk production from 13 to 15.5 litres. The cows also showed better body score (Table 1).

e) **Urea molasses multinutrient blocks (UMMB)** : Supplementation with UMMB can increase digestibility of fibrous feeds by upto 20%, increase the nutrients the animal receives and can increase feed intake by 25 to 30%. UMMB was introduced for feeding of dairy milch animals and it was found that milk production increased by 27 % (7 litres to 8.9 litres) (Table 1).

Table 1: Impact of KVK's Interventions

Technology	* Average Milk production before intervention(l)	*Average Milk production after intervention(l)	Percentage increase (%)
Use of probiotics	6.72	8.0	19%
Use of forage	6.0	6.25	4.16 %
Use of probiotics-yeast mix	6.5	8.0	23%
Bypass protein	13	15.5	19.23%
Urea molasses multi nutrient block	7.0	8.9	27 %
	39.22	46.65	19 %

*Average of 10 cows

It is evident from (Table 1) that KVK's interventions through GOFAMS led to increase in milk production from 39.22 litres to 46.65 litres (19%). An effort was also made by KVK to evaluate GOFAMS impact on reducing the GHG emission. Emission intensity is the quantity of emissions produced per unit of agricultural product such as carcass, milk solids or fleece weight. Thus we attempted to quantify the emission intensity of methane using the formula of Lines-Kelly, 2014.

Lines-Kelly Formula

Emission intensity (EI) : Total methane production (Kg)/ Total milk production (L)

With annual enteric methane production per cow being taken as 165 kg milch cow population in Pathanamthitta district : 39701 nos.

Table 2: Reduction in emission intensity through GOFAMS

Technology	Emission intensity before intervention (g /litre)	Emission intensity after intervention (g /litre)	Reduction in emission intensity (g /litre)
Use of probiotics	24.55	20.62	3.93
Use of forage	27.50	26.40	1.1
Use of probiotics-yeast mix	25.38	20.62	4.76
Bypass protein	12.69	10.64	2.05
Urea molasses multi nutrient block	23.57	18.53	5.04
	22.74	19.36	3.38

Table 2. and Fig. 3 clearly depicts the cumulative impact of technologies disseminated through GOFAMS, which resulted in contributing to emission intensity (EI) reduction by 3.38 g/litre of milk, while total increase in milk production was 19 % (Table 1).

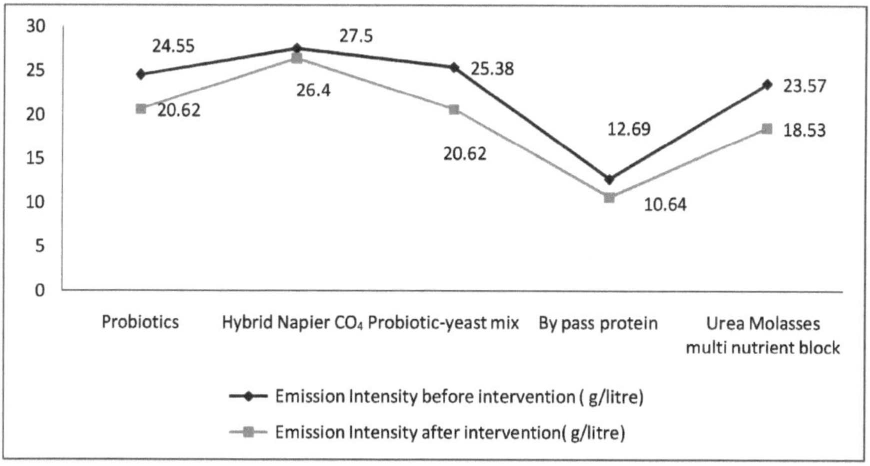

Fig. 3: GOFAMS Impact

Although there was varied positive influence of different technologies on milk production and emission intensity (EI) reduction, feeding milch cows with urea molasses multi nutrient block (UMMB) resulted in maximum EI reduction by 30% (Fig. 4).

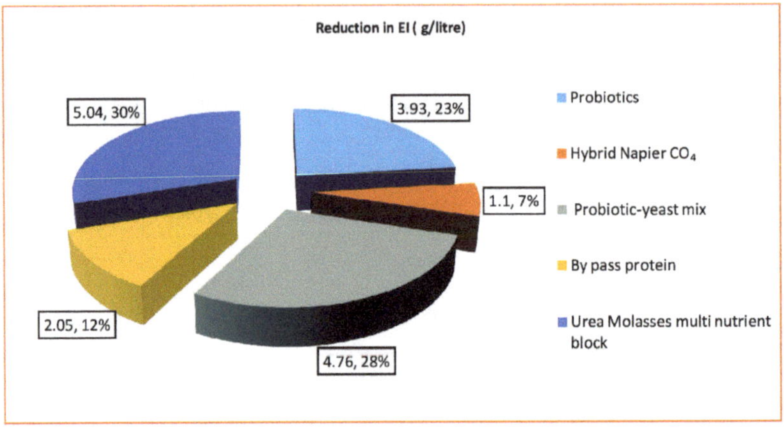

Fig. 4: Technology impacts through GOFAMS

Conclusion

Agriculture in general, and livestock production in particular, contribute to global warming through emissions of the non-CO_2 GHGs CH_4 and N_2O. Most CH_4 is emitted from ruminants (animal + manure), whereas, N_2O is mainly emitted from fertilized land. Methane mitigation options from ruminants focus on increasing production per animal, modifying diet, decreasing numbers of methanogens and methanogen activity. Thus mitigation of CH_4 emissions can be effectively achieved by GOFAMS by selectively adopting the potential technologies such as probiotics, forages, probiotic-yeast mix, by pass protein and urea molasses multi nutrient block (UMMB) that improve the efficiency of animal production. GOFAMS is a unique strategy to promote climate resilient dairy farming.

References

Annual Report (2013-2014). CARD-Krishi Vigyan Kendra, Pathanamthitta.

Chabra, A., Manjunath, K.R, Panigrahy, S. (2010). Assessing The Role of Indian Livestock in Climate Change. ISPRS Archives XXXVIII-8/W3 Workshop Proceedings: Impact of Climate Change on Agriculture.

Crutzen, P.J., Aselmann and I. Seiler, W. (1986). Methane production by domestic animals,wild animals, other herbivorous fauna and humans. *Tellus,* 38B:271-284.

Dastagiri, M.B. (2004). Demand and supply projections for livestock products in India. Policy paper, 21, National Centre for Agricultural Economics and Policy Research (ICAR) New Delhi, India, 58p.

FAOSTAT. Agricultural Production Database. Food and Agricultural Organisation. http://www.apps.fao.org

FAO (2006). Steinfeld H, Gerber P, Wassenaar T, Castel V, Rosales M, de Haan C (eds.) Livestock's Long Shadow. Environmental issues and options.*http://fao.org/docrep.*

Global Methane Initiative (2011). Global Methane emissions and mitigation opportunities .Viewed online at http://www. global methane .org/documents/analysis_fs_en.pdf

MOA (2003). 17th Livestock Census of India, Dept. of Animal Husbandry and Dairying, Ministry of Agriculture, Govt. of India, available online at *http://www.indiastat.com.*

Patra, A.K. (2014). Trends and Projected Estimates of GHG Emissions from Indian Livestock in Comparison with GHG Emissions from World and Developing Countries. *Asian-Australias J Anim Sci.* Apr 2014; 27(4):592-599.

Rebecca Lines-Kelly (2014). Enteric Methane Research: A summary of Current Knowledge and Research. Published by NSW Department of Primary Industries. pp.7.

Swamy, M., Bhattacharya, S. (2006). Budgeting anthropogenic green house gas emission from Indian livestock using country specific emission coefficients. *Current Science*, 91(10): 1340-1353. 17th Quinquennial Livestock Census 2003, Directorate of Animal Husbandry, Kerala.

Extension Management Strategies for Sustainable Agriculture, pp 479-485

New India Publishing Agency, New Delhi, India

53

Management of Agricultural Risks – Farmer's Perspective

Sendilkumar, R. and V. Ravichandran

The agricultural production is inherently a risky business characterized by seasonality; perishability and variability which make it susceptible to vagaries of climate and markets and thus farmers face a variety of agricultural risks. In response to these risks over a period of time, several risk management mechanisms have been evolved. Risk management strategies refer to *ex-ante* measures taken before such an event has taken place, whereas risk coping strategies are, *ex-post* strategies dealing with the consequences of the event (World Bank Report, 2005). Invariably, farmers use both *ex-ante* and *ex-post* strategies. It is interesting to note that so far agriculture risk management has been in terms of risk avoidance and risk reduction and mostly these risks are addressed at farm or production level. Currently there has been a shift in agriculture risk management from risk avoidance and risk reduction strategies to risk mitigation strategies. Similarly, there exist shifts from production to market orientation in agriculture. To understand this phenomenon involved, it is felt necessary to study the agricultural risk management behaviour farmers. Keeping this discussion as focal point, the study was conducted with an objective to assess risk management behaviour of farmers.

Methodology

Six blocks of Nagapattinam district of Tamil nadu was selected with a sample size of 180, comprised of 120 insured and 60 non insured. The 180 samples were distributed to 12 villages by adopting simple proportionate random sampling based on the secondary data served by the Primary agricultural credit banks, commercial banks and block level agricultural office. The major components identified for the computation of agricultural risk management behaviour were risk awareness behaviour, risk aversion behaviour, risk reduction behaviour, risk coping behaviour, and risk transfer behaviour. The selections of sub-components were done based on the consensus approach and item analysis approach. The Relevancy Co-efficient Index (RCI) for each major component and subcomponents were worked out by using the formula devised by Ramana (1999). The selected major components and subcomponents formed the constituents of the index to measure the agricultural risk management behaviour of paddy farmers. All the subcomponents were measured in terms of degree of symbolic agreement on three point continuum scale (Agree, Undecided, Disagree) and with the scores of 3, 2 and 1 respectively. Mean scores were computed for each of the subcomponents and average mean scores were arrived. The constituents of scale to construct index was tested for its reliability and validity for the standardization.

Finding and Discussion

Component wise Agricultural Risk Management Behaviour (Symbolic agreement)

The component wise agricultural risk management behaviour was computed and presented in the Table 1.

Table 1: Mean scores on the agricultural risk management behaviour of paddy farmers (symbolic agreement)

S. No. Components	Mean scores		
	Insured **(n=120)**	Non insured **(n=60)**	Total **(n=180)**
I. Risk Awareness Behaviour	**1.6990**	**1.6310**	**1.6768**
1. Systematic inventory of all types of risk	1.7945	1.9200	1.8360
2. Planning for hazard monitoring and forecasting	1.6510	1.5470	1.6163
3. Hazard evaluation and mental mapping	1.3996	1.3816	1.3940
4. Vulnerability and risk assessment	1.1924	1.1446	1.1765
5. Knowing Risk information through mass media	1.9714	1.8456	1.9295
6. Risk management education and training	2.1855	1.9543	2.1080
II. Risk Aversion Behaviour	**1.6809**	**1.6836**	**1.6656**
1. Planning well in advance to leave the land in fallow condition	1.7208	1.7503	1.7306
2. Deciding not to grow single crop/ monocrop	1.7820	1.6358	1.7333
3. Changing cropping pattern by foreseeing the risk	1.8670	1.4641	1.7327
4. Contingency decision to adjust cultivation aspects (*eg.* Shifting of sowing time)	2.1600	2.1640	2.1613
5. Diversifying income sources by anticipating the loss	1.9115	2.2458	2.0230
6. Migrating towards other profession from agriculture	1.2520	1.3308	1.2783
7. Shifting from own farm cultivation to labour market	1.0731	1.1945	1.1136

S. No.	Components	Mean scores		
		Insured (n=120)	Non insured (n=60)	Total (n=180)
III.	**Risk Reduction Behaviour**	**1.8515**	**1.7257**	**1.8076**
1.	Selecting the crop and variety based on available choices	2.4531	2.4326	2.4460
2.	Temporarily taking cultivation in small portion of land area to avoid huge loss	1.7159	1.6126	1.6815
3.	Diversifying the farm in to different allied enterprises	2.2680	1.6670	2.0676
4.	Diversified cropping	1.5136	1.4896	1.5056
5.	Diversifying the activities towards non-farm resources	1.3292	1.3378	1.3321
6.	Adopting strategies to minimizing external inputs & maximizing internal inputs	1.6986	1.5880	1.6617
7.	Adopting risk reduction technology (IPM, INM, Soil and water conservation *etc.*,)	1.9823	1.9253	1.9633
IV.	**Risk Coping Behaviour**	**1.7387**	**1.6857**	**1.7210**
1.	Rely on new credit facility	1.7496	1.6815	1.7269
2.	Selling the assets/live stocks/trees/stored produces *etc.*, to compensate the loss occurred	1.0842	0.9720	1.0468
3.	Accommodating to temporary off-farm employment	1.1819	1.1878	1.1838
4.	Contracting with traders and merchants for credit and marketing	1.869	1.8270	1.8550
5.	Making arrangement for interlocked contracts (sub lease sub contracting)	1.9107	1.8193	1.8802
6.	Utilizing the reserve savings at the time of contingencies	2.3583	2.3066	2.3411
7.	Availing loan from the formal/informal institutions	2.0173	2.0060	2.0135
V.	**Risk Transfer Behaviour**	**2.1294**	**1.4108**	**1.8837**
1.	Availing the Government's price support policies	2.6625	2.5365	2.6205
2.	Switch over to crop insurance programme as one of the risk transfer device	2.8262	1.0756	2.2427
3.	Storing the Produce in Warehouse	0.8997	0.6205	0.8066

Source: Primary data

Risk Awareness Behaviour

The average mean score worked out for risk awareness behaviour was 1.6768. Risk management education (2.1080), knowing risk information through mass media (1.9295) and systematic inventory of all types of risk (1.8360) contribute to the risk awareness behaviour greatly. Risk related information either on production or market risks were disseminated mainly through risk management awareness programmes through bulletins from formal institutions like TNAU, Indian Meterological Department and Mass Media. Mostly the farmers were aware of such risks from these formal sources and get reinforced through informal means also *viz.* peer group, and other fellow farmers in their area.

Risk Aversion Behaviour

The average mean score for the risk aversion behaviour worked out was 1.6656, whereas the mean scores for the sub components were taking up contingency decision to adjust cultivation aspects (2.1613), diversifying income sources by anticipating the loss (2.0230), deciding not to grow single crop (1.7333), changing cropping pattern by foreseeing the risk (1.7327) and planning well in advance to leave the land in fallow condition (1.7306) have been observed in the descending order. The study area was very often prone to all kinds of risks especially the production risks and as well as market risks. For these reasons only, the paddy farmers had developed their own strategies as identified in the subcomponents of risk aversion behaviour to combat the risks. The farmers with more of risk aversion behaviour pattern were found to be very cautious, not innovative and even do not have a mind set to taste the risk. A kind of self defense mechanism vested with individual decision making pattern prevails in the study area.

Risk Reduction Behaviour

The average mean score for the risk reduction behaviour was 1.8076; while as the mean scores for the subcomponents were found between 2.4460 and 1.3321. Among the seven subcomponents, three scored high mean score *viz*., selecting the crop and variety based on available choices (2.4460), diversifying the farm into different allied enterprises (2.0676) and adopting risk reduction technology (1.9633).

It could be inferred from the findings that risk reduction behaviour showed an optimum blend of both formal and informal mechanisms. Selecting the crop and variety based on available choices and diversifying the farm into different allied enterprises were normally denoted as the informal mechanisms. On the other hand, adopting risk reduction technologies, including IPM, INM, soil and water conservation practices etc. were referred as formal mechanisms. The informal mechanisms were developed by the farmers themselves by virtue of their experience and their time tested results, while the formal mechanism was lab tested. Hence the farmers would have symbolically exhibited the risk reduction behaviour with a balance of both informal and formal mechanisms. However, the other subcomponents *viz.* diversified cropping (1.5056) strategies to minimizing external inputs and maximizing internal inputs (1.6617) diversifying towards non-farm resources (1.3321) had emerged as the constituents of building risk reduction behaviour. It was quite important to note that these subcomponents were the long term strategies aiming at risk reduction behaviour especially to address the production risks.

Temporarily taking cultivation in small portion of land area to avoid huge loss (1.6815) was the short term strategy with flexibility and zero cost aimed at to reflect the risk reduction behaviour concerned to production risks. Hence, it is suggested that risk reduction behaviour should have the equilibrium of both formal and informal mechanisms to address the agricultural risks optimally.

Risk Coping Behaviour

The average mean score for the risk coping behaviour was 1.7210. A range of 2.3411 to 1.0468 was obtained for subcomponents.Among the seven, five sub components scored more than the average mean score. They were, utilizing the reserve savings at the time of contingencies (2.3411), availing loan from the formal or informal sources (2.0135), making arrangements for interlocked contracts (1.8802), contracting with traders and merchants for credit and marketing (1.8550) and relay on new credit facility (1.7269). It was quite interesting to note that utilizing the reserve savings at the time of contingencies scored the top priority. The obvious reason would be that, it was an instantaneous coping mechanism exhibited by farmer to meet contingencies and all other means would come only after it. It is a front runner to address most of the financial risks. Hence, the reason why it had shown high mean score.

The other subcomponents of risk coping behaviour clearly indicate the alliance of interventions from institutions. The paddy farmers adhere to adapt the risk coping behaviour, mainly to manage the financial and market-risks. All the subcomponents tailored aims at temporary relief and settlement from huge loss. Thus the paddy farmers resorted to exhibit their risk coping behaviour; jointly inter wovened with institutional and non-institutional sources of intervention to address financial and market risks.

Risk Transfer Behaviour

It is seen that the average mean score worked out for risk transfer behaviour was 1.8837. Out of three subcomponents identified, two secured high mean score, availing the governments' price support policies (2.6205) and switching over to crop insurance programme (2.2427). The former one being mostly fabricated to address the market risks and the later addressed the production risks. Thus it was apparent that risk transfer behaviour mostly demands the Government interventions from the farmer's side. It could be attributed that paddy was the major stable crop and the announcement of Minimum Support Price being the major intervention by the Government could have made the farmers resorted to use of such benefits.

Switching over to crop insurance programme as one of the risk transfer device was yet another subcomponent. This intervention has been designed by the Government to address the risks pattern in crop production and farming. The awareness on crop insurance scheme and the promotional efforts made by the Government machinery would have influenced the farmer to use this risk transfer behaviour.

'Storing the produce in warehouse' was seen with very low mean score. This might be due to fact that this intervention is yet to get exposured by the paddy farmers and only very few farmers have the privilege of including this intervention as one of the subcomponents of risk transfer behaviour. Hence, it could be concluded that there is a scope for government interventions to knock the individual at the door step.

Agricultural Risk Management Behaviour of Paddy Farmers (Symbolic agreement)

The major as well as subcomponents identified were considered for the study to arrive at Agricultural Risk Management Behavioural Index (ARMBI). The ARMBI was used to analyse the risk management phenomenon in their totality by combining subcomponents of phenomenon into one composite index. The data appropriate for constructing ARMBI were carefully collected and presented in Table 2 to fabricate index.

Table 2: Composite agricultural risk management behaviour of paddy farmers

S.No.	Risk management components	Index		
		(n=120)	**(n=60)**	**(n=180)**
1.	Risk awareness behaviour	77.05	74.05	76.04
2.	Risk aversion behaviour	82.59	83.84	83.01
3.	Risk reduction behaviour	84.21	78.22	82.21
4.	Risk coping behaviour	90.76	87.99	89.84
5.	Risk transfer behaviour	86.91	56.83	76.88
	Composite Agricultural Risk Management Behavioural Index	84.30	76.18	81.59

A glance at the Table 2 revealed that the composite agricultural risk management behavioural index worked out was 81.59. The index range for various subcomponents, (76.04) risk awareness behaviour, (83.01) risk aversion behaviour, (82.10) risk reduction behaviour, (89.84) risk coping behaviour and (76.88) risk transfer behaviour.

Conclusion

To comprehend, the agricultural risk management behavioural index was higher with respect to all the subcomponents. Even though the risk aversion behaviour of paddy farmers was seen with high index, it had provided an adequate balance to exhibit the composite index. Based on the light of findings the following policy suggestions are recommended:

1. Adequate empowerment mechanism may be taken up towards risk preparedness plan following ex-ante strategies and as a policy measure developing individual database supporting the issue at the micro level.
2. The intensive efforts are needed to popularize the advantages of futuristic risk management tools like crop insurance, contract farming, and ware house receipts.
3. Efforts may be made to establish a risk management data base extension centre. The existing provisions under ATMA at block level may be utilized.
4. The cost effective strategies which reflects the risk aversion behaviour such as contingency decision to adjust cultivation practices, planning well in advance to leave the land not put under cultivation *etc*., to be validated scientifically.
5. Appropriate policy measures have to be devised to encourage maximum of risk reduction behaviour than risk coping behaviour. On these lines, large scale demonstrations on risk reduction technology may be arranged.
6. As the study area is very often subjected to natural disasters like storm, cyclone, torrential rain, (tsunami, nisha cyclone) there is a need for establishing permanent research centre with a focus to develop separate package for risk coping mechanisms.
7. To enhance the reach effect, the advantage of risk management tools has to be popularized through appropriate crop insurance educational programmes.

References

Ramanna, K.N. (1999). People's Participation in Planning and Implementation of Integrated Watershed Development Programme – A Comparative Study of Watersheds of Government vis-à-vis NGO. Unpub. Ph.D. Thesis. TNAU, Coimbatore.

Sendilkumar, R. (2009). Risk management behavioural pattern of farmers - An empirical Study. UnPub. Ph.D thesis, Dept of Agricultural Extension and Rural sociology, TNAU, Coimbatore.

World Bank (2005). Managing Agricultural Production Risk Innovations in Developing Countries. The World Bank Agriculture and Rural Development Department.

Extension Management Strategies for Sustainable Agriculture, pp 487-495

New India Publishing Agency, New Delhi, India

54

Factoring of Attributes Influencing Consumption of Greens Among Family Members in Selected Rural and Urban Families of Madurai in Tamil Nadu

Sheela, P. and M. Ramasubramanian

Rationale of the Study

Greens are the cheapest of all the vegetables within the reach of poor man, being richest in their nutritional value. The problem of malnutrition is assuming seriousness in the vulnerable groups viz., adolescents, pregnancy, lactation and school children, not because of poverty but because of ignorance, illiteracy and callousness of the people. The lack of knowledge especially on the nutritive value of these green leafy vegetables among the public in general is the main drawback in their lower consumption. (Jyoti, 2008)

Consumption pattern of greens tends to vary among different type of families. Families with working women may have a different consumption pattern when compared to families with house wives. The consumption of greens bound to vary with respect to differences in families in terms of number of members, number of children, income status, working status of family members, and social networking status of family member's information availability about greens.

Moreover, a plethora of factors would influence a family and its members to develop a liking to consume greens. It starts from accessibility, availability and to affordability. These factors, either solitarily or in combination would influence the consumption pattern of greens in a family. Previous studies have associated food consumption with behavioral attitudes such as health consciousness, environmental consciousness, trust of food claims and desirability of food attributes such as taste, texture, freshness (Arlappa *et al.,* 2010, Gan *et al.,* 2008). Many researchers agree to say that determinants of the consumption of green products could be due to cultural orientation like value, belief or norm, psychological, economical or socio-demographic factors (Honkanen *et al.,* 2006, Kamphuis, 2007).

Price could also be seen as a determinant for many researchers. In fact, price, quality and convenience can have competitive advantage in buying decision of green product (Ottoman, 1994). But, according to Loureiro and Hine (2002), the price is not an obstacle of the buying decision of green product which was endorsed by Olumakaiye *et al.* (2010) who found that the sales of green products in Europe were predicted to double by 2015 according to perceived benefits of those products, consumers are not paying attention to thc price.

The foregoing discussion substantiated that despite the importance of consumption of greens, it has been influenced by several factors. More micro level studies are to be conducted to analyse the determinants of consumption of greens. Hence, a study has been contemplated with the following objectives

1. To study the determinants of consumption of greens among family members.
2. To analyse the underlying pattern among the determinants of consumption of greens.

Research Methodology

Since this study is attempted to unearth the determinants of greens consumption, which has already occurred, the *ex-post facto* design, is selected for study. Madurai district was purposively selected for the study as it is one of the backward rural district in southern Tamil Nadu with more number of people living below poverty line and ultimately with poor nutrition.

Madurai district comprises of seven taluks, thirteen blocks and six hundred and sixty four revenue villages. Since, the respondents for this study are to be selected from rural and urban background, the secondary data collected from ICDS office, Madurai, pertaining to nutritional indicators in different blocks were analyzed. Three rural blocks namely Usilampatti, Melur and Thirumangalam were selected for rural respondents based on the criteria of more number of moderately and severely underweight children in these three blocks.

In each of the selected block, one village was purposively selected based on the information obtained from Child Development Project Officer's and Agriculture Officers. In each of the selected village, 20 respondents were selected using snow ball sampling method wherein the word of mouth of one respondent to indicate another prospective respondent was considered.

For the selection of urban respondents for the study the researcher discussed with officials in Madurai Collectorate and Madurai Corporation. In order to ensure representativeness of the sample, the following yardsticks were used

1. The respondents should represent all the socio-economic status namely upper, middle and lower class.
2. The representation of both of the housewives and working women should be ensured.

In accordance with the above yardsticks the areas namely Anna Nagar, Goripalayam and Alagappa Nagar in Madurai city were selected. The same procedure of snow ball sampling techniques was followed for selecting sixty urban respondents, twenty each from selected three areas.

The factors influencing consumption of greens were listed and measured on a three point continuum. Since, the data was amenable for factor analysis and from the magnitude of factor loadings one can interpret with greater precision, factor analysis was run for this study.

Findings and Discussion

The factors influencing greens consumption may be associated with qualities of greens, influence of accessibility, affordability, availability, influence of family members, value system etc. In order to find out the pattern of grouping among these factors and to identity underlying relationship between them, factor analysis was run with the help of SPSS version 16.00 and the results are presented below. Table 1 gave an account of number of factors extracted using principle component analysis with varimax rotation. It also indicated the total variance explained by the factors put together and the amount of variance explained by each factors.

Closer observation of the Table 1 revealed that 51.48 per cent of variation in the consumption of greens was explained cumulatively by the extracted four factors. The first factor accounted for 18.74 per cent variation in consumption of greens followed by 12.77 per cent, 10.44 per cent and 9.54 per cent of variation explained by second, third and fourth factor respectively. The following Table 2 showed the factors loaded into four components and the magnitude of factor loadings.

Table 1: Percentage of variance explained by extracted factors in Principal Component Analysis (PCA)

Component	Initial Eigen values			Extracted sum of squared Loadings			Rotation Sums of Squared Loadings		
	Total	% of Var	Cum %	Total	% of Var	Cum %	Total	% of Var	Cum %
1.	2.999	18.743	18.743	2.999	18.743	18.743	2.780	17.373	17.373
2.	2.043	12.767	31.510	2.043	12.767	31.510	1.866	11.662	29.035
3.	1.670	10.436	41.946	1.670	10.436	41.946	1.805	11.282	40.317
4.	1.526	9.536	51.482	1.526	9.536	51.482	1.786	11.165	51.482
5.	1.368	8.547	60.030						
6.	1.075	6.717	66.747						
7.	0.962	6.011	72.758						
8.	0.952	5.951	78.710						
9.	0.757	4.730	83.439						
10.	0.580	3.627	87.066						
11.	0.517	3.229	90.294						
12.	0.467	2.917	93.211						
13.	0.339	2.119	95.330						
14.	0.318	1.984	97.315						
15.	0.270	1.690	99.005						

% Var – Percentage Variation   Cum % - Cumulative percentage

Table 2: Rotated component matrix and factor loadings

S. No	Factors influencing consumption of greens	Component			
		1	2	3	4
1.	Drudgery involved in preparing greens for cooking	0.735			
2.	Taste of greens	-0.715			
3.	Time taken for cooking	0.704			
4.	Variety of dishes prepared from greens	0.617			
5.	Information about the good qualities of greens given in mass media				
6.	Children's liking towards junk foods				
7.	Availability at an accessible place		0.663		
8.	Seasonal availability of greens		0.649		
9.	Availability of favourite greens		0.587		
10.	Information given to children in schools		-0.550		
11.	Prejudices/ beliefs that prevent consumption			0.846	
12.	Expenditure incurred for greens			-.554	
13.	Consumption of greens as a value inherited from forefathers				0.531
14.	Ability of greens to cure specific health ailments				0.822
15.	Availability of variety of greens at a time				0.526
16.	Motivated by neighbours cooking greens				

Factor I

The highest factor loading in the 'Component I' was for the factor "drudgery involved in preparing greens for cooking" with 0.735. This was closely followed by a high negative loading (-0.715) for the factor 'taste of greens'. A factor loading of 0.704 was attributed to the factor "time taken for cooking certain types of greens". The factor "variety of dishes prepared from greens" got a factor loading of 0.671 under component one.

The examination of factor loadings clearly painted a picture that drudgery and time taken was a concern which has already been reported in the findings in the previous section. Though there existed scope for variety of dishes from greens endorsed by majority of urban respondents reported elsewhere due to the increasing exposure to recipes shown in satellite television, the drudgery involved and time taken for preparation of dishes, given the attitude of people that they wanted everything quickly would have affected the consumption pattern.

Taste of greens wasn't found to be a predominant influencing factor among rural respondents, but it did influence an appreciable number of rural respondents as has been evident from Table 2. Most of the rural respondents opined that rather than actual taste, be it good or bad, the custom that they inherited and the local availability was the more influencing factor. Hence, the three factors namely, drudgery, time taken and variety whose loadings are high and positive might have suppressed the loading for taste and hence the negative loading was recorded for taste of greens as an influencing factor. Since this component involves on the one side drudgery and time taken which were negative points for consumption and on the other side positive points like taste and variety on the other side, assigning a name 'Drudgery and Variety' factor, would be appropriate.

Factor II

Four factors were extracted into the 'component II'. Among the four, the factor 'availability at an accessible place' got the highest positive loading of 0.663. This was followed by the positive loading of 0.643 for the factor 'seasonal availability of greens'. Availability of favourite greens as a factor also influenced consumption with a positive loading of 0.587. The information given to children in schools was loaded with a negative loading of -0.550.

A closer examination of component II revealed that it was the availability of greens that overwhelmed the component. Availability in easily accessible place, seasonal availability and availability of favourite greens were obviously and logically would influence the consumption. No doubt their contribution would definitely have a say in the consumption pattern. But the negative loading of information given to children in school could be attributed to, the reason of more children either failed to pass on the information to their parents or not given with such information, especially in rural areas. Though it is a factor to reckon with, the high positive loadings of availability factors suppressed this factor and projected it to be negative. This indicated that despite the influence of information given in schools to children it was the availability that matters.

Since this component was dominated by availability of greens, be it seasonal availability or availability of favourite greens or accessibility, this component could be appropriately named as 'availability of greens factors'.

Factor III

The third component extracted three factors. The first factor which got highest factor loadings (0.846) was "beliefs that prevent consumption". This was followed by impressive negative loading (-0.554) for the factor expenditure incurred for the greens. A factor loading of 0.531 was obtained by the factor "consumption of greens as a value inherited from forefathers".

The analysis of factor loading for the component III revealed that the beliefs and values did have an influence on consumption of greens. Since greens consumption is an age old practice, the people got to know several beliefs and values from their forefathers which might have influenced their consumption pattern of greens. The discussion on how beliefs prevent consumption of greens elsewhere could be corroborated here. Similarly, values could have an indelible impact on how people behave including consumption. Since the beliefs got a higher loading and it prevents consumption, which is a negative factor it accompanied another negative factor called 'expenditure incurred due to greens'. Though the expenditure incurred due to greens was meagre, the common attitude of the people is to get everything free. This attitude has been reflected in the negative factor loading for this factor.

Since, beliefs and values did occupy a predominant position under component III it is pretty appropriate to name this component as 'beliefs and values factor for greens consumption'.

Factor IV

The fourth component extracted two factors namely 'curing specific health ailments' and 'availability of variety of greens at a time' with positive factor loadings of 0.822 and 0.526 respectively.

Greens are known for their medicinal properties and people irrespective of rural and urban domicile have been eating greens for specific health ailments, the details of which has already been discussed elsewhere in this report. Further this factor depends on availability of different types of greens at a time, since certain ailments requires a consortium of greens to be consumed. Hence, these two factors were extracted together under component IV.

Since this factor is specifically meant for curing health ailments, it is appropriate to name this factor as 'Health benefits'.

Conclusions and Recommendations

A study was conducted in Madurai district among sixty rural and sixty urban families to identify the factors influencing consumption of greens and to identify the underlying pattern among the factors through factor analysis. The results revealed that 51.48 per cent of variation in the consumption of greens was explained cumulatively by the extracted four factors. The first factor account for 18.74 per cent variation in consumption of greens followed by 12.77 per cent, 10.44 per cent and 9.54 per cent of variation explained by second, third and fourth factor respectively. The analysis of pattern clearly revealed that in the first factor, the attributes related to drudgery involved in cooking and variety of greens were grouped together. The attributes pertaining to availability of greens dominated the second factor whereas the beliefs and values related factors were concentrated in the third factor. The fourth factor had grouping of attributes related to health benefits.

From the findings of the study, the following recommendations were suggested.

1. The respondents were concerned about availability of variety of greens at a time and hence the planners at helm of affairs may think of establishing 'green shops' which can fully made functional using Self Help Group women. Already in Tamil Nadu, Government has initiated 'Kaikani angadi' in selected cities. In such already established centres, a separate place for greens may be set up.
2. Availability and accessibility has been the factors which were predominantly endorsed by many respondents and hence the policy makers and planners need to evolve strategies to make it available for consumers.
3. Drudgery involved in cooking had been indicated by many respondents as one of the factor determining the consumption and hence the food scientists along with processing engineers may evolve some technologies/tools to reduce the drudgery involved in cooking of greens.
4. Similarly time taken for cooking was also considered as one of the factor influencing consumption and hence the food scientists should evolve technologies to reduce the cooking time for certain greens.
5. Taste and appearance are the foremost factors which were found to influence children to take greens. Hence food scientists should try new products from greens with good taste and appearance without affecting the nutritional qualities of greens.
6. Value of eating greens was one of the influencing factor indicated by respondents which got eroded day by day and hence the parents are suitably sensitized to inculcate the value of eating greens among their children.
7. Several beliefs inhibit the eating of greens and hence from the school itself children need to be informed about getting rid of such beliefs. Mass media can be used for this purpose also.

References

Arlappa, N., Laxmaiah, A., Balakrishna, N. and G.N.V. Brahmam (2010). Consumption pattern of pulses, vegetables and nutrients among rural population in India. *African Journal of Food Science,* 4(10):668-675.

Gan, C., Wee, H. Y., Ozanne, L. and Kao, T. H. (2008). Consumer's purchasing behaviour towards green products in New Zealand, *Innovative Marketing.* 4(1): 93-102.

Honkanen, P., B. Verplanken and S.O. Olsen (2006). Ethical values and motives driving organic food choice. *Journal of Consumer Behaviour.* 5(5):420–430.

Huang, C.H.L. (1996). Consumer Preferences and Attitudes towards Organically Grown Produce. *European Review of Agricultural Economics.* 23:331-342.

Jyoti, Sajjan (2008). Consumption pattern of green leafy Vegetables and impact of Nutrition Education on haemoglobin status of rural Adolescent girls. *Ms.c thesis.* Department of food science and nutrition. College of Rural Home Science, University of Agricultural Sciences, Dharwad.

Kamphuis, C.B.M., van Lenthe, F.J., Giskes, K., Brug, J. and Mackenbach, J.P. (2007). Perceived environmental determinants of physical activity and fruit and vegetable consumption among high and low socioeconomic groups in the Netherlands. *Health and Place,* 13: 493-503.

Loureiro, M. and S. Hine. (2002). "Discovering Niche Markets: A Comparison of Consumer Willingness to Pay for Local, Organic, and GMO-Free Products. *Journal of Agricultural and Applied Economics,* 34:477-488.

Olumakaiye, M.F., Atinmo, T., Olubayo-Fatiregun, M.A. (2010). Food consumption patterns of Nigerian adolescents and effect on body weight. *Journal of Nutrition Education Behaviour.* 42(3): 144-151.

Ottoman, J. (1994). Green Marketing: challenges and opportunities for the new marketing age. London: *McGraw-Hill/Contemporary.* 88-100.

NIPA Publications on Extension Education

S.No.	Title	Author	ISBN	Year
1.	Agricultural Extension: Worldwide Innovations	Saravanan, R.	9788189422967	2008
2.	Agricultural Marketing: Perspectives and Potentials	Bhat, Anil	9789385516153	2016
3.	Agrobiodiversity and Sustainable Rural Development	S.K. Soam	9789383305896	2015
4.	Agro-enterprises for Rural Development and Livelihood Security	Sharma J.P. & S.K.Dubey	9789380235851	2011
5.	Agroforestry Systems for Resource Conservation and Livelihood Security in Lower Himalays	Panwar,P & Dadhwal	9789381450215	2012
6.	Agroforestry: Principles and Practices	Patra, Alok Kumar	9789381450765	2013
7.	Agroforestry: Systems and Practices	Puri, Sunil & Pankaj Panwar	9788189422622	2007
8.	Agroforestry: Systems and Prospects	C.B.Pandey & O.P.Chaturvedi	9789381450970	2014
9.	Ancestral Knowledge in Agri-Allied Science	S.R. Kumar	9789383305216	2014
10.	Biotechnology in India: Initiatives and Accomplishments	Niladri Bag	9789385516252	2016
11.	Computers in Agriculture: Fundamentals and Applications	Sharma Manish & Anil Bhatt	9789385516160	2015
12.	Conducting An Effective and Successful Training Programme	Sontakki,Bharat S.	9789383305223	2015
13.	Dimensions of Extension Education	Mohapatra, B.P.	9789381450987	2014
14.	Disasters: Strengthening Community Mitigation and Preparedness	Khanna, B.K. & Neena Khanna	9789380235455	2011
15.	Economics,Marketing and Sales of Agricultural Products	T. S. Nath	9789381450659	2013
16.	Evaluation and Impact Assessment of Technologies and Developmental Activities in Agriculture, Fisheries and Allied Fields	Roy, A.K.	9789380235400	2011
17.	Extension Education Management in Veterinary Sciences and Animal Husbandry	Singh, Ruchi	9789385516054	2016
18.	Extension Management in the Information Age Initiatives and Impacts	Philip, H. & T.Rathakrishnan	9789381450543	2013
19.	Extension of Technologies: From Labs to Farms	Anandaraja, N.	9788189422837	2008
20.	Extension Techniques for Livestock Development	Singh S.K. & S.D.Singh	9789380235233	2011
21.	Family Farming and Rural Economic Development	M.L.Choudhary & Aditya	9789383305858	2015
22.	Farm Women Empowerment Through Dairy Co-operative Societies	Swain, Pitambar	9789383305186	2014
23.	ICTs for Agricultural Extension: Global Experiments, Innovations and Experiences	Saravanan, R. ed.	9789380235240	2010

S.No.	Title	Author	ISBN	Year
24.	ICTs for Transfer of Technologies: Tools & Techniques	Verma, S.R.	9788193014479	2015
25.	Information and Communication Technology for Agriculture and Rural Development	Saravanan, R.	9789380235882	2011
26.	Information and Knowledge Management: Tools, Techniques and Practices	Roy, A.K.	9789381450628	2013
27.	Information Technology in Veterinary Science	Patil, Vivek	9788190851244	2009
28.	Mobile Phones for Agricultural Extension	Saravanan,R.	9789383305230	2014
29.	Question Bank in Extension Education	Mohammad,Asif	9788193014493	2015
30.	Rural Livelihood and Food Security	Wani, M.H.	9789380235936	2012
31.	Samekit Krishi Pranali	Kumar, Sanjeev	9789381450154	2012
32.	SHGs in Techno-Economic Empowerment of Tribal Women	Satapathy, C. & Sabita Mishra	9789380235875	2011
33.	Social-Ecological Diversity and Traditional Food Systems	Singh,Ranjay, Nancy J.Turner	9789383305360	2014
34.	Technologies for Livelihood Enhancement	V.L. Chopra	9789383305810	2015
35.	Traditional Agricultural Practices: Applications and Technical Implementations	Rathakrishnan,T.	9789380235028	2009
36.	Women in Agriculture and Rural Development	S., Shakuntala: eds.	9788189422998	2009
37.	Women in Sustainable Agriculture	C.Satapathy & Sabita Mishra	9789383305056	2014
38.	Youth in Agriculture and Rural Development	K.Narayana Gowda:Ed.	9789381450758	2013

FSC
www.fsc.org
MIX
Papier | Fördert
gute Waldnutzung
FSC® C083411

Zeitfracht Medien GmbH
Ferdinand-Jühlke-Straße 7
99095 Erfurt, Deutschland
produktsicherheit@kolibri360.de